Piezoelectricity in Classical and Modern Systems

Online at: https://doi.org/10.1088/978-0-7503-5557-5

Piezoelectricity in Classical and Modern Systems

Morten Willatzen
Beijing Institute of Nanoenergy and Nanosystems, Chinese Academy of Sciences, Beijing, People's Republic of China

IOP Publishing, Bristol, UK

ISBN   978-0-7503-5557-5 (ebook)
ISBN   978-0-7503-5555-1 (print)
ISBN   978-0-7503-5558-2 (myPrint)
ISBN   978-0-7503-5556-8 (mobi)

DOI   10.1088/978-0-7503-5557-5

Version: 20240601

IOP ebooks

British Library Cataloguing-in-Publication Data: A catalogue record for this book is available from the British Library.

Published by IOP Publishing, wholly owned by The Institute of Physics, London

IOP Publishing, No.2 The Distillery, Glassfields, Avon Street, Bristol, BS2 0GR, UK

US Office: IOP Publishing, Inc., 190 North Independence Mall West, Suite 601, Philadelphia, PA 19106, USA

This book is dedicated to my wife Marianna and sons Jeppe and Søren for their constant encouragement, motivation, and support.

Contents

Part II   Dynamic deformation

4   Simple oscillator systems   4-1

5   Transverse vibrations of strings   5-1

6   Vibrations of bars   6-1

Part IV Appendices

Preface

With the increasing range of applications of ultrasonics in medicine, navigation, industry, energy harvesting, etc, the importance of piezoelectricity has followed alongside because it is the main mechanism for the technological use of ultrasound. Piezoelectricity, as a phenomenon, plays a substantial role in understanding transport, photonics, and strain phenomena in both classical devices and nanodevices. It is the hope that the present book will shine light on the fundamental theory of piezoelectricity as well as its many classical and nanotechnology applications. A central topic for the study of piezoelectricity is symmetry properties of solids. Piezoelectricity only exists in non-centrosymmetric structures. Out of the 32 crystal classes in three dimensions, 21 have a non-centrosymmetric unit cell, and 20 of them are piezoelectric. The lack of a centrosymmetric unit cell is a prerequisite for the generation of an electric field in the presence of a mechanical stress or strain, and the reverse effect: the generation of a mechanical stress or strain in the presence of an electric field. The group-symmetry properties of crystal structures dictate whether the structure can be piezoelectric and the generated electric field components subject to an applied strain or stress component or vice versa. Evidently, the theory of electrostatics (and to some extent electromagnetics), mechanical deformations, and vibrations must be understood before discussing how piezoelectricity, and other electromechanical effects, can modify electric and mechanical fields.

The present book is inspired and guided by the excellent textbook accounts due to Landau and Lifshitz [1], Auld [2], and Kinsler, Frey, Coppens, and Sanders [3] on the introduction to elasticity theory, symmetry analysis, piezoelectricity, and mechanical vibrations.

The general theory of strain, stress, wave propagation in solids, and elementary mathematical methods are presented in the first part of the book. In chapter 1, the concepts of mechanical strain and stress in solids are introduced. Chapter 2 is devoted to the theory of how scalars, vectors, and tensors of rank 2 transform under coordinate system transformations. This analysis is essential for the discussions in later chapters on the general form of the mechanical, electrical, and piezoelectric tensors associated with the different crystal classes. Chapter 3 discusses propagation of mechanical waves in anisotropic (crystal) structures and includes examples from some of the technologically important cubic and hexagonal crystal classes.

The second part of the book starts with a repetition of the basic theory of zero-dimensional oscillators (chapter 4) and mechanical vibrations of transverse strings (chapter 5), bars (chapter 6), membranes (chapter 7), and cylindrical rods (chapter 8). These first five chapters of Part II are all standard undergraduate expositions of the theory of mechanical vibrations and can be skipped by the reader who is familiar with these topics. Some of the general concepts and mathematical methods used in solving problems in vibrations and piezoelectricity are presented in chapters 4–8.

The third part of the book provides a thorough analysis of piezoelectricity, group-symmetry properties and piezoelectric applications. Chapter 9 presents a toy model of piezoelectricity in a one-dimensional system, following largely the presentation of Auld

[2], but including extra details and exercises. Piezoelectric properties are discussed in chapter 10 for real three-dimensional systems. Chapter 10 contains a group-symmetry analysis of piezoelectricity and piezoelectric tensors in solids. An alternative, yet more formalized, method to obtain the general form of stiffness, permittivity, and piezoelectric tensors is given in chapter 11. The next three chapters 12–14 briefly touch upon three important classical and nanotechnology applications of piezoelectricity.

Chapter 12 briefly discusses the piezoelectric cantilever sensor used to measure small masses in nanotechnology. Chapter 13 contains a detailed one-dimensional model framework of a reciprocal piezoelectric transducer system coupled to fluids. In this chapter, many of the concepts from previous chapters are used. In chapter 14, a three-dimensional axisymmetric piezoelectric transducer model is presented with calculations in Matlab. Chapter 15 discusses flexoelectricity, the coupling of strain gradients to the electric field and vice versa, which is an important phenomenon in many two-dimensional materials that can undergo large flexing. Flexoelectricity is an electro-mechanical phenomenon like piezoelectricity but the former distinguishes itself from the latter in being allowed in both centrosymmetric and non-centrosymmetric systems. Chapter 16 contains a discussion about the modern theory of polarization used in atomistic calculations to determine piezoelectric properties of bulk and nanoscale systems and follows to a large extent the presentation given by Vanderbilt [4]. Chapter 17 addresses the optical properties of semiconductors considering both bulk and quantum-confined systems. A large section of the chapter discusses the influence of piezoelectric fields on electronic eigenstates and the consequences for photonic properties. Finally, chapter 18 presents an exotic angle on ultrasonics applications by analyzing the interesting phenomenon of single-bubble sonoluminescence.

I am grateful to Professor Zhong Lin Wang, Beijing Institute for Nanoenergy and Nanosystems, Chinese Academy of Sciences for encouraging me to write a book on piezoelectricity with an emphasis on classical and modern applications. Much of the work presented was carried out in Denmark during the COVID-19 years when I was unable to work in China due to travel restrictions. Support from the Beijing Institute of Nanoenergy and Nanosystems, Chinese Academy of Sciences, while working outside China is acknowledged with gratitude. I am indebted to PhD student Zhiwei Zhang at the Beijing Institute of Nanoenergy and Nanosystems for preparing figures and assisting me with innumerous practicalities during my time in China. I thank Emil Villekjær Dinesen, a former student at Rosborg Gymnasium and now a physics student at the University of Southern Denmark, and Peter Kjeldsen, senior lecturer at Rosborg Gymnasium, for stimulating discussions and inspiring collaboration on the material presented in the chapter on sonoluminescence. Thanks also go to Dr Lars Duggen at the University of Southern Denmark, Dr Jiajia Shao at the Beijing Institute of Nanoenergy and Nanosystems, and my long-time collaborator, Professor Lok C Lew Yan Voon, Findlay University, for several years of interesting discussions on many of the topics presented in the book.

Morten Willatzen
Beijing, China and Egtved, Denmark
November 1, 2023

References

[1] Landau L D and Lifshitz E M 1986 *Theory of elasticity Course of Theoretical Physics* **vol 7** 3rd edn (Oxford: Heinemann)

[2] Auld B A 1990 *Acoustic Fields and Waves in Solids* **vol I** 2nd edn (Malabar, FL: Krieger)

[3] Kinsler L E, Frey A R, Coppens A B and Sanders J V 2000 *Fundamentals of Acoustics* (New York: Wiley) 4th edn

[4] Vanderbilt D 2018 *Berry Phases in Electronic Structure Theory* (Cambridge: Cambridge University Press)

Author biography

Morten Willatzen

Dr. Morten Willatzen is a Senior Full Professor at the Beijing Institute of Nanoenergy and Nanosystems, Chinese Academy of Sciences. He was previously employed as a Full Professor at the Department of Photonics Engineering, Technical University of Denmark (2012–2017) and the Mads Clausen Institute, University of Southern Denmark (2004–2012). From 2000–2004 he was employed as Associate Professor at the Mads Clausen Institute, University of Southern Denmark. From 1995–2000 he was employed as a senior scientist at Danfoss A/S working on applications in thermo-fluid systems and flow measurement. From 1994–1995, he was a postdoctoral fellow at the Max-Planck Institute for Solid-State Research, Stuttgart working in the group of Professor Manuel Cardona. The author holds an Honorary Professorship at the Mads Clausen Institute, University of Southern Denmark. He received his MSc degree in physics and mathematics in 1989 from Aarhus University (MSc dissertation in quantum-field theory on the topic "Massive Neutrinos") and the PhD degree in theoretical physics from the Niels Bohr Institute, University of Copenhagen (PhD dissertation in theoretical physics entitled "Theory of Gain in Bulk and Quantum-Well Semiconductor Lasers"). Morten Willatzen's research interests include solid-state physics, mathematical physics, piezoelectricity, electronic bandstructures, and, more recently, triboelectric nanogenerators and contact electrification. In addition to the present "Piezoelectricity in Classical and Modern Systems" published by IOP, he is the author of two books "The k.p Method" published by Springer in 2009 and "Separable Boundary-Value Problems in Physics" published by Wiley in 2011 both co-authored with Lok C. Lew Yan Voon. Morten Willatzen has authored more than 10 book chapters and 350 papers in international journals and proceedings.

Other Books by the Author:
Lew Yan Voon L C and Willatzen M 2009 The k.p Method – Electronic Properties of Semiconductors (Berlin: Springer)
Willatzen M and Lew Yan Voon L C 2011 Separable Boundary-Value Problems in Physics (Weinheim: Wiley)

Part I

Strain and stress in solids

Chapter 1

Definition of strain and stress

The theory of vibrations in crystals is based on the introduction of mechanical strain and stress. In chapter 1, strain is defined as a second-rank tensor specifying the change in the vector displacement between two points of a solid body separated by an infinitesimal distance. The vector force required to strain a material then defines the stress tensor from Newton's second law. The free energy density of a crystal is at a minimum in the absence of strain corresponding to thermal equilibrium, and therefore a quadratic function of the strain. The coupling between the free energy and strain specifies the elastic coefficients of the crystal.

1.1 Strain

When a solid body is deformed, the particle originally at a position $\mathbf{r}$ is moved to the position $\mathbf{r}'$ (figure 1.1). The components of the displacement $\mathbf{u} = \mathbf{r}' - \mathbf{r}$ are given by,

$$u_i = x_i' - x_i, \quad i = 1, 2, 3, \tag{1.1}$$

where $\mathbf{r} = (x_1, x_2, x_3) \equiv (x, y, z)$. Consider two points in the solid separated by an infinitesimal distance $d\mathbf{r}$ *before* the deformation and $d\mathbf{r}'$ *after* the deformation. The distance between the points before the deformation is,

$$dL = \sqrt{dx_1^2 + dx_2^2 + dx_3^2}. \tag{1.2}$$

After the deformation the distance between the points is,

$$dL' = \sqrt{dx_1'^2 + dx_2'^2 + dx_3'^2}. \tag{1.3}$$

Since,

$$du_i = dx_i' - dx_i, \tag{1.4}$$

doi:10.1088/978-0-7503-5557-5ch1

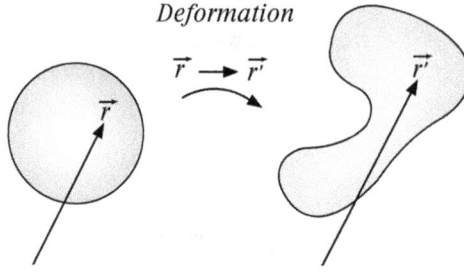

Figure 1.1. Deformation of a solid body.

and,

$$du_i = \frac{\partial u_i}{\partial x_1}dx_1 + \frac{\partial u_i}{\partial x_2}dx_2 + \frac{\partial u_i}{\partial x_3}dx_3 \equiv \frac{\partial u_i}{\partial x_k}dx_k, \tag{1.5}$$

where repeated indices are to be summed over (Einstein notation), i.e.,

$$dx_i'^2 = dx_1'^2 + dx_2'^2 + dx_3'^2,$$
$$\frac{\partial u_i}{\partial x_k}dx_k = \frac{\partial u_i}{\partial x_1}dx_1 + \frac{\partial u_i}{\partial x_2}dx_2 + \frac{\partial u_i}{\partial x_3}dx_3, \tag{1.6}$$

it follows that,

$$dL'^2 = dx_i'^2 = (dx_i + du_i)^2 = \left(dx_i + \frac{\partial u_i}{\partial x_k}dx_k\right)^2$$
$$= dx_i^2 + 2\frac{\partial u_i}{\partial x_k}dx_k dx_i + \frac{\partial u_i}{\partial x_k}\frac{\partial u_i}{\partial x_l}dx_l \tag{1.7}$$
$$= dL^2 + 2 \cdot \frac{1}{2}\left(\frac{\partial u_i}{\partial x_k} + \frac{\partial u_k}{\partial x_i}\right)dx_i dx_k + \frac{\partial u_l}{\partial x_i}\frac{\partial u_l}{\partial x_k}dx_i dx_k,$$

where, after the last equality, i and l were interchanged in the last term. In the second term, it was also used that,

$$\frac{\partial u_i}{\partial x_k}dx_k dx_i = \frac{\partial u_k}{\partial x_i}dx_k dx_i \tag{1.8}$$

since both k and i subscripts are summed from 1 to 3.

We are now ready to define the strain tensor S_{ik},

$$dL'^2 - dL^2 \equiv 2S_{ik}dx_i dx_k, \tag{1.9}$$

and it follows from equation (1.7) that,

$$S_{ik} = \frac{1}{2}\left(\frac{\partial u_i}{\partial x_k} + \frac{\partial u_k}{\partial x_i} + \frac{\partial u_l}{\partial x_i}\frac{\partial u_l}{\partial x_k}\right). \tag{1.10}$$

It should be observed that the strain tensor is symmetrical, i.e.,

$$S_{ik} = S_{ki}.$$ (1.11)

For small deformations, it is reasonable to discard the last term in equation (1.10), thus,

$$S_{ik} = \frac{1}{2}\left(\frac{\partial u_i}{\partial x_k} + \frac{\partial u_k}{\partial x_i}\right).$$ (1.12)

1.2 Stress

Consider a volume V as shown in figure 1.2. The total force on the volume is given by,

$$\mathcal{F}_i = \int_V F_i \, dV = \int_V \frac{\partial T_{ik}}{\partial x_k} \, dV = \int_S T_{ik} \, df_k,$$ (1.13)

where F_i are the components of the force density (units Nm^{-3}), and S is the surface area. We have used that a vector (first-rank tensor $[F_i]$) can be written as the divergence of a second-rank tensor T_{ik} defined as the *stress tensor*. This result is similar to the standard case in vector calculus where a scalar is replaced by the divergence of a vector. In the last equality we employed Gauss's theorem and introduced the surface element vector df_k directed along the outward normal.

Before continuing let us restate the above relation between force density and stress as it is important for later use:

$$F_i = \frac{\partial T_{ik}}{\partial x_k} = \rho \frac{\partial^2 u_i}{\partial t^2},$$ (1.14)

where the second equality is Newton's Second law for the general dynamic elastic problem and ρ denotes the mass density. Note also, that if an external force on a unit

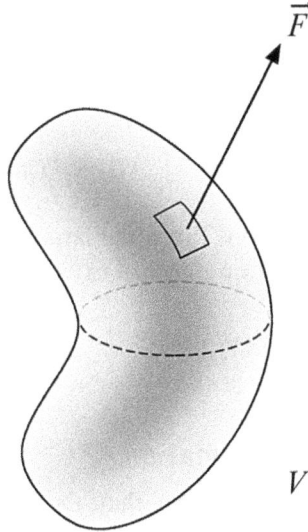

Figure 1.2. Force acting on a solid body.

area of the surface of the solid is $\mathbf{P}$ then the force that acts on a surface element df is $\mathbf{P}df = P_i df$. In equilibrium, this force is equal to the force of the stresses $T_{ik} df_k$ acting on that element,

$$P_i df = T_{ik} df_k. \tag{1.15}$$

Hence, since,

$$df_k = n_k df, \tag{1.16}$$

where the normal vector along the outward normal is $\mathbf{n}$, the surface condition

$$T_{ik} n_k = P_i, \tag{1.17}$$

is obtained. For a solid body in vacuum, the latter condition becomes,

$$T_{ik} n_k = 0. \tag{1.18}$$

Note that in the absence of shear stresses, equation (1.17) becomes,

$$T_{ij} = -P_i \delta_{ij}, \tag{1.19}$$

since $\mathbf{P}$ is positive when the force $\mathbf{F} = \mathbf{P}df$ points opposite to the normal vector $\mathbf{n}$. Consider now the moment of the forces $\mathbf{M} = \int_V \mathbf{F} \times \mathbf{r} \, dV$ on the volume V,

$$\mathbf{M} = \mathbf{e}_1 \int_V (F_2 x_3 - F_3 x_2) dV + \mathbf{e}_2 \int_V (F_3 x_1 - F_1 x_3) dV + \mathbf{e}_3 \int_V (F_1 x_2 - F_2 x_1) dV \\ \equiv (M_{23}, M_{31}, M_{12}), \tag{1.20}$$

where $\mathbf{e}_i$ are unit vectors along the x_i directions. Then M_{ik} becomes,

$$\begin{aligned} M_{ik} &= \int_V (F_i x_k - F_k x_i) dV = \int_V \left(\frac{\partial T_{il}}{\partial x_l} x_k - \frac{\partial T_{kl}}{\partial x_l} x_i \right) dV \\ &= \int_V \frac{\partial (T_{il} x_k - T_{kl} x_i)}{\partial x_l} dV - \int_V \left(T_{il} \frac{\partial x_k}{\partial x_l} - T_{kl} \frac{\partial x_i}{\partial x_l} \right) dV \\ &= \int_V \frac{\partial (T_{il} x_k - T_{kl} x_i)}{\partial x_l} dV - \int_V (T_{il} \delta_{kl} - T_{kl} \delta_{il}) dV \\ &= \int_S (T_{il} x_k - T_{kl} x_i) df_l + \int_V (T_{ki} - T_{ik}) dV, \end{aligned} \tag{1.21}$$

where δ_{ij} denotes a Kronecker delta. From the latter result, we conclude that since the moment of force vanishes in the absence of external surface forces, then the stress tensor must be symmetrical,

$$T_{ki} = T_{ik}. \tag{1.22}$$

1.3 Thermodynamics of deformation

Consider a deformed body and assume that the deformation implies that the displacement vector $\mathbf{u}$ changes by $\delta\mathbf{u}$. The work *done by* the internal stresses is,

$$\int_V \delta R \, dV = \int_V F_i \delta u_i dV = \int_V \frac{\partial T_{ik}}{\partial x_k} \delta u_i dV$$
$$= \int_S T_{ik} \delta u_i df_k - \int_V T_{ik} \frac{\partial \delta u_i}{\partial x_k} dV, \quad (1.23)$$

where the latter equality follows from integration by parts.

If we consider an infinite medium which is not deformed at infinity, the surface integral vanishes, and,

$$\int_V \delta R \, dV = -\frac{1}{2} \int_V T_{ik} \left(\frac{\partial \delta u_i}{\partial x_k} + \frac{\partial \delta u_k}{\partial x_i} \right) dV = -\frac{1}{2} \int_V T_{ik} \delta \left(\frac{\partial u_i}{\partial x_k} + \frac{\partial u_k}{\partial x_i} \right) dV$$
$$= -\int_V T_{ik} \delta S_{ik} dV, \quad (1.24)$$

so,

$$\delta R = -T_{ik} \delta S_{ik}. \quad (1.25)$$

The infinitesimal change in the internal energy dE is from the first law of thermodynamics,

$$dE = \Theta d\sigma - dR, \quad (1.26)$$

where Θ and σ denote the absolute temperature and the entropy, respectively. The free energy F is defined by,

$$F = E - \Theta\sigma, \quad (1.27)$$

thus,

$$dF = dE - \Theta d\sigma - \sigma d\Theta = -dR - \sigma d\Theta = T_{ik} dS_{ik} - \sigma d\Theta, \quad (1.28)$$

and an important relation between stress and strain is obtained,

$$T_{ik} = \left(\frac{\partial F}{\partial S_{ik}} \right)_\Theta. \quad (1.29)$$

This relation is a central result in elasticity theory. We shall derive expressions for the free energy in terms of the strain tensor applicable to different solid (crystal) structures. Once this expression is obtained, we can immediately determine the stress–strain relationship for the solid also known as the generalized Hooke's law.

1.4 Free energy of a solid crystal

The free energy of a solid is at a minimum (thermal equilibrium) in the absence of strain. Hence, the free energy can only contain terms of second order or higher in the strain tensor. To lowest order,

$$F = \frac{1}{2} \lambda_{iklm} S_{ik} S_{lm}. \quad (1.30)$$

Since S_{ik} is symmetric,

$$\lambda_{iklm} = \lambda_{kilm} = \lambda_{ikml} = \lambda_{lmik}, \tag{1.31}$$

and the number of independent elastic coefficients reduces from 81 to 21. This is true for all solid crystals.

1.4.1 Voigt notation

Due to the symmetry of the strain and stress tensors, it is convenient to introduce a more compact notation known as the Voigt notation. Here, double subscripts are substituted according to,

$$
\begin{aligned}
xx &= 1, \quad yy = 2, \quad zz = 3, \\
yz &= 4, \quad xz = 5, \quad xy = 6,
\end{aligned} \tag{1.32}
$$

such that,

$$
\begin{aligned}
\lambda_{xxxx} &= \lambda_{11}, \\
\lambda_{xxyy} &= \lambda_{12}, \\
\lambda_{xyxy} &= \lambda_{66}, \\
\lambda_{yzyz} &= \lambda_{44}.
\end{aligned} \tag{1.33}
$$

Similarly,

$$
\begin{aligned}
S_{xx} &= S_1, \quad S_{yy} = S_2, \quad S_{zz} = S_3, \\
2S_{yz} &= S_4, \quad 2S_{xz} = S_5, \quad 2S_{xy} = S_6, \\
T_{xx} &= T_1, \quad T_{yy} = T_2, \quad T_{zz} = T_3, \\
T_{yz} &= T_4, \quad T_{xz} = T_5, \quad T_{xy} = T_6.
\end{aligned} \tag{1.34}
$$

It is left as an exercise to show that,

$$T_{ik}\,dS_{ik} = T_I\,dS_I, \tag{1.35}$$

where repeated indices involving capital letters, here I, are to be summed from 1 to 6, refer to equation (1.32). Hence, equation (1.28) can be written as,

$$dF = T_I\,dS_I - \sigma d\Theta, \tag{1.36}$$

so,

$$T_I = \left(\frac{\partial F}{\partial S_I}\right)_\Theta, \tag{1.37}$$

which will be used in the following.

1.4.2 Stiffness tensor for the cubic crystal O_h $(m\bar{3}m)$ (diamond)

Since the following operations,

$$
\begin{aligned}
x &\rightarrow x, \quad y \rightarrow -z, \quad z \rightarrow y, \\
x &\rightarrow y, \quad y \rightarrow -x, \quad z \rightarrow z, \\
x &\rightarrow z, \quad y \rightarrow y, \quad z \rightarrow -x,
\end{aligned} \tag{1.38}
$$

are symmetries for the cubic point group O_h ($m\bar{3}m$),

$$
\begin{aligned}
\lambda_{xxxx} &= \lambda_{zzzz} = \lambda_{yyyy}, \\
\lambda_{xxyy} &= \lambda_{zzyy} = \lambda_{zzxx}, \\
\lambda_{xyxy} &= \lambda_{xzxz} = \lambda_{yzyz},
\end{aligned}
\tag{1.39}
$$

and only 3 independent elastic coefficients exist.

It is clear that the free energy of a cubic crystal with point group O_h ($m\bar{3}m$) is,

$$
\begin{aligned}
F &= \frac{1}{2}\lambda_{xxxx}\left(S_{xx}^2 + S_{yy}^2 + S_{zz}^2\right) + \lambda_{xxyy}(S_{xx}S_{yy} + S_{xx}S_{zz} + S_{yy}S_{zz}) \\
&\quad + 2\lambda_{xyxy}\left(S_{xy}^2 + S_{xz}^2 + S_{yz}^2\right)
\end{aligned}
\tag{1.40}
$$

$$
\begin{aligned}
&= \frac{1}{2}\lambda_{xxxx}\left(S_1^2 + S_2^2 + S_3^2\right) + \lambda_{xxyy}(S_1S_2 + S_1S_3 + S_2S_3) \\
&\quad + \frac{1}{2}\lambda_{xyxy}\left(S_6^2 + S_5^2 + S_4^2\right).
\end{aligned}
\tag{1.41}
$$

Here, it is used that,

$$
\begin{aligned}
\lambda_{xxyy} &= \lambda_{yyxx}, \\
\lambda_{xyxy} &= \lambda_{yxxy} = \lambda_{xyyx} = \lambda_{yxyx},
\end{aligned}
\tag{1.42}
$$

to get the pre-factors $\frac{1}{2}$, 1, and 2 to the three elastic coefficients in equation (1.40) for the free energy.

Hence, it follows from equation (1.37) that,

$$
\begin{aligned}
T_1 &= \frac{\partial F}{\partial S_1} = \lambda_{xxxx}S_1 + \lambda_{xxyy}(S_2 + S_3) \equiv c_{11}S_1 + c_{12}S_2 + c_{12}S_3, \\
T_2 &= \frac{\partial F}{\partial S_2} = \lambda_{xxxx}S_2 + \lambda_{xxyy}(S_1 + S_3) \equiv c_{12}S_1 + c_{11}S_2 + c_{12}S_3, \\
T_3 &= \frac{\partial F}{\partial S_3} = \lambda_{xxxx}S_3 + \lambda_{xxyy}(S_1 + S_2) \equiv c_{12}S_1 + c_{12}S_2 + c_{11}S_3, \\
T_4 &= \frac{\partial F}{\partial S_4} = \lambda_{xyxy}S_4 \equiv c_{44}S_4, \\
T_5 &= \frac{\partial F}{\partial S_5} = \lambda_{xyxy}S_5 \equiv c_{44}S_5, \\
T_6 &= \frac{\partial F}{\partial S_6} = \lambda_{xyxy}S_6 \equiv c_{44}S_6,
\end{aligned}
\tag{1.43}
$$

and the Voigt representation of the stress–strain relations of a cubic crystal with point group O_h ($m\bar{3}m$) becomes,

$$
\begin{bmatrix} T_1 \\ T_2 \\ T_3 \\ T_4 \\ T_5 \\ T_6 \end{bmatrix} = \begin{bmatrix} c_{11} & c_{12} & c_{12} & 0 & 0 & 0 \\ c_{12} & c_{11} & c_{12} & 0 & 0 & 0 \\ c_{12} & c_{12} & c_{11} & 0 & 0 & 0 \\ 0 & 0 & 0 & c_{44} & 0 & 0 \\ 0 & 0 & 0 & 0 & c_{44} & 0 \\ 0 & 0 & 0 & 0 & 0 & c_{44} \end{bmatrix} \begin{bmatrix} S_1 \\ S_2 \\ S_3 \\ S_4 \\ S_5 \\ S_6 \end{bmatrix}. \tag{1.44}
$$

The stiffness tensor found above applies to all cubic structures (in the next subsection we provide another example for the zincblende structure with the point group T_d ($\bar{4}3m$)). Hence, we conclude that only 3 independent elastic coefficients exist in a cubic crystal.

1.4.3 Stiffness tensor for the cubic crystal T_d ($\bar{4}3m$) (zincblende)

The following operations are symmetry operations for the zincblende crystal T_d ($\bar{4}3m$),

$$
\begin{aligned}
x &\to -x, & y &\to -y, & z &\to z, \\
x &\to -x, & y &\to y, & z &\to -z, \\
x &\to x, & y &\to -y, & z &\to -z \\
x &\to y, & y &\to -x, & z &\to -z \\
y &\to z, & z &\to -y, & x &\to -x \\
x &\to z, & z &\to -x, & y &\to -y,
\end{aligned} \tag{1.45}
$$

and it follows that the free energy has the general form,

$$
\begin{aligned}
F &= \frac{1}{2}\lambda_{xxxx}\left(S_{xx}^2 + S_{yy}^2 + S_{zz}^2\right) + \lambda_{xxyy}(S_{xx}S_{yy} + S_{xx}S_{zz} + S_{yy}S_{zz}) \\
&\quad + 2\lambda_{xyxy}\left(S_{xy}^2 + S_{xz}^2 + S_{yz}^2\right)
\end{aligned} \tag{1.46}
$$

$$
\begin{aligned}
&= \frac{1}{2}\lambda_{xxxx}\left(S_1^2 + S_2^2 + S_3^2\right) + \lambda_{xxyy}(S_1S_2 + S_1S_3 + S_2S_3) \\
&\quad + \frac{1}{2}\lambda_{xyxy}\left(S_6^2 + S_5^2 + S_4^2\right),
\end{aligned} \tag{1.47}
$$

and the same stiffness tensor as for the point group O_h ($m\bar{3}m$) applies.

Chapter 2

Transformation properties of strain and stress

The use of symmetry transformations in solid state physics and elasticity theory is indispensable. In chapter 2, the coordinate transformation matrix (**a**) and the transformation matrices for stress (**M**) and strain (**N**) are introduced. Applications on how the stiffness tensor transforms in crystals are given. From the derived transformation rules, it is shown that the equations-of-motion for wave propagation take the same form in any coordinate system.

2.1 Orthogonal transformations

We shall follow Auld [1] in deriving the transformation properties of strain and stress tensors. Suppose that a given vector **v** is represented by components v_x, v_y, v_z relative to one set of coordinate axes x, y, z and v_x', v_y', v_z' relative to another set of axes x', y', z'. It is clear from figure 2.1 that the connection between the vector representations in the two coordinate systems is,

$$v_x' = \cos\theta_{x'x}v_x + \cos\theta_{x'y}v_y = a_{xx}v_x + a_{xy}v_y. \tag{2.1}$$

Evidently,

$$v_i' = a_{ij}v_j, \tag{2.2}$$

for all three components $i, j = x, y, z$, where the coefficients a_{ij} define a transformation matrix **a**. Since the transformation does not change the length of **v**,

$$\mathbf{v} \cdot \mathbf{v} = v_x^2 + v_y^2 + v_z^2 = v_x'^2 + v_y'^2 + v_z'^2. \tag{2.3}$$

Writing **v** as a column matrix,

$$\mathbf{v} = \begin{bmatrix} v_x \\ v_y \\ v_z \end{bmatrix}, \tag{2.4}$$

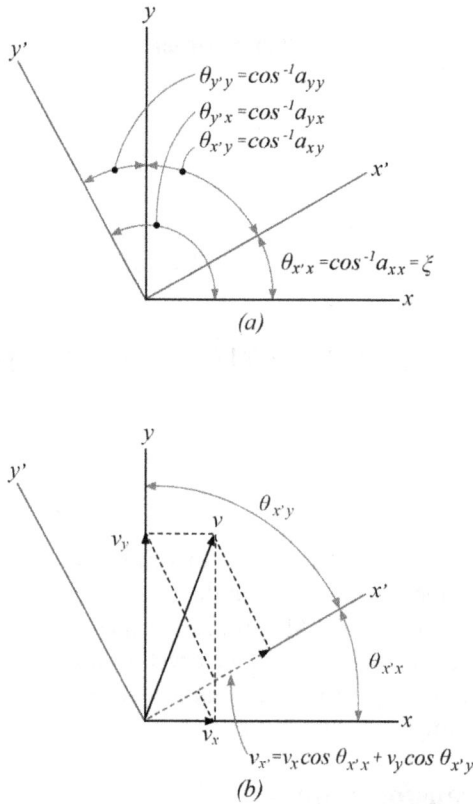

$$\theta_{y'y} = cos^{-1}a_{yy}$$
$$\theta_{y'x} = cos^{-1}a_{yx}$$
$$\theta_{x'y} = cos^{-1}a_{xy}$$

$$\theta_{x'x} = cos^{-1}a_{xx} = \zeta$$

(a)

$$\theta_{x'y}$$

$$\theta_{x'x}$$

$$v_{x'} = v_x cos\,\theta_{x'x} + v_y cos\,\theta_{x'y}$$

(b)

Figure 2.1. Rotation of coordinate systems. Adapted from reference [1] with permission from Krieger Publishing Company. All rights reserved.

equation (2.2) reads,

$$\mathbf{v}' = \mathbf{a}\mathbf{v}, \tag{2.5}$$

and equation (2.3) is expressed as,

$$\mathbf{v}^T\mathbf{v} = \mathbf{v}'^T\mathbf{v}' = \mathbf{v}^T\mathbf{a}^T\mathbf{a}\mathbf{v}, \tag{2.6}$$

where $\mathbf{v}^T$ denotes the transpose of $\mathbf{v}$, etc. Hence, the transformation matrix $\mathbf{a}$ must satisfy,

$$\mathbf{a}^T\mathbf{a} = \mathbf{I}, \tag{2.7}$$

where $\mathbf{I}$ is the identity matrix, or,

$$\mathbf{a}^{-1} = \mathbf{a}^T. \tag{2.8}$$

Denoting the particle displacement gradient vector as,

$$\mathcal{E}_{ij} = \frac{\partial u_i}{\partial x_j}, \tag{2.9}$$

it is clear that,

$$\mathbf{du} = \mathcal{E}\mathbf{dr}. \tag{2.10}$$

Since $d\mathbf{u}$ is a vector, it transforms according to,

$$\mathbf{du'} = \mathbf{au} = \mathbf{a}\mathcal{E}\mathbf{dr}. \tag{2.11}$$

Similarly,

$$\mathbf{dr'} = \mathbf{adr}, \tag{2.12}$$

or,

$$\mathbf{dr} = \mathbf{a}^{-1}\mathbf{dr'}, \tag{2.13}$$

so,

$$\mathbf{du'} = \mathbf{a}\mathcal{E}\mathbf{a}^{-1}\mathbf{dr'}. \tag{2.14}$$

This shows that the displacement gradient matrix in transformed coordinates is,

$$\mathcal{E}' = \mathbf{a}\mathcal{E}\mathbf{a}^{-1} = \mathbf{a}\mathcal{E}\mathbf{a}^{T}. \tag{2.15}$$

Since the strain tensor is the symmetric part of the particle displacement gradient,

$$\mathbf{S} = \frac{1}{2}(\mathcal{E} + \mathcal{E}^{T}), \tag{2.16}$$

it follows that the strain tensor also transforms according to,

$$\mathbf{S'} = \mathbf{aSa}^{T}, \tag{2.17}$$

or in terms of components,

$$S'_{ij} = a_{ik} S_{kl} a_{lj}^{T} = a_{ik} S_{kl} a_{jl} = a_{ik} a_{jl} S_{kl}. \tag{2.18}$$

2.1.1 Example

Consider a two-dimensional static displacement,

$$\mathbf{u}(\mathbf{r}) = \hat{x}Cy + \hat{y}Cx. \tag{2.19}$$

Since,

$$S_{xy} = S_{yx} = C, \tag{2.20}$$

the strain tensor has the form,

$$\mathbf{S} = \begin{bmatrix} 0 & C & 0 \\ C & 0 & 0 \\ 0 & 0 & 0 \end{bmatrix}. \tag{2.21}$$

Let us determine the form of the strain tensor in a coordinate system rotated an angle ξ around the z axis corresponding to the **a** matrix,

$$\mathbf{a} = \begin{bmatrix} \cos\xi & \sin\xi & 0 \\ -\sin\xi & \cos\xi & 0 \\ 0 & 0 & 1 \end{bmatrix}. \tag{2.22}$$

Then,

$$\begin{bmatrix} u'_x \\ u'_y \\ u'_z \end{bmatrix} = \begin{bmatrix} \cos\xi & \sin\xi & 0 \\ -\sin\xi & \cos\xi & 0 \\ 0 & 0 & 1 \end{bmatrix}\begin{bmatrix} u_x \\ u_y \\ u_z \end{bmatrix}, \tag{2.23}$$

that is,

$$u'_x = \cos\xi \, u_x + \sin\xi \, u_y, \tag{2.24}$$

$$u'_y = -\sin\xi \, u_x + \cos\xi \, u_y, \tag{2.25}$$

$$u'_z = u_z. \tag{2.26}$$

The strain tensor in rotated coordinates now reads,

$$\begin{aligned} \mathbf{S}' &= \begin{bmatrix} \cos\xi & \sin\xi & 0 \\ -\sin\xi & \cos\xi & 0 \\ 0 & 0 & 1 \end{bmatrix}\begin{bmatrix} 0 & C & 0 \\ C & 0 & 0 \\ 0 & 0 & 0 \end{bmatrix}\begin{bmatrix} \cos\xi & -\sin\xi & 0 \\ \sin\xi & \cos\xi & 0 \\ 0 & 0 & 1 \end{bmatrix} \\ &= \begin{bmatrix} C\sin 2\xi & C\cos 2\xi & 0 \\ C\cos 2\xi & -C\sin 2\xi & 0 \\ 0 & 0 & 0 \end{bmatrix}. \end{aligned} \tag{2.27}$$

It is seen that the strain components change radically when referred to a new coordinate system. In the original coordinate system only the shear strain components $S_{xy} = S_{yx}$ were present. Rotation of the coordinates adds an extension along the x' axis and a compression along the y' axis. When $\xi = \pi/4$ the shear strain components vanish completely, leaving only an extension along the x' and a compression along y'. This is clearly illustrated in figure 2.2.

2.1.2 Example

Inversion of the coordinate axes,

$$\mathbf{x}' = -\mathbf{x}, \tag{2.28}$$

$$\mathbf{y}' = -\mathbf{y}, \tag{2.29}$$

$$\mathbf{z}' = -\mathbf{z}, \tag{2.30}$$

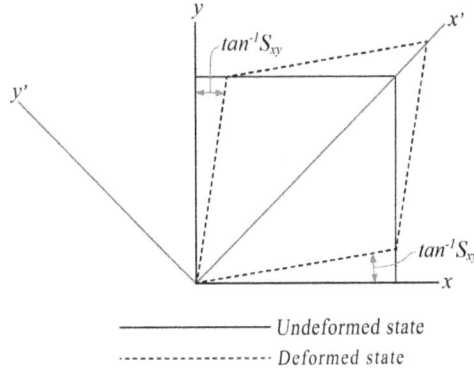

Figure 2.2. Shear deformation. Adapted from reference [1] with permission from Krieger Publishing Company. All rights reserved.

is described by the transformation matrix,

$$\mathbf{a} = \begin{bmatrix} -1 & 0 & 0 \\ 0 & -1 & 0 \\ 0 & 0 & -1 \end{bmatrix} \tag{2.31}$$

such that all vector components change sign,

$$u_x' = -u_x, \tag{2.32}$$

$$u_y' = -u_y, \tag{2.33}$$

$$u_z' = -u_z, \tag{2.34}$$

but the strain tensor is unchanged,

$$\mathbf{S}' = \begin{bmatrix} -1 & 0 & 0 \\ 0 & -1 & 0 \\ 0 & 0 & -1 \end{bmatrix} \begin{bmatrix} S_{xx} & S_{xy} & S_{xz} \\ S_{yx} & S_{yy} & S_{yz} \\ S_{zx} & S_{zy} & S_{zz} \end{bmatrix} \begin{bmatrix} -1 & 0 & 0 \\ 0 & -1 & 0 \\ 0 & 0 & -1 \end{bmatrix} = \mathbf{S}. \tag{2.35}$$

This result will be shown to have important implications when we discuss piezoelectric properties.

Since the stress tensor is a second-rank tensor, it transforms in the same way as the strain tensor, i.e.,

$$T_{ij}' = a_{ik} a_{jl} T_{kl}, \tag{2.36}$$

and,

$$\begin{aligned} T_{xx}' &= a_{xx}^2 T_{xx} + a_{xx} a_{xy} T_{xy} + a_{xx} a_{xz} T_{xz} \\ &\quad + a_{xy} a_{xx} T_{yx} + a_{xy}^2 T_{yy} + a_{xy} a_{xz} T_{yz} \\ &\quad + a_{xz} a_{xx} T_{zx} + a_{xz} a_{xy} T_{zy} + a_{xz}^2 T_{zz} \\ &= a_{xx}^2 T_{xx} + a_{xy}^2 T_{yy} + a_{xz}^2 T_{zz} + 2a_{xy} a_{xz} T_{yz} + 2a_{xx} a_{xz} T_{xz} + 2a_{xx} a_{xy} T_{xy}. \end{aligned} \tag{2.37}$$

Writing this using Voigt notation,

$$T_1' = a_{xx}^2 T_1 + a_{xy}^2 T_2 + a_{xz}^2 T_3 + 2a_{xy}a_{xz} T_4 + 2a_{xx}a_{xz} T_5 + 2a_{xx}a_{xy} T_6. \qquad (2.38)$$

and repeating similar steps for the other stress components leads to the transformation law,

$$T_I' = M_{IJ} T_J, \quad I, J = 1, 2, 3, 4, 5, 6, \qquad (2.39)$$

where,

$$\mathbf{M} = \begin{bmatrix} a_{xx}^2 & a_{xy}^2 & a_{xz}^2 & 2a_{xy}a_{xz} & 2a_{xz}a_{xx} & 2a_{xx}a_{xy} \\ a_{yx}^2 & a_{yy}^2 & a_{yz}^2 & 2a_{yy}a_{yz} & 2a_{yz}a_{yx} & 2a_{yx}a_{yy} \\ a_{zx}^2 & a_{zy}^2 & a_{zz}^2 & 2a_{zy}a_{zz} & 2a_{zz}a_{zx} & 2a_{zx}a_{zy} \\ a_{yx}a_{zx} & a_{yy}a_{zy} & a_{yz}a_{zz} & a_{yy}a_{zz} + a_{yz}a_{zy} & a_{yx}a_{zz} + a_{yz}a_{zx} & a_{yy}a_{zx} + a_{yx}a_{zy} \\ a_{zx}a_{xx} & a_{zy}a_{xy} & a_{zz}a_{xz} & a_{xy}a_{zz} + a_{xz}a_{zy} & a_{xz}a_{zx} + a_{xx}a_{zz} & a_{xx}a_{zy} + a_{xy}a_{zx} \\ a_{xx}a_{yx} & a_{xy}a_{yy} & a_{xz}a_{yz} & a_{xy}a_{yz} + a_{xz}a_{yy} & a_{xz}a_{yx} + a_{xx}a_{yz} & a_{xx}a_{yy} + a_{xy}a_{yx} \end{bmatrix}. \qquad (2.40)$$

Starting from the strain transformation law,

$$S_{ij}' = a_{ik}a_{jl} S_{kl}, \qquad (2.41)$$

and following the same line of argument, one finds,

$$S_I' = N_{IJ} S_J, \quad I, J = 1, 2, 3, 4, 5, 6, \qquad (2.42)$$

where,

$$\mathbf{N} = \begin{bmatrix} a_{xx}^2 & a_{xy}^2 & a_{xz}^2 & a_{xy}a_{xz} & a_{xz}a_{xx} & a_{xx}a_{xy} \\ a_{yx}^2 & a_{yy}^2 & a_{yz}^2 & a_{yy}a_{yz} & a_{yz}a_{yx} & a_{yx}a_{yy} \\ a_{zx}^2 & a_{zy}^2 & a_{zz}^2 & a_{zy}a_{zz} & a_{zz}a_{zx} & a_{zx}a_{zy} \\ 2a_{yx}a_{zx} & 2a_{yy}a_{zy} & 2a_{yz}a_{zz} & a_{yy}a_{zz} + a_{yz}a_{zy} & a_{yx}a_{zz} + a_{yz}a_{zx} & a_{yy}a_{zx} + a_{yx}a_{zy} \\ 2a_{zx}a_{xx} & 2a_{zy}a_{xy} & 2a_{zz}a_{xz} & a_{xy}a_{zz} + a_{xz}a_{zy} & a_{xz}a_{zx} + a_{xx}a_{zz} & a_{xx}a_{zy} + a_{xy}a_{zx} \\ 2a_{xx}a_{yx} & 2a_{xy}a_{yy} & 2a_{xz}a_{yz} & a_{xy}a_{yz} + a_{xz}a_{yy} & a_{xz}a_{yx} + a_{xx}a_{yz} & a_{xx}a_{yy} + a_{xy}a_{yx} \end{bmatrix}. \qquad (2.43)$$

2.2 Transformation law for the stiffness matrix

The above transformation laws allow us to obtain the transformation law of the stiffness matrix. Indeed, from,

$$\mathbf{T} = \mathbf{cS}, \qquad (2.44)$$

it follows that,

$$\mathbf{T'} = \mathbf{McS}, \qquad (2.45)$$

and since,

$$\mathbf{S} = \mathbf{N}^{-1}\mathbf{S}', \qquad (2.46)$$

the transformation law is,

$$\mathbf{T}' = \mathbf{McN}^{-1}\mathbf{S}', \qquad (2.47)$$

or,

$$\mathbf{c}' = \mathbf{McN}^{-1}. \qquad (2.48)$$

The above relation involves taking the inverse of a 6×6 matrix $\mathbf{N}^{-1}$. We can avoid the complex operation of taking the inverse of $\mathbf{N}$ by noting that,

$$\mathbf{S}' = \mathbf{aSa}^T = \mathbf{NS}, \text{ or}$$
$$S'_{ij} = a_{ik} a_{jl} S_{kl}, \qquad (2.49)$$

and,

$$\mathbf{S} = \mathbf{a}^T\mathbf{S}'\mathbf{a} = \mathbf{N}^{-1}\mathbf{S}', \text{ or}$$
$$S_{ij} = a_{ki} a_{lj} S'_{kl}. \qquad (2.50)$$

A comparison of equations (2.49) and (2.50) shows that the matrix $\mathbf{N}^{-1}$ is obtained from $\mathbf{N}$ by interchanging all subscripts, which is simply the transpose of the matrix $\mathbf{M}$,

$$\mathbf{N}^{-1} = \mathbf{M}^T. \qquad (2.51)$$

Thus,

$$\mathbf{c}' = \mathbf{McM}^T, \qquad (2.52)$$

which is easy to evaluate.

2.3 Stiffness tensor of cubic crystal in rotated coordinates

Consider a cubic crystal for which we derived the stiffness tensor to be,

$$\mathbf{c} = \begin{bmatrix} c_{11} & c_{12} & c_{12} & 0 & 0 & 0 \\ c_{12} & c_{11} & c_{12} & 0 & 0 & 0 \\ c_{12} & c_{12} & c_{11} & 0 & 0 & 0 \\ 0 & 0 & 0 & c_{44} & 0 & 0 \\ 0 & 0 & 0 & 0 & c_{44} & 0 \\ 0 & 0 & 0 & 0 & 0 & c_{44} \end{bmatrix}, \qquad (2.53)$$

assuming the cubic crystal axes X, Y, Z correspond to the x, y, z axes, respectively. We shall derive the stiffness tensor in rotated coordinates defined by the rotation angle ξ around the z axis. The transformation matrix $\mathbf{a}$ is,

$$\mathbf{a} = \begin{bmatrix} \cos \xi & \sin \xi & 0 \\ -\sin \xi & \cos \xi & 0 \\ 0 & 0 & 1 \end{bmatrix}, \tag{2.54}$$

and the 6×6 transformation matrix $\mathbf{M}$ becomes,

$$\mathbf{M} = \begin{bmatrix} \cos^2 \xi & \sin^2 \xi & 0 & 0 & 0 & \sin 2\xi \\ \sin^2 \xi & \cos^2 \xi & 0 & 0 & 0 & -\sin 2\xi \\ 0 & 0 & 1 & 0 & 0 & 0 \\ 0 & 0 & 0 & \cos \xi & -\sin \xi & 0 \\ 0 & 0 & 0 & \sin \xi & \cos \xi & 0 \\ -\dfrac{\sin 2\xi}{2} & \dfrac{\sin 2\xi}{2} & 0 & 0 & 0 & \cos 2\xi \end{bmatrix}. \tag{2.55}$$

The transformed stiffness tensor is easily evaluated from equation (2.52),

$$\mathbf{c} = \begin{bmatrix} c'_{11} & c'_{12} & c'_{13} & 0 & 0 & c'_{16} \\ c'_{12} & c'_{11} & c'_{13} & 0 & 0 & -c'_{16} \\ c'_{13} & c'_{13} & c'_{33} & 0 & 0 & 0 \\ 0 & 0 & 0 & c'_{44} & 0 & 0 \\ 0 & 0 & 0 & 0 & c'_{44} & 0 \\ c'_{16} & -c'_{16} & 0 & 0 & 0 & c'_{66} \end{bmatrix}, \tag{2.56}$$

where,

$$
\begin{aligned}
c'_{11} &= c_{11} - \left(\frac{c_{11} - c_{12}}{2} - c_{44} \right) \sin^2 2\xi, \\
c'_{12} &= c_{12} + \left(\frac{c_{11} - c_{12}}{2} - c_{44} \right) \sin^2 2\xi, \\
c'_{13} &= c_{12}, \\
c'_{16} &= -\left(\frac{c_{11} - c_{12}}{2} - c_{44} \right) \sin 2\xi \cos 2\xi, \\
c'_{33} &= c_{11}, \\
c'_{44} &= c_{44}, \\
c'_{66} &= c_{44} + \left(\frac{c_{11} - c_{12}}{2} - c_{44} \right) \sin^2 2\xi.
\end{aligned}
\tag{2.57}
$$

It is clear that the matrix form is different compared to the original stiffness tensor as some entries are now nonzero and others that were equal are now different.

Consider the case where $\xi = 45°$ that is the x' axis is along the [110] direction and the y' axis is in the [$\bar{1}$10] direction (refer to figure 2.3). Then the transformed stiffness tensor becomes,

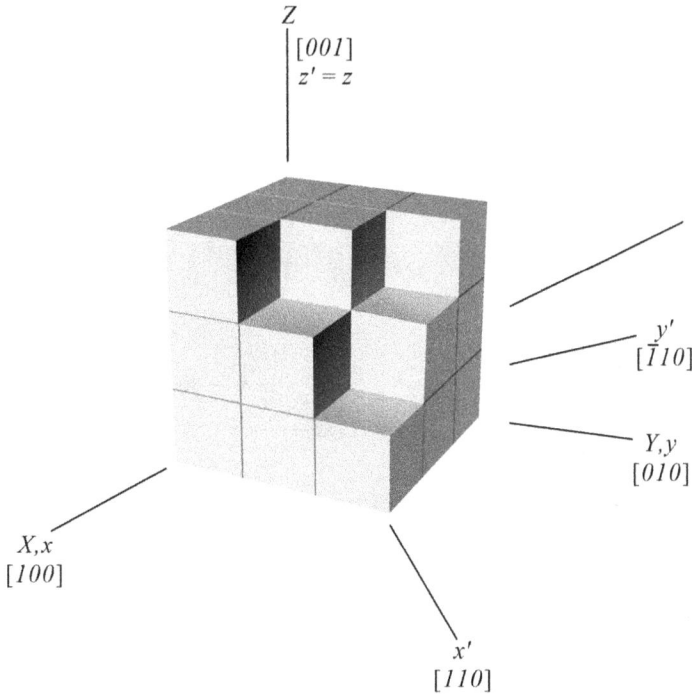

Figure 2.3. Transformation of a crystal coordinate system by 45 degrees about the z axis. Adapted from reference [1] with permission from Krieger Publishing Company. All rights reserved.

$$\mathbf{c} = \begin{bmatrix} c_{11}' & c_{12}' & c_{12} & 0 & 0 & 0 \\ c_{12}' & c_{11}' & c_{12} & 0 & 0 & 0 \\ c_{12} & c_{12} & c_{11} & 0 & 0 & 0 \\ 0 & 0 & 0 & c_{44} & 0 & 0 \\ 0 & 0 & 0 & 0 & c_{44} & 0 \\ 0 & 0 & 0 & 0 & 0 & \dfrac{c_{11} - c_{12}}{2} \end{bmatrix}, \tag{2.58}$$

with,

$$c_{11}' = \frac{c_{11} + c_{12} + 2c_{44}}{2},$$

$$c_{12}' = \frac{c_{11} + c_{12} - 2c_{44}}{2}.$$

2.4 equations-of-motion in rotated coordinates

We will next derive the equations-of-motion in rotated coordinates. Since the gradient operator is a vector, it transforms as,

$$\frac{\partial}{\partial x_j'} = a_{jo}\frac{\partial}{\partial x_o},$$

(2.59)

and the stiffness tensor transforms as,

$$\mathbf{T}' = \mathbf{aTa}^T, \text{ or}$$
$$T_{ij}' = a_{ik}T_{kl}a_{lj}^T = a_{ik}T_{kl}a_{jl},$$

(2.60)

so,

$$\frac{\partial T_{ij}'}{\partial x_j'} = a_{jo}\frac{\partial}{\partial x_o}(a_{ik}a_{jl}T_{kl}) = a_{jo}a_{jl}a_{ik}\frac{\partial T_{kl}}{\partial x_o}$$
$$= a_{oj}^T a_{jl}a_{ik}\frac{\partial T_{kl}}{\partial x_o} = \delta_{ol}a_{ik}\frac{\partial T_{kl}}{\partial x_o} = a_{ik}\frac{\partial T_{kl}}{\partial x_l}.$$

(2.61)

Further, since the particle displacement **u** is a vector,

$$u_i' = a_{ik}u_k,$$

(2.62)

it follows that the original equations-of-motion,

$$\frac{\partial T_{ij}}{\partial x_j} = -\rho\omega^2 u_i,$$

(2.63)

are equivalent to the transformed form,

$$\frac{\partial T_{ij}'}{\partial x_j'} = -\rho\omega^2 u_i'.$$

(2.64)

In conclusion, the equations-of-motion has the same form in any rotated coordinate system.

Reference

Auld B A 1990 *Acoustic Fields and Waves in Solids* **vol I** 2nd edn (Malabar, FL: Krieger)

IOP Publishing

Piezoelectricity in Classical and Modern Systems

Morten Willatzen

Chapter 3

Wave propagation in solids

In chapter 3, wave propagation in cubic, isotropic, and hexagonal crystals is discussed. It is used that the free energy must be invariant subject to any symmetry operation. The Christoffel equation is determined, providing an effective way to calculate wave speeds along different crystal directions from the Christoffel tensor. Newton's second law for an unbounded solid in terms of displacement vectors and vibration frequencies is discussed. This chapter is rounded off by deriving the basic theory of surface wave propagation.

3.1 Compressional wave propagation along the [110] direction in a cubic crystal

Let us consider a uniform plane wave propagating along the [110] direction (x' axis). In this case we would need to use particle displacements u_x, u_y and coordinate variables x, y to describe the mechanical field if reference is made to the cubic crystal axes. That is, the non-zero strain components are,

$$S_{xx} = \frac{\partial u_x}{\partial x},$$

$$S_{yy} = \frac{\partial u_y}{\partial y}, \tag{3.1}$$

$$S_{xy} = \frac{1}{2}\left(\frac{\partial u_x}{\partial y} + \frac{\partial u_y}{\partial x}\right).$$

It is obviously easier to use rotated coordinates x', y', z' where the particle displacement is,

$$\mathbf{u} = \hat{x}' \cos(\omega t - kx'), \tag{3.2}$$

and there is only one strain component,

$$S'_{x'x'} = \frac{\partial u_{x'}}{\partial x'} = k \sin(\omega t - kx'), \tag{3.3}$$

or,

$$\mathbf{S}' = \begin{bmatrix} k\sin(\omega t - kx') \\ 0 \\ 0 \\ 0 \\ 0 \\ 0 \end{bmatrix}. \tag{3.4}$$

The stress field $\mathbf{T}'$ is computed from equation (2.58) as $\mathbf{c}'\mathbf{S}'$,

$$T_1' = T_{x'x'}' = c_{11}'k\sin(\omega t - kx'),$$
$$T_2' = T_{y'y'}' = c_{12}'k\sin(\omega t - kx'),$$
$$T_3' = T_{z'z'}' = c_{12}k\sin(\omega t - kx'),$$

and the equations-of-motion are,

$$\frac{\partial T_{x'x'}'}{\partial x'} = -c_{11}'k^2\cos(\omega t - kx') = -\rho\omega^2 u_{x'},$$

$$\frac{\partial T_{y'x'}'}{\partial x'} = 0 = -\rho\omega^2 u_{y'},$$

$$\frac{\partial T_{z'x'}'}{\partial x'} = 0 = -\rho\omega^2 u_{z'}.$$

It now follows from equation (3.2) that the dispersion is,

$$c_{11}'k^2 = \rho\omega^2, \tag{3.5}$$

and the velocity of a compressional wave propagating along the [110] direction in a cubic crystal is thus,

$$V_l^{[110]} = \sqrt{\frac{c_{11}'}{\rho}} = \sqrt{\frac{c_{11} + c_{12} + 2c_{44}}{2\rho}}. \tag{3.6}$$

Evidently, the wave velocity for propagation along the [110] direction is different from the wave speed along the X, Y, Z crystal axes which can be shown (left as an exercise) to be,

$$V_l^{[100]} = V_l^{[010]} = V_l^{[001]} = \sqrt{\frac{c_{11}}{\rho}}. \tag{3.7}$$

3.1.1 Isotropic crystals

For an isotropic crystal, the stress–strain relations have a similar structure as for the cubic structure. However, a relation exists between the stiffness coefficients in the isotropic crystal, as we shall now prove. This relation implies that the three independent elastic coefficients for a cubic crystal reduces to two independent elastic coefficients for an isotropic solid.

If the crystal is isotropic, the free energy must be invariant subject to a symmetry operation, for example a 45 degrees clockwise rotation about the z axis,

$$x \rightarrow \frac{1}{\sqrt{2}}(x + y),$$

$$y \rightarrow \frac{1}{\sqrt{2}}(y - x), \tag{3.8}$$

$$z \rightarrow z.$$

The strain tensor transforms as (prove it using the transformation rules derived in the preceding sections),

$$S_{xx} \rightarrow \frac{1}{2}(S_{xx} + S_{yy} + 2S_{xy}),$$

$$S_{yy} \rightarrow \frac{1}{2}(S_{xx} + S_{yy} - 2S_{xy}),$$

$$S_{zz} \rightarrow S_{zz},$$

$$S_{yz} \rightarrow \frac{1}{\sqrt{2}}(S_{yz} - S_{xz}), \tag{3.9}$$

$$S_{xz} \rightarrow \frac{1}{\sqrt{2}}(S_{xz} + S_{yz}),$$

$$S_{xy} \rightarrow \frac{1}{2}(S_{yy} - S_{xx}),$$

i.e.,

$$S_1 \rightarrow \frac{1}{2}(S_1 + S_2 + S_6),$$

$$S_2 \rightarrow \frac{1}{2}(S_1 + S_2 - S_6),$$

$$S_3 \rightarrow S_3,$$

$$S_4 \rightarrow \frac{1}{\sqrt{2}}(S_4 - S_5), \tag{3.10}$$

$$S_5 \rightarrow \frac{1}{\sqrt{2}}(S_5 + S_4),$$

$$S_6 \rightarrow S_2 - S_1.$$

Thus, the free energy in equation (1.41) transforms as,

$$F = \frac{1}{2}\lambda_{xxxx}\left(\frac{1}{2}S_1^2 + \frac{1}{2}S_2^2 + S_1 S_2 + \frac{1}{2}S_6^2 + S_3^2\right)$$

$$+ \lambda_{xxyy}\left(\frac{1}{4}S_1^2 + \frac{1}{4}S_2^2 + \frac{1}{2}S_1 S_2 - \frac{1}{4}S_6^2 + S_1 S_3 + S_2 S_3\right) \tag{3.11}$$

$$+ \frac{1}{2}\lambda_{xyxy}\left(S_1^2 + S_2^2 - 2S_1 S_2 + S_4^2 + S_5^2\right).$$

Now, collecting all terms proportional to S_1^2 in the latter expression and equating to the same term in equation (1.41) yields,

$$\frac{1}{4}\lambda_{xxxx}S_1^2 + \frac{1}{4}\lambda_{xxyy}S_1^2 + \frac{1}{2}\lambda_{xyxy}S_1^2 = \frac{1}{2}\lambda_{xxxx}S_1^2, \tag{3.12}$$

i.e.,

$$\frac{1}{4}\lambda_{xxxx}S_1^2 = \frac{1}{4}\lambda_{xxyy}S_1^2 + \frac{1}{2}\lambda_{xyxy}S_1^2 \Longrightarrow$$

$$\lambda_{xyxy} = \frac{1}{2}(\lambda_{xxxx} - \lambda_{xxyy}), \tag{3.13}$$

or, in terms of c_{IJ},

$$c_{44} = \frac{1}{2}(c_{11} - c_{12}), \tag{3.14}$$

which is the sought isotropic relation.

3.2 Hexagonal crystals

Now, the general form of the elastic tensor λ_{iklm} will be derived. Examples for the two hexagonal crystal classes: C_6 and $D_{3h}(\bar{6}m2)$ are considered.

3.2.1 Crystal class C_6

Consider the hexagonal crystal class C_6. If the sixth-order axis is chosen as the z axis and the complex coordinates are introduced,

$$\begin{aligned}\xi &= x + iy, \\ \eta &= x - iy,\end{aligned} \tag{3.15}$$

it is found for a rotation of an angle $\pi/3$ about the z axis, which is a symmetry, that ξ and η undergo the transformation,

$$\begin{aligned}\xi &\rightarrow \xi e^{i\pi/3}, \\ \eta &\rightarrow \eta e^{-i\pi/3}.\end{aligned} \tag{3.16}$$

Since the free energy must be unchanged by a symmetry transformation, only λ_{iklm} coefficients that contain the same number of ξ's and η's can be non-zero. Hence, only the following 5 non-zero independent elasticity coefficients exist,

$$\lambda_{zzzz}, \quad \lambda_{\xi\eta\xi\eta}, \quad \lambda_{\xi\xi\eta\eta}, \quad \lambda_{\xi\eta zz}, \quad \lambda_{\xi z\eta z}. \tag{3.17}$$

In actual fact, these correspond to the 5 independent elastic stiffness coefficients c_{11}, c_{12}, c_{13}, c_{33}, c_{44} as we will demonstrate.

3.2.2 Stiffness tensor of hexagonal structures

The energy of a hexagonal structure is,

$$U = \frac{1}{2}\lambda_{zzzz}S_{zz}^2 + \frac{1}{2}(\lambda_{zz\xi\eta} + \lambda_{zz\eta\xi} + \lambda_{\xi\eta zz} + \lambda_{\eta\xi zz})S_{zz}S_{\xi\eta}$$

$$+ \frac{1}{2}(\lambda_{\xi\eta\xi\eta} + \lambda_{\eta\xi\xi\eta} + \lambda_{\xi\eta\eta\xi} + \lambda_{\eta\xi\eta\xi})S_{\xi\eta}^2 + \frac{1}{2}(\lambda_{\xi\xi\eta\eta} + \lambda_{\eta\eta\xi\xi})S_{\xi\xi}S_{\eta\eta}$$

$$+ \frac{1}{2}(\lambda_{\xi z\eta z} + \lambda_{z\xi\eta z} + \lambda_{\xi z z\eta} + \lambda_{z\xi z\eta})S_{\xi z}S_{\eta z}$$

$$= \frac{1}{2}\lambda_{zzzz}S_{zz}^2 + 2\lambda_{zz\xi\eta}S_{zz}(S_{xx} + S_{yy})$$

$$+ 2\lambda_{\xi\eta\xi\eta}(S_{xx} + S_{yy})^2 + \lambda_{\xi\xi\eta\eta}(S_{xx} - S_{yy} + 2iS_{xy})(S_{xx} - S_{yy} - 2iS_{xy}) \qquad (3.18)$$

$$+ 2\lambda_{\xi z\eta z}(S_{xz} + iS_{yz})(S_{xz} - iS_{yz})$$

$$= \frac{1}{2}\lambda_{zzzz}S_3^2 + 2\lambda_{zz\xi\eta}S_3(S_1 + S_2)$$

$$+ 2\lambda_{\xi\eta\xi\eta}(S_1 + S_2)^2 + \lambda_{\xi\xi\eta\eta}(S_1 - S_2 + iS_6)(S_1 - S_2 - iS_6)$$

$$+ \frac{1}{2}\lambda_{\xi z\eta z}(S_5 + iS_4)(S_5 - iS_4).$$

In deriving the second equality, the expressions,

$$\xi\xi = (x + iy)^2 = x^2 - y^2 + 2ixy,$$
$$\xi\eta = (x + iy)(x - iy) = x^2 + y^2, \qquad (3.19)$$
$$\xi z = (x + iy)z = xz + iyz,$$

etc, were used, and components of a second-rank tensor such as strain fulfill,

$$S_{\xi\xi} = S_{xx} - S_{yy} + 2iS_{xy},$$
$$S_{\xi\eta} = S_{xx} + S_{yy}, \qquad (3.20)$$
$$S_{\xi z} = S_{xz} + iS_{yz},$$

etc.

The stress–strain relations can now be obtained,

$$T_1 = \frac{\partial U}{\partial S_1} = (4\lambda_{\xi\eta\xi\eta} + 2\lambda_{\xi\xi\eta\eta})S_1 + (4\lambda_{\xi\eta\xi\eta} - 2\lambda_{\xi\xi\eta\eta})S_2 + 2\lambda_{zz\xi\eta}S_3,$$

$$T_2 = \frac{\partial U}{\partial S_2} = (4\lambda_{\xi\eta\xi\eta} - 2\lambda_{\xi\xi\eta\eta})S_1 + (4\lambda_{\xi\eta\xi\eta} + 2\lambda_{\xi\xi\eta\eta})S_2 + 2\lambda_{zz\xi\eta}S_3,$$

$$T_3 = \frac{\partial U}{\partial S_3} = 2\lambda_{zz\xi\eta}S_1 + 2\lambda_{zz\xi\eta}S_2 + \lambda_{zzzz}S_3,$$

$$T_4 = \frac{\partial U}{\partial S_4} = \lambda_{\xi z\eta z}S_4, \qquad (3.21)$$

$$T_5 = \frac{\partial U}{\partial S_5} = \lambda_{\xi z\eta z}S_5,$$

$$T_6 = \frac{\partial U}{\partial S_6} = 2\lambda_{\xi\xi\eta\eta}S_6,$$

which, written in matrix form, reads,

$$
\mathbf{c} = \begin{bmatrix}
(4\lambda_{\xi\eta\xi\eta} + 2\lambda_{\xi\xi\eta\eta}) & (4\lambda_{\xi\eta\xi\eta} - 2\lambda_{\xi\xi\eta\eta}) & 2\lambda_{zz\xi\eta} & 0 & 0 & 0 \\
(4\lambda_{\xi\eta\xi\eta} - 2\lambda_{\xi\xi\eta\eta}) & (4\lambda_{\xi\eta\xi\eta} + 2\lambda_{\xi\xi\eta\eta}) & 2\lambda_{zz\xi\eta} & 0 & 0 & 0 \\
2\lambda_{zz\xi\eta} & 2\lambda_{zz\xi\eta} & \lambda_{zzzz} & 0 & 0 & 0 \\
0 & 0 & 0 & \lambda_{\xi z\eta z} & 0 & 0 \\
0 & 0 & 0 & 0 & \lambda_{\xi z\eta z} & 0 \\
0 & 0 & 0 & 0 & 0 & 2\lambda_{\xi\xi\eta\eta}
\end{bmatrix}, \tag{3.22}
$$

and by renaming the coefficients above,

$$
\mathbf{c} = \begin{bmatrix}
c_{11} & c_{12} & c_{13} & 0 & 0 & 0 \\
c_{12} & c_{11} & c_{13} & 0 & 0 & 0 \\
c_{13} & c_{13} & c_{33} & 0 & 0 & 0 \\
0 & 0 & 0 & c_{44} & 0 & 0 \\
0 & 0 & 0 & 0 & c_{44} & 0 \\
0 & 0 & 0 & 0 & 0 & \frac{1}{2}(c_{11} - c_{12})
\end{bmatrix}. \tag{3.23}
$$

It should be pointed out that $c_{66} = \frac{1}{2}(c_{11} - c_{12})$ reflects the isotropy of hexagonal crystals in the plane perpendicular to the sixth-order axis.

3.2.3 Crystal class D_{3h} ($\bar{6}m2$)

In the case of the crystal class D_{3h} ($\bar{6}m2$), we first note that a rotation of $\pi/3$ about the z axis is no longer a symmetry. However, a rotation of $2\pi/3$ about the z axis is a symmetry. In this case, ξ and η undergo the transformations,

$$
\begin{aligned}
\xi &\to \xi e^{i2\pi/3}, \\
\eta &\to \eta e^{-i2\pi/3}.
\end{aligned} \tag{3.24}
$$

At a first glance, one could then suggest elastic coefficients of the type $\lambda_{\xi\xi\xi z}$, $\lambda_{\eta\eta\eta z}$, etc contribute to the free energy. They do not, however, since a reflection in the x–y plane ($z \to -z$ is also a symmetry of D_{3h} ($\bar{6}m2$) thus λ coefficients involving an odd number of z indices are not allowed. With these considerations in mind, the general form of the free energy expression in equation (3.18) can be seen to still apply and the stiffness tensor takes the form in equation (3.23).

3.3 Unbounded isotropic media

Before we proceed to study piezoelectric media, it is instructive to analyze wave propagation in isotropic solid media. This is important as piezoelectric sensors or actuators are often in direct contact with isotropic solid media and the ultrasonic signal, detected or transmitted by a piezoelectric element, is shaped in form and time by the media it propagated through.

Let us therefore recast the elastic equations of an isotropic medium in the form,

$$(\lambda + \mu)\frac{\partial}{\partial x}\left(\frac{\partial u_x}{\partial x} + \frac{\partial u_y}{\partial y} + \frac{\partial u_z}{\partial z}\right) + \mu\nabla^2 u_x + \rho f_x = \rho\frac{\partial^2 u_x}{\partial t^2}, \qquad (3.25)$$

$$(\lambda + \mu)\frac{\partial}{\partial y}\left(\frac{\partial u_x}{\partial x} + \frac{\partial u_y}{\partial y} + \frac{\partial u_z}{\partial z}\right) + \mu\nabla^2 u_y + \rho f_y = \rho\frac{\partial^2 u_y}{\partial t^2}, \qquad (3.26)$$

$$(\lambda + \mu)\frac{\partial}{\partial z}\left(\frac{\partial u_x}{\partial x} + \frac{\partial u_y}{\partial y} + \frac{\partial u_z}{\partial z}\right) + \mu\nabla^2 u_z + \rho f_z = \rho\frac{\partial^2 u_z}{\partial t^2}. \qquad (3.27)$$

Note that the elastic equations above can be written in vector form,

$$(\lambda + \mu)\nabla\nabla \cdot \mathbf{u} + \mu\nabla^2\mathbf{u} = \rho\frac{\partial^2\mathbf{u}}{\partial t^2}. \qquad (3.28)$$

3.3.1 Exercise

For an isotropic medium derive equations (3.25)–(3.27) from equation (1.14) using the relations between the Lamé coefficients and the stiffness tensor,

$$
\begin{aligned}
c_{11} &= \lambda + 2\mu, \\
c_{44} &= \frac{1}{2}(c_{11} - c_{12}) = \mu, \\
c_{12} &= \lambda.
\end{aligned}
\qquad (3.29)
$$

From the Helmholtz decomposition rule [1] it is possible, for a general displacement vector $\mathbf{u}$, to write,

$$\mathbf{u} = \nabla\Phi + \nabla \times \mathbf{H}, \qquad (3.30)$$

where $\mathbf{H}$ is a zero divergence vector,

$$\nabla \cdot \mathbf{H} = 0. \qquad (3.31)$$

The fields Φ and $\mathbf{H}$ are scalar and vector potentials, respectively.

Substituting equation (3.30) in equation (3.28),

$$(\lambda + \mu)\nabla\nabla \cdot (\nabla\Phi + \nabla \times \mathbf{H}) + \mu\nabla^2(\nabla\Phi + \nabla \times \mathbf{H}) = \rho\left(\nabla\frac{\partial^2\Phi}{\partial t^2} + \nabla \times \frac{\partial^2\mathbf{H}}{\partial t^2}\right). \quad (3.32)$$

From the vector identity,

$$\nabla^2\mathbf{u} = \nabla\nabla \cdot \mathbf{u} - \nabla \times \nabla \times \mathbf{u}, \qquad (3.33)$$

it is found that,

$$\left[(\lambda + \mu) \boldsymbol{\nabla}\boldsymbol{\nabla} \cdot \left(\boldsymbol{\nabla}\Phi \right) - \rho\boldsymbol{\nabla}\frac{\partial^2 \Phi}{\partial t^2} \right] + \mu\boldsymbol{\nabla}\boldsymbol{\nabla} \cdot (\boldsymbol{\nabla}\Phi) - \mu\boldsymbol{\nabla} \times \boldsymbol{\nabla} \times \boldsymbol{\nabla}\Phi$$

$$+ (\lambda + \mu)\boldsymbol{\nabla}\boldsymbol{\nabla} \cdot \boldsymbol{\nabla} \times \mathbf{H} + \left[\mu\boldsymbol{\nabla}^2 \boldsymbol{\nabla} \times \mathbf{H} - \rho\boldsymbol{\nabla} \times \frac{\partial^2 \mathbf{H}}{\partial t^2} \right] = 0. \tag{3.34}$$

From the identities,

$$\boldsymbol{\nabla} \cdot \boldsymbol{\nabla}\Phi = \boldsymbol{\nabla}^2\Phi,$$
$$\boldsymbol{\nabla} \times \boldsymbol{\nabla}\Phi = 0, \tag{3.35}$$
$$\boldsymbol{\nabla} \cdot \boldsymbol{\nabla} \times \mathbf{H} = 0,$$

the following equation is obtained,

$$\boldsymbol{\nabla}\left[(\lambda + 2\mu)\boldsymbol{\nabla}^2\Phi - \rho\frac{\partial^2 \Phi}{\partial t^2} \right] + \boldsymbol{\nabla} \times \left[\mu\boldsymbol{\nabla}^2\mathbf{H} - \rho\frac{\partial^2 \mathbf{H}}{\partial t^2} \right] = 0, \tag{3.36}$$

which can only be satisfied if both terms vanish, i.e.,

$$\boldsymbol{\nabla}^2\Phi = \frac{1}{c_L^2}\frac{\partial^2 \Phi}{\partial t^2}, \text{ and} \tag{3.37}$$

$$\boldsymbol{\nabla}^2\mathbf{H} = \frac{1}{c_T^2}\frac{\partial^2 \mathbf{H}}{\partial t^2}, \tag{3.38}$$

where,

$$c_L^2 = \frac{\lambda + 2\mu}{\rho},$$
$$c_T^2 = \frac{\mu}{\rho}. \tag{3.39}$$

The subscripts L and T refer to that the waves associated with the potentials Φ and $\mathbf{H}$ are longitudinal and transverse, respectively, as the Helmholtz decomposition reveals. Evidently the two wave speeds are different but for general boundary-confined systems in elasticity, the solution consists of a mixture of longitudinal and transverse waves. Only when simple boundary conditions apply can the wave solution be of either longitudinal or transverse form.

3.4 Christoffel equation

Consider elastic waves in an infinite crystal medium and recall the elastic equations,

$$\frac{\partial T_{ik}}{\partial x_k} = \rho\frac{\partial^2 u_i}{\partial t^2}. \tag{3.40}$$

Since the constitutive law for a non-piezoelectric medium is,

$$T_{ik} = c_{iklm}S_{lm}, \tag{3.41}$$

and the strain tensor is,

$$S_{lm} = \frac{1}{2}\left(\frac{\partial u_l}{\partial x_m} + \frac{\partial u_m}{\partial x_l}\right). \tag{3.42}$$

Hence,

$$\rho\frac{\partial^2 u_i}{\partial t^2} = \frac{1}{2}c_{iklm}\left(\frac{\partial^2 u_l}{\partial x_k \partial x_m} + \frac{\partial^2 u_m}{\partial x_k \partial x_l}\right). \tag{3.43}$$

In the following, repeated use of the symmetry of the elastic stiffness tensor is made,

$$c_{iklm} = c_{ikml} = c_{kilm}. \tag{3.44}$$

Consider a plane wave,

$$u_i = A_i \exp\left[i\left(k_j x_j - \omega t\right)\right], \tag{3.45}$$

where k_j are the wavenumber components. Equation (3.43) then yields,

$$\rho\omega^2 u_i = \frac{1}{2}c_{iklm}(k_k k_m u_l + k_k k_l u_m) = c_{iklm}k_k k_l u_m. \tag{3.46}$$

Since,

$$u_i = u_m \delta_{im}, \tag{3.47}$$

the latter expression is identical to,

$$(\rho\omega^2 \delta_{im} - c_{iklm}k_k k_l)u_m = 0, \tag{3.48}$$

known as the Christoffel equation for anisotropic media. Defining the Christoffel tensor as,

$$\Gamma_{im} = c_{iklm}n_k n_l, \tag{3.49}$$

and noticing that,

$$k_k = k n_k, \tag{3.50}$$

where **n** is a unit vector along the wavefront normal, equation (3.48) can be written as,

$$(\Gamma_{im}k^2 - \rho\omega^2\delta_{im})u_m = 0, \tag{3.51}$$

or,

$$(\Gamma_{im} - \rho c^2\delta_{im})u_m = 0, \tag{3.52}$$

where,

$$c^2 = \frac{\omega^2}{k^2}. \tag{3.53}$$

For nontrivial solutions to equation (3.52) the determinant of the coefficients must vanish,

$$|\Gamma_{im} - \rho c^2 \delta_{im}| = 0. \tag{3.54}$$

The latter equation is known as the Christoffel determinantal equation and is convenient for numerical solution. Upon solving the Christoffel determinantal equation, three solutions for c_i^2 ($i = 1, 2, 3$) are found giving the three wave speeds for a certain but arbitrary propagation direction $\mathbf{n}$.

Let us show how to obtain Γ_{11}. It follows from equation (3.49) that,

$$\begin{aligned}
\Gamma_{11} = {}& c_{1111}n_1 n_1 + c_{1211}n_2 n_1 + c_{1311}n_3 n_1 \\
& + c_{1121}n_1 n_2 + c_{1221}n_2 n_2 + c_{1321}n_3 n_2 \\
& + c_{1131}n_1 n_3 + c_{1231}n_2 n_3 + c_{1331}n_3 n_3,
\end{aligned} \tag{3.55}$$

which can be written as,

$$\Gamma_{11} = c_{1111}n_1^2 + c_{1221}n_2^2 + c_{1331}n_3^2 + 2c_{1211}n_2 n_1 + 2c_{1311}n_3 n_1 + 2c_{1321}n_3 n_2, \tag{3.56}$$

or, using Voigt notation,

$$\Gamma_{11} = c_{11}n_1^2 + c_{66}n_2^2 + c_{55}n_3^2 + 2c_{16}n_2 n_1 + 2c_{15}n_3 n_1 + 2c_{56}n_3 n_2. \tag{3.57}$$

All the other Christoffel tensor components can be found in a similar way.

3.4.1 Exercise

Show that

$$\begin{aligned}
\Gamma_{12} = {}& c_{16}n_1^2 + c_{26}n_2^2 + c_{45}n_3^2 + (c_{12} + c_{66})n_1 n_2 \\
& + (c_{14} + c_{56})n_1 n_3 + (c_{46} + c_{25})n_2 n_3.
\end{aligned} \tag{3.58}$$

3.4.2 Example

Consider a cubic crystal for which the stiffness tensor is,

$$\mathbf{c} = \begin{bmatrix}
c_{11} & c_{12} & c_{12} & 0 & 0 & 0 \\
c_{12} & c_{11} & c_{12} & 0 & 0 & 0 \\
c_{12} & c_{12} & c_{11} & 0 & 0 & 0 \\
0 & 0 & 0 & c_{44} & 0 & 0 \\
0 & 0 & 0 & 0 & c_{44} & 0 \\
0 & 0 & 0 & 0 & 0 & c_{44}
\end{bmatrix}. \tag{3.59}$$

and determine the wave speeds along the [110] direction, i.e., when,

$$n_1 = n_2 = \frac{1}{\sqrt{2}}, \tag{3.60}$$
$$n_3 = 0.$$

Show that the Christoffel determinantal equation becomes,

$$\begin{vmatrix} \frac{1}{2}(c_{11} + c_{44}) - \rho c^2 & \frac{1}{2}(c_{12} + c_{44}) & 0 \\ \frac{1}{2}(c_{12} + c_{44}) & \frac{1}{2}(c_{11} + c_{44}) - \rho c^2 & 0 \\ 0 & 0 & c_{44} - \rho c^2 \end{vmatrix} = 0, \quad (3.61)$$

and that the three wave speed solutions are,

$$c_1 = \sqrt{\frac{c_{44}}{\rho}},$$

$$c_2 = \sqrt{\frac{c_{11} + c_{12} + 2c_{44}}{2\rho}}, \quad (3.62)$$

$$c_3 = \sqrt{\frac{c_{11} - c_{12}}{2\rho}}.$$

Previously, it was obtained that only two wave speeds, c_L and c_T, are possible. Is there a conflict with the result in equation (3.62) for which three wave speeds are found?

3.5 Surface waves

An important area of wave physics is the understanding of wave propagation near a surface between, for example, a crystal solid and air/vacuum or between different solids. We know that the equation-of-motion for an isotropic solid takes the form,

$$\frac{\partial^2 \mathbf{u}}{\partial t^2} - c^2 \nabla^2 \mathbf{u} = 0, \quad (3.63)$$

where $\mathbf{u}$ can be a longitudinal or transverse displacement associated with c_L or c_T, respectively. Consider the surface of a medium to be plane and equivalent to the x–y plane. The medium is assumed to occupy the half space $z < 0$ and the material is embedded in vacuum. Let us consider a plane monochromatic wave propagating along the x axis. Then,

$$\mathbf{u} = \hat{\mathbf{u}} f(z) \exp\left[i(kx - \omega t)\right], \quad (3.64)$$

where $\hat{\mathbf{u}}$ is a unit vector and substitution in to equation (3.63) yields,

$$\frac{d^2 f}{dz^2} = \left(k^2 - \frac{\omega^2}{c^2}\right) f. \quad (3.65)$$

A surface wave that decays in to the solid is only possible if $k^2 - \frac{\omega^2}{c^2} > 0$. Then the solution is,

$$f(z) = constant \times \exp\left[+\sqrt{k^2 - \frac{\omega^2}{c^2}}\,z\right] \equiv constant \times \exp\left[+\kappa z\right]. \qquad (3.66)$$

To determine the combination of a longitudinal wave $\mathbf{u}_L$ and a transverse wave $\mathbf{u}_T$ that constitutes the total wave solution,

$$\mathbf{u} = \mathbf{u}_L + \mathbf{u}_T, \qquad (3.67)$$

the boundary conditions,

$$T_{ik}n_k = 0, \qquad (3.68)$$

must be imposed, i.e., with $\mathbf{n}$ along the z direction,

$$T_{xz} = T_{yz} = T_{zz} = 0, \qquad (3.69)$$

or by use of the relations between the Lamé coefficients and Young's modulus E and the Poisson ratio σ,

$$\lambda = \frac{E\sigma}{(1 + \sigma)(1 - 2\sigma)},$$
$$\mu = \frac{E}{2(1 + \sigma)}, \qquad (3.70)$$

i.e.,

$$S_{xz} = 0,$$
$$S_{yz} = 0,$$
$$\sigma(S_{xx} + S_{yy}) + (1 - \sigma)S_{zz} = 0. \qquad (3.71)$$

From the second equation,

$$S_{yz} = \frac{1}{2}\left(\frac{\partial u_y}{\partial z} + \frac{\partial u_z}{\partial y}\right) = \frac{1}{2}\frac{\partial u_y}{\partial z} = 0, \qquad (3.72)$$

and, therefore, from equations (3.64)–(3.65),

$$u_y = 0. \qquad (3.73)$$

Since the divergence of the transverse part $\mathbf{u}_T$ vanishes ($\mathbf{u}_T$ is the curl of another vector),

$$\frac{\partial u_{T,x}}{\partial x} + \frac{\partial u_{T,z}}{\partial z} = 0, \qquad (3.74)$$

it follows from equations (3.64) and (3.66) that,

$$iku_{T,x} + \kappa_T u_{T,z} = 0, \qquad (3.75)$$

where,

$$\kappa_T = \sqrt{k^2 - \frac{\omega^2}{c_T^2}} \,. \tag{3.76}$$

This implies,

$$\begin{aligned} u_{T,x} &= \kappa_T \alpha \exp\left[ikx + \kappa_T z - i\omega t\right], \\ u_{T,z} &= -ik\alpha \exp\left[ikx + \kappa_T z - i\omega t\right], \end{aligned} \tag{3.77}$$

where α is some constant.

The longitudinal part, $\mathbf{u}_L$ has a zero curl, or,

$$\frac{\partial u_{L,x}}{\partial z} - \frac{\partial u_{L,z}}{\partial x} = 0, \tag{3.78}$$

whence,

$$iku_{L,z} - \kappa_L u_{l,x} = 0, \tag{3.79}$$

where,

$$\kappa_L = \sqrt{k^2 - \frac{\omega^2}{c_L^2}} \,. \tag{3.80}$$

Thus,

$$\begin{aligned} u_{L,x} &= k\beta \exp\left[ikx + \kappa_L z - i\omega t\right], \\ u_{L,z} &= -i\kappa_L \beta \exp\left[ikx + \kappa_L z - i\omega t\right], \end{aligned} \tag{3.81}$$

where β is a constant.

Now from the first and third of the equations in equation (3.71),

$$\begin{aligned} \frac{\partial u_x}{\partial z} + \frac{\partial u_z}{\partial x} &= 0, \\ c_L^2 \frac{\partial u_z}{\partial z} + \left(c_L^2 - 2c_T^2\right)\frac{\partial u_x}{\partial x} &= 0, \end{aligned} \tag{3.82}$$

where use was made of equations (3.29) and (3.39). Substituting $u_x = u_{L,x} + u_{T,x}$, $u_z = u_{L,z} + u_{T,z}$ in to the first of the latter equations and employing the expressions derived above for $u_{L,x}$, $u_{T,x}$, $u_{L,z}$, $u_{T,z}$ yield,

$$\alpha\left(k^2 + \kappa_T^2\right) + 2\beta k\kappa_L = 0. \tag{3.83}$$

Using the same substitution in the second of the latter equations,

$$2\alpha c_T^2 \kappa_T k + \beta\left[c_L^2\left(\kappa_L^2 - k^2\right) + 2c_T^2 k^2\right] = 0. \tag{3.84}$$

Dividing this expression by c_T^2 and substituting,

$$\kappa_L^2 - k^2 = -\frac{\omega^2}{c_L^2} = -\left(k^2 - \kappa_T^2\right)\frac{c_T^2}{c_L^2}, \tag{3.85}$$

it follows that,

$$2\alpha\kappa_T k + \beta\left(k^2 + \kappa_T^2\right) = 0. \tag{3.86}$$

For equations (3.83) and (3.86) to be compatible, the following must hold,

$$\left(k^2 + \kappa_T^2\right)^2 = 4k^2\kappa_T\kappa_L, \tag{3.87}$$

or squaring plus substituting the values of κ_T^2 and κ_L^2,

$$\left(2k^2 - \frac{\omega^2}{c_T^2}\right)^4 = 16k^4\left(k^2 - \frac{\omega^2}{c_T^2}\right)\left(k^2 - \frac{\omega^2}{c_L^2}\right). \tag{3.88}$$

This expression gives us the relation between ω and k. It is convenient to write,

$$\omega = c_T k\xi, \tag{3.89}$$

and rewrite equation (3.88) as,

$$\xi^6 - 8\xi^4 + 8\xi^2\left(3 - 2\frac{c_T^2}{c_L^2}\right) - 16\left(1 - \frac{c_T^2}{c_L^2}\right) = 0. \tag{3.90}$$

Evidently, ξ depends only on the ratio $\frac{c_T}{c_L}$ which is a constant characteristic of the medium and determined uniquely by the Poisson ratio as,

$$\frac{c_T}{c_L} = \sqrt{\frac{1 - 2\sigma}{2(1 - \sigma)}}. \tag{3.91}$$

Reference

[1] Morse P M and Feshbach H 1953 *Methods of Theoretical Physics* **vol 1** 1st edn (Minneapolis, MN: Feshbach Publishing)

Part II

Dynamic deformation

Chapter 4

Simple oscillator systems

To address more complicated systems, it is instructive and necessary to first derive the theory of mechanical vibrations of simple linear oscillator systems. Depending on the values of the three oscillator parameters (stiffness, mass, and the damping coefficient), three types of vibrations (overdamped, critical damped, and under-damped) can exist. For any frequency of vibration, the general solution to any external force input can be found. General time-dependent problems are then solved using Fourier theory. The analysis is presented in chapter 4 and the analytical calculations are supplemented by Matlab program codes.

We start out by writing down the equation-of-motion for a mass m connected to a damper R_m and a spring s (figure 4.1),

$$m\frac{d^2x}{dt^2} + R_m\frac{dx}{dt} + sx = 0, \tag{4.1}$$

where x is the displacement of the mass m from equilibrium and t is time.

A simple exponential ansatz is attempted,

$$x(t) = e^{\gamma t}, \tag{4.2}$$

which, upon insertion to equation (4.1), yields

$$\gamma = \frac{-R_m \pm \sqrt{R_m^2 - 4ms}}{2m}. \tag{4.3}$$

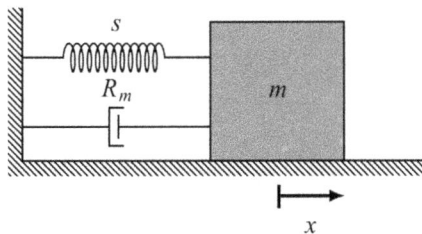

Figure 4.1. Schematic drawing of a damped oscillator system.

4-1

It is now convenient to address three cases separately,

- Case 1: $R_m^2 < 4ms$ (under damping)
- Case 2: $R_m^2 > 4ms$ (over damping)
- Case 3: $R_m^2 = 4ms$ (critical damping)

4.1 Case 1: underdamping

If $R_m^2 < 4ms$, underdamping results. By use of the parameters,

$$\gamma = -\beta \pm i\omega_d, \tag{4.4}$$

$$\beta = \frac{R_m}{2m}, \tag{4.5}$$

$$\omega_0 = \sqrt{\frac{s}{m}}, \tag{4.6}$$

$$\omega_d = \sqrt{\omega_0^2 - \beta^2} = \sqrt{\frac{s}{m} - \left(\frac{R_m}{2m}\right)^2} > 0, \tag{4.7}$$

the general complex solution to equation (4.1) can be written as,

$$x(t) = Ae^{\gamma_1 t} + Be^{\gamma_2 t} = e^{-\beta t}(A_1 e^{i\omega_d t} + A_2 e^{-i\omega_d t}). \tag{4.8}$$

The real part of the complex solution can be written as,

$$x(t) = e^{-\beta t}(A\cos(\omega_d t) + B\sin(\omega_d t)) = Ce^{-\beta t}\cos(\omega_d t + \phi), \tag{4.9}$$

where A, B or C, ϕ are constants determined by the initial conditions.

4.1.1 Example

Consider the following parameters: $R_m = 2\,\mathrm{N\,s\,m^{-1}}$, $s = 2\,\mathrm{N\,m^{-1}}$, and $m = 1\,\mathrm{kg}$ and initial conditions,

$$x(0) = 1 \text{ m}, \tag{4.10}$$

$$\frac{dx}{dt}(0) = 0. \tag{4.11}$$

We find,

$$R_m^2 - 4ms = 2^2 - 4 \cdot 2 \cdot 1 = -4,$$

$$\beta = \frac{R_m}{2m} = \frac{2}{2 \cdot 1} = 1,$$

$$\omega_0 = \sqrt{\frac{s}{m}} = \sqrt{\frac{2}{1}} = \sqrt{2}, \tag{4.12}$$

$$\omega_d = \sqrt{\omega_0^2 - \beta^2} = \sqrt{\sqrt{2}^2 - 1^2} = 1.$$

The initial conditions give,

$$x(0) = A = 1,$$

$$\frac{dx}{dt}(0) = -\beta A + \omega_d B = 0 \Longrightarrow$$

$$B = \frac{\beta}{\omega_d} A = \frac{1}{1} \cdot 1 = 1,$$

(4.13)

so,

$$x(t) = e^{-t}(\cos(t) + \sin(t)).$$

(4.14)

4.1.2 Exercise

Consider the following parameters: $R_m = 2 \text{ N s m}^{-1}$, $s = 2 \text{ N m}^{-1}$, and $m = 1$ kg and initial conditions,

$$x(0) = 0,$$

(4.15)

$$\frac{dx}{dt}(0) = 1 \text{ m s}^{-1}.$$

(4.16)

Find the solution $x(t)$ and make a plot $(t, x(t))$.

4.2 Case 2: overdamping

In the case of heavily damped systems, $R_m^2 > 4ms$, the γ solutions are,

$$\gamma_1 = \frac{-R_m + \sqrt{R_m^2 - 4ms}}{2m} < 0,$$

(4.17)

$$\gamma_2 = \frac{-R_m - \sqrt{R_m^2 - 4ms}}{2m} < 0,$$

(4.18)

and the general real solution becomes,

$$x(t) = Ae^{\gamma_1 t} + Be^{\gamma_2 t} = Ae^{\frac{-R_m + \sqrt{R_m^2 - 4ms}}{2m} t} + Be^{\frac{-R_m - \sqrt{R_m^2 - 4ms}}{2m} t}.$$

(4.19)

Observe that effective damping of disturbances is governed by the slowest time constant γ_1.

4.2.1 Example

Consider the following parameters: $R_m = 6 \text{ N s m}^{-1}$, $s = 2.75 \text{ N m}^{-1}$, and $m = 1$ kg and initial conditions,

$$x(0) = 1 \text{ m},$$

(4.20)

$$\frac{dx}{dt}(0) = 0. \tag{4.21}$$

We find,

$$R_m^2 - 4ms = 6^2 - 4 \cdot 2.75 \cdot 1 = 25,$$

$$\gamma_1 = \frac{-6 + \sqrt{6^2 - 4 \cdot 2.75}}{2 \cdot 1} = \frac{-6 + 5}{2} = -\frac{1}{2},$$

$$\gamma_2 = \frac{-6 - \sqrt{6^2 - 4 \cdot 2.75}}{2 \cdot 1} = \frac{-6 - 5}{2} = -\frac{11}{2}. \tag{4.22}$$

The initial conditions give,

$$x(0) = A + B = 1 \Longrightarrow B = 1 - A \Longrightarrow$$

$$\frac{dx}{dt}(0) = A\gamma_1 + B\gamma_2 = A(\gamma_1 - \gamma_2) + \gamma_2 = 0 \Longrightarrow A = \frac{\gamma_2}{\gamma_2 - \gamma_1} \Longrightarrow$$

$$B = 1 - A = -\frac{\gamma_1}{\gamma_2 - \gamma_1} \Longrightarrow \tag{4.23}$$

$$x(t) = \frac{11}{10} e^{-\frac{1}{2}t} - \frac{1}{10} e^{-\frac{11}{2}t},$$

where t is in seconds and x is in meters.

4.2.2 Exercise

Consider the following parameters: $R_m = 6 \text{ N s m}^{-1}$, $s = 2.75 \text{ N m}^{-1}$, and $m = 1 \text{ kg}$ and initial conditions,

$$x(0) = 0, \tag{4.24}$$

$$\frac{dx}{dt}(0) = 1 \text{ m s}^{-1}. \tag{4.25}$$

Find the solution $x(t)$ and make a plot $(t, x(t))$.

4.3 Case 3: critical damping

If $R_m^2 = 4ms$, a special situation occurs known as critical damping, for which the γ solutions are,

$$\gamma_1 = \gamma_2 = -\frac{R_m}{2m}, \tag{4.26}$$

and the general real solution becomes,

$$x(t) = Ae^{-\frac{Rm}{2m}t} + Bte^{-\frac{Rm}{2m}t}. \tag{4.27}$$

4.3.1 Example

Consider the following parameters: $R_m = 2\,\text{N}\,\text{s}\,\text{m}^{-1}$, $s = 1\,\text{N}\,\text{m}^{-1}$, and $m = 1\,\text{kg}$ and initial conditions,

$$x(0) = 1\ \text{m}, \tag{4.28}$$

$$\frac{dx}{dt}(0) = 0. \tag{4.29}$$

We find,

$$R_m^2 - 4ms = 2^2 - 4 \cdot 1 \cdot 1 = 0,$$

$$\gamma_1 = \frac{-2 + \sqrt{2^2 - 4 \cdot 1}}{2 \cdot 1} = \frac{-2 + 0}{2} = -1,$$

$$\gamma_2 = \frac{-2 - \sqrt{2^2 - 4 \cdot 1}}{2 \cdot 1} = \frac{-2 - 0}{2} = -1. \tag{4.30}$$

The initial conditions give,

$$x(0) = A = 1 \Longrightarrow$$

$$\frac{dx}{dt}(0) = A\gamma_1 + B = B - A = 0 \Longrightarrow B = A = 1, \tag{4.31}$$

and the solution is,

$$x(t) = e^{-t} + te^{-t}, \tag{4.32}$$

where t is in seconds and x is in meters.

4.3.2 Exercise

Consider the following parameters: $R_m = 2\,\text{N}\,\text{s}\,\text{m}^{-1}$, $s = 1\,\text{N}\,\text{m}^{-1}$, and $m = 1\,\text{kg}$ and initial conditions,

$$x(0) = 0, \tag{4.33}$$

$$\frac{dx}{dt}(0) = 1\ \text{m}\ \text{s}^{-1}. \tag{4.34}$$

Find the solution $x(t)$ and make a plot $(t, x(t))$.

4.3.3 Program for computing the displacement as a function of time for an underdamped, overdamped, or critically damped oscillator

In figure 4.2, a program to compute vibrations of a general damped oscillator system in Matlab is shown.

```
close all
clear all

% underdamped system

R_m = 2e0;                          % damping constant
ss = 2e0;                           % spring constant
mm = 1e0;                           % mass

% initial conditions
% x(0) = 1; dx/dt(0) = 0

beta = R_m/(2e0*mm);                % beta constant
omega_0 = sqrt(ss/mm);              % natural frequency omega_0
omega_d = sqrt(omega_0^2-beta^2);   % damped frequency omega_d

% solution x(t)
numb_ii = 100;                      % number of time values

t_vec = zeros(numb_ii,1);           % initialization of time vector
x_vec = zeros(numb_ii,1);           % initialization of solution vector
t_start = 0e0;                      % plot starting time
t_final = 10e0;                     % plot final time
for ii=1:numb_ii
  tt = t_start + (ii-1)/(numb_ii-1)*t_final;
  t_vec(ii) = tt;
  xx = exp(-beta*tt)*( 1e0*cos(omega_d*tt) + 1e0*sin(omega_d*tt) );
  x_vec(ii) = xx;
end

plot(t_vec,x_vec)
grid
xlabel('time t [s]')
ylabel('x(t) [m]')

print('underdamped_plot','-dpdf','-fillpage')

clear all

% overdamped system

R_m = 6e0;                          % damping constant
ss = 2.75e0;                        % spring constant
mm = 1e0;                           % mass

% initial conditions
% x(0) = 1; dx/dt(0) = 0

gamma_1 = ( -R_m + sqrt(R_m^2 - 4e0*mm*ss) )/(2e0*mm);
gamma_2 = ( -R_m - sqrt(R_m^2 - 4e0*mm*ss) )/(2e0*mm);

% solution x(t)
numb_ii = 100;                          % number of time values
```

Figure 4.2. Program for computing the displacement as a function of time for an underdamped, overdamped, or critical damped oscillator.

4.4 Forced oscillations

Consider next a forced oscillator system (figure 4.3) for which the equation-of-motion is,

$$m\frac{d^2x}{dt^2} + R_m\frac{dx}{dt} + sx = f(t), \tag{4.35}$$

where f is an external force.

Assume the external force is a monofrequency force,

$$f(t) = Fe^{i\omega t}, \tag{4.36}$$

where ω is the angular frequency of the force and F is its amplitude. Searching for solutions of the same frequency,

$$x(t) = Ae^{i\omega t}, \tag{4.37}$$

we find,

$$(-Am\omega^2 + iAR_m\omega + As)e^{i\omega t} = Fe^{i\omega t}, \tag{4.38}$$

and the complex solution becomes,

$$x(t) = \frac{1}{i\omega}\frac{Fe^{i\omega t}}{R_m + i(m\omega - s/\omega)}. \tag{4.39}$$

The complex oscillator speed is,

$$u(t) = \frac{dx}{dt} = i\omega x = \frac{Fe^{i\omega t}}{R_m + i(m\omega - s/\omega)}. \tag{4.40}$$

It is convenient to introduce the mechanical impedance defined by,

$$Z = \frac{f}{u} = |Z|e^{i\theta}. \tag{4.41}$$

Equation (4.40) now yields,

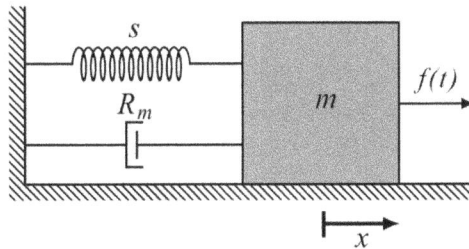

Figure 4.3. Schematic drawing of a forced oscillator system.

$$|Z| = \sqrt{R_m^2 + (m\omega - s/\omega)^2}\,,$$

$$\theta = \arctan\left(\frac{m\omega - s/\omega}{R_m}\right),$$

(4.42)

and simple manipulations lead to expressions for the real displacement,

$$x(t) = \frac{F}{\omega|Z|}\sin(\omega t - \theta),$$

(4.43)

and the real speed,

$$u(t) = \frac{F}{|Z|}\cos(\omega t - \theta).$$

(4.44)

Equation (4.43) is a particular solution to the inhomogeneous differential equation, equation (4.35).

The complete general solution is a particular solution added to the general homogeneous solution.

4.4.1 Complete general solution: underdamping

The complete general solution for an underdamped oscillator is therefore,

$$x(t) = e^{-\beta t}(A\cos(\omega_d t) + B\sin(\omega_d t)) + \frac{F}{\omega|Z|}\sin(\omega t - \theta).$$

(4.45)

4.4.2 Complete general solution: overdamping

The complete general solution for an overdamped oscillator is,

$$x(t) = Ae^{-\frac{R_m + \sqrt{R_m^2 - 4ms}}{2m}t} + Be^{-\frac{R_m - \sqrt{R_m^2 - 4ms}}{2m}t} + \frac{F}{\omega|Z|}\sin(\omega t - \theta).$$

(4.46)

4.4.3 Complete general solution: Critical damping

The complete general solution for a critically damped oscillator is,

$$x(t) = Ae^{-\frac{Rm}{2m}t} + Bte^{-\frac{Rm}{2m}t} + \frac{F}{\omega|Z|}\sin(\omega t - \theta).$$

(4.47)

In all cases, underdamped, overdamped, or critically damped systems, two constants A and B are specified by the initial conditions. All other quantities (ω_0, ω_d, β, $|Z|$, θ) are uniquely determined by the system parameters m, R_m, s. The angular frequency ω is determined by the angular frequency of the external force $f = Fe^{i\omega t}$.

Chapter 5

Transverse vibrations of strings

The simplest geometric complexity in vibration theory above the simple oscillator systems presented in chapter 4 is the theory of strings. In chapter 5, the wave equation for strings in addition to various relevant boundary conditions, eigenfrequencies, and eigenstates are presented. Examples of analytical solutions are given for different boundary conditions and external force inputs. The material and methods presented in chapter 5 are essential for the later analysis of beams and membrane vibrations.

5.1 Wave equation of strings

Consider a string defined as a thin rod with an infinitesimal cross-section. In this case, vibrations are functions of the length coordinate x and time (refer to figure 5.1) and for small displacements vibrations can be considered transverse (along the y direction) [1]. We assume the tension of the string is uniform and equal to T and the mass density per length is ρ_L. For the y-component net force on the string segment between x and $x + dx$, the expression,

$$df_y = (T \sin \theta)_{x+dx} - (T \sin \theta)_x, \tag{5.1}$$

applies, where θ is the angle between the tangent of the string and the x axis. Since small vibrations are considered, θ is small, i.e.,

$$\sin \theta \approx \tan \theta = \frac{\partial y}{\partial x}, \tag{5.2}$$

so,

$$df_y = (T \sin \theta)_{x+dx} - (T \sin \theta)_x = \frac{\left(T\frac{\partial y}{\partial x}\right)_{x+dx} - \left(T\frac{\partial y}{\partial x}\right)_x}{dx}dx = T\frac{\partial^2 y}{\partial x^2}dx, \tag{5.3}$$

where the assumption that the tension T is uniform was used in obtaining the last equality. Applying next Newton's Second Law,

doi:10.1088/978-0-7503-5557-5ch5

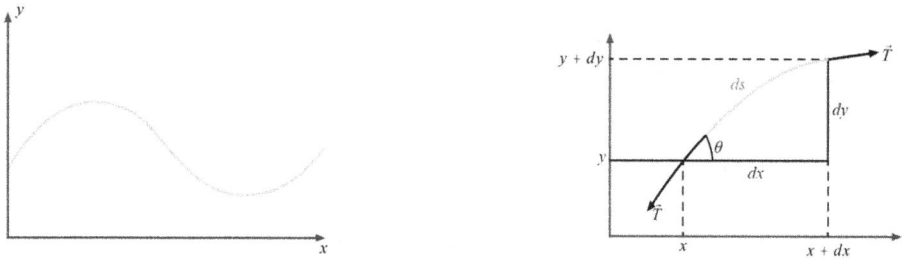

Figure 5.1. Transverse string vibrations. Left (a) a vibrating string and (b) forces on an infinitesimal string segment.

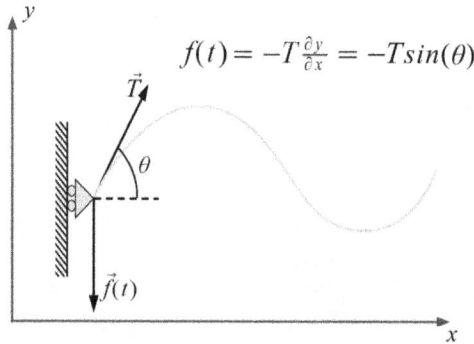

$$f(t) = -T\frac{\partial y}{\partial x} = -T sin(\theta)$$

Figure 5.2. Forced string of semi-infinite extension.

$$df_y = \rho_L dx \frac{\partial^2 y}{\partial t^2}, \tag{5.4}$$

yields,

$$\frac{\partial^2 y}{\partial x^2} = \frac{1}{c^2}\frac{\partial^2 y}{\partial t^2}, \tag{5.5}$$

where the constant c^2 is defined by,

$$c^2 = \frac{T}{\rho_L}. \tag{5.6}$$

The general solution to equation (5.5) is,

$$y(x, t) = y_1(ct - x) + y_2(ct + x), \tag{5.7}$$

where y_1, y_2 are unspecified functions. Evidently, y_1 and y_2 correspond to forward- and backward-traveling wave components, respectively, and c is the wave speed.

5.2 Forced vibration of a semi-infinite string

Consider next a semi-infinite string extending to the right from $x = 0$ (figure 5.2). The string is assumed stretched to a tension T and a force $f(t) = F \cos \omega t$ drives the string to move along the y direction. In this case, forward-traveling waves are generated only, so,

$$y(x, t) = y_1(ct - x). \tag{5.8}$$

The string vibration at $x = 0$ can be written as,

$$y(x, 0) = A \exp(i\omega t) = Ae^{ik(ct)}, \tag{5.9}$$

where the wavenumber k is ω/c. The solution at any time t and position x is then,

$$y(x, t) = Ae^{i(\omega t - kx)}, \tag{5.10}$$

and the wavelength λ is $2\pi/k$. Hence, the well-known wave condition,

$$c = \lambda f, \tag{5.11}$$

is obtained where $f = \omega/(2\pi)$ is the frequency of the force.

At the left end ($x = 0$) the total force is,

$$f + T \sin \theta = \rho_L dx \frac{\partial^2 y}{\partial t^2}, \tag{5.12}$$

so, in the limit $dx \to 0$ we get the boundary condition,

$$f = -T \left(\frac{\partial y}{\partial x} \right)_{x=0}. \tag{5.13}$$

Thus,

$$Fe^{i\omega t} = -T(-ik)Ae^{i\omega t}, \tag{5.14}$$

and,

$$y(x, t) = \frac{F}{ikT} e^{i(\omega t - kx)},$$
$$u(x, t) = \frac{\partial y}{\partial t} = \frac{Fc}{T} e^{i(\omega t - kx)} = \frac{F}{\rho_L c} e^{i(\omega t - kx)}, \tag{5.15}$$

where $u(x, t)$ is the transverse speed of the wave.

The input mechanical impedance Z_{m0} is defined as the ratio between the driving force and the transverse wave speed,

$$Z_{m0} = \frac{f}{u(0, t)} = \rho_L c. \tag{5.16}$$

For a semi-infinite string with a free end (no driving force) it follows from the above considerations that the string slope satisfies $\frac{\partial y}{\partial x} = 0$ at the free end.

5.2.1 Exercise

Show that if the string is semi-infinite extending to the left from $x = 0$, then the mechanical impedance satisfies at the right end,

$$Z_{m0} = \frac{f}{u(0,\ t)} = -\rho_L c. \tag{5.17}$$

5.3 Normal modes of the fixed-fixed string

Consider a string of length L fixed at both ends. From the equation-of-motion, equation (5.5), the general solution is,

$$y(x,\ t) = A e^{i(\omega t - kx)} + B e^{i(\omega t + kx)}, \tag{5.18}$$

where $k = \omega/c$ (and ω is unspecified at this point). Application of the boundary conditions that the string is fixed at both ends yields,

$$A + B = 0, \tag{5.19}$$

$$A e^{-ikL} + B e^{ikL} = 0. \tag{5.20}$$

The solution is,

$$B = -A, \tag{5.21}$$

$$2iA \sin(kL) = 0. \tag{5.22}$$

There are two solutions. Either $A = 0$ or $kL = n\pi$ where n is an integer. Thus, a true vibration ($A \neq 0$) forces n to be a nonzero integer (convince yourself it is enough to consider positive n values). Then, a discrete set of solutions for the wavenumber and frequency exists. We assign the different solutions by the subscript n, i.e.,

$$k_n = \frac{n\pi}{L}, \tag{5.23}$$

$$f_n = \omega_n/(2\pi) = \frac{nc}{2L}. \tag{5.24}$$

Thus, the solutions are of the form,

$$y(x,\ t) = \sum_{n=1}^{\infty} y_n(x,\ t), \tag{5.25}$$

$$y_n(x,\ t) = A_n \sin(k_n x) e^{i\omega_n t}, \tag{5.26}$$

where A_n are coefficients to be determined by initial conditions. A real representation of the solution is found by setting,

$$A_n = \alpha_n - i\beta_n, \tag{5.27}$$

where α_n, β_n are real numbers. Thus, the real part of equation (5.25) is

$$y(x,\ t) = \sum_{n=1}^{\infty}[\alpha_n \cos{(\omega_n t)} + \beta_n \sin{(\omega_n t)}] \sin{(k_n x)}. \tag{5.28}$$

5.3.1 Exercise

Assume that at time $t = 0$, the position-dependent string displacement $y(x, 0)$ and speed $u(x, 0) = \frac{\partial y}{\partial t}|_{t=0}$ are known. Derive,

$$y(x,\ 0) = \sum_{n=1}^{\infty}\alpha_n \sin{(k_n x)}, \tag{5.29}$$

$$u(x,\ 0) = \sum_{n=1}^{\infty}\omega_n\beta_n \sin{(k_n x)}, \tag{5.30}$$

5.3.2 Exercise

Derive,

$$\alpha_n = \frac{2}{L}\int_0^L y(x,\ 0)\sin{(k_n x)}dx, \tag{5.31}$$

$$\beta_n = \frac{2}{\omega_n L}\int_0^L u(x,\ 0)\sin{(k_n x)}dx. \tag{5.32}$$

5.3.3 Exercise

Assume that initially a string is pulled aside a distance h at $x = 2L/3$ and then released. Find the solution $y(x,\ t)$.

5.3.4 Exercise

Assume that a string at rest is struck a blow such that the initial speed distribution is $u(x,\ 0) = U\delta(x - 2L/3)$. Find the solution $y(x,\ t)$.

Reference

[1] Kinsler L E, Frey A R, Coppens A B and Sanders J V 2000 *Fundamentals of Acoustics* (New York: Wiley) 4th edn

IOP Publishing

Piezoelectricity in Classical and Modern Systems

Morten Willatzen

Chapter 6

Vibrations of bars

Bars are different compared to strings in that bending and twisting phenomena can take place due to their finite cross-sectional dimensions. The theory of string vibrations in chapter 5 will be adapted to address longitudinal bar vibrations in chapter 6. Transverse vibrations are handled using the Euler-Lagrange formalism. The governing differential equation of transverse bar vibrations is a fourth-order equation, which is solved for the possible eigenfrequencies subject to different boundary conditions. This chapter concludes by deriving the wave equation for torsional waves.

In the following, three types of bar vibrations are discussed: longitudinal, transverse, and torsional vibrations. Longitudinal vibrations of bars assume deformations of a bar along its main axis. Consider an element of a bar of length dx as shown in figure 6.1. The displacement $u(x, t)$ is a function of position x and time t. In the presence of strain, the left side of the element in the unstrained bar, x, is moved to $x + u(x, t)$. Similarly, the right side of the element at position $x + dx$ is moved to $x + dx + u(x + dx, t)$. At any time t the displacement $u(x + dx)$ satisfies,

$$u(x + dx, t) = u(x, t) + \frac{\partial u}{\partial x}\bigg|_t dx, \qquad (6.1)$$

and the increase in length of the element is,

$$u(x + dx, t) - u(x, t) = \frac{\partial u}{\partial x}\bigg|_t dx. \qquad (6.2)$$

Strain is defined as the change in element length compared to the original element length,

$$S = \frac{u(x + dx, t) - u(x, t)}{dx} = \frac{\partial u}{\partial x}\bigg|_t. \qquad (6.3)$$

If a bar is strained, longitudinal elastic forces are induced that seek to counteract the deformation, see figure 6.1. On both sides of an element, x and $x + dx$, internal forces F_x and F_{x+dx} act. The direction of force is chosen to be positive (negative) if it is compressive (tensile). The bar is then subject to a stress,

doi:10.1088/978-0-7503-5557-5ch6

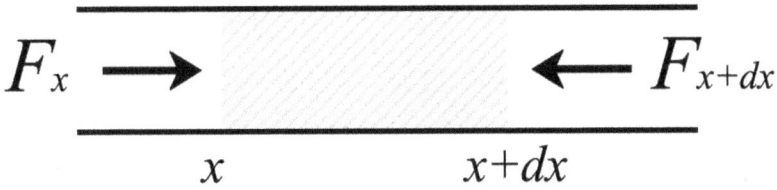

Figure 6.1. Deformations of and forces on an element of a bar subject to longitudinal vibrations.

$$\sigma = \frac{F}{A}, \tag{6.4}$$

where A is the cross-sectional area of the bar. For most materials subject to small deformations, the stress is proportional to the strain, a rule known as Hooke's law,

$$\sigma = -Y\frac{\partial u}{\partial x}, \tag{6.5}$$

where Y is Young's modulus of elasticity which is a material property. Thus,

$$F = -AY\frac{\partial u}{\partial x}. \tag{6.6}$$

With reference to figure 6.1b, the net element force to the right is,

$$dF = F_x - F_{x+dx} = F_x - \left(F_x + \left.\frac{\partial F}{\partial x}\right|_t dx\right) = -\left.\frac{\partial F}{\partial x}\right|_t dx, \tag{6.7}$$

so,

$$dF = AY\frac{\partial^2 u}{\partial x^2}dx. \tag{6.8}$$

It now follows from Newton's second law that,

$$dF = dm\frac{\partial^2 u}{\partial t^2},$$

$$AY\frac{\partial^2 u}{\partial x^2}dx = \rho A dx\frac{\partial^2 u}{\partial t^2}, \tag{6.9}$$

where ρ is the mass density of the bar. From the last equality, the wave equation for longitudinal vibrations of a bar is obtained,

$$\frac{\partial^2 u}{\partial x^2} - \frac{\rho}{Y}\frac{\partial^2 u}{\partial t^2} = 0, \tag{6.10}$$

or,

$$\frac{\partial^2 u}{\partial x^2} - \frac{1}{c^2}\frac{\partial^2 u}{\partial t^2} = 0, \tag{6.11}$$

where the wave speed c is introduced,

$$c = \sqrt{\frac{Y}{\rho}}. \tag{6.12}$$

The wave equation for longitudinal vibrations is similar to the wave equation of transverse string vibrations and therefore shares the same mathematical solution. The general solution to equation (6.11) is,

$$u(x, t) = u_R(ct - x) + u_L(ct + x), \tag{6.13}$$

where u_R is any function of $ct - x$ which represents waves propagating to the right. Similarly, u_L is any function of $ct + x$ representing waves propagating to the left. Complex harmonic solutions to equation (6.11) are,

$$u(x, t) = Ae^{i(\omega t - kx)} + Be^{i(\omega t + kx)}, \tag{6.14}$$

where A, B are complex constants, ω is the angular frequency, and $k = \omega/c$ is the wavevector.

6.1 Simple boundary conditions

Let us next give the solutions to some of the simple boundary conditions.

6.1.1 Fixed boundaries at both ends

If both ends of the bar are subject to fixed boundaries, i.e., $u(0, t) = u(L, t) = 0$, the coefficients of the general harmonic solution must obey,

$$A + B = 0,$$
$$Ae^{-ikL} + Be^{ikL} = 0, \tag{6.15}$$

where L is the length of the bar. The general complex solution is then,

$$u(x, t) = \sum_{n=1}^{\infty} - 2iA_n e^{i\omega_n t} \sin k_n x, \tag{6.16}$$

where,

$$\omega_n = \frac{n\pi c}{L},$$
$$k_n = \frac{\omega_n}{c}, \tag{6.17}$$

and A_n is the complex amplitude of the nth mode of vibration. Introducing the real amplitude constants, α_n and β_n,

$$2A_n = \beta_n + i\alpha_n, \tag{6.18}$$

the complete real solution becomes,

$$u(x, t) = \sum_{n=1}^{\infty} u_n(x, t) = \sum_{n=1}^{\infty} (\alpha_n \cos \omega_n t + \beta_n \sin \omega_n t) \sin k_n x. \tag{6.19}$$

The values of α_n and β_n can finally be determined if initial conditions $u(x, t = 0)$ and $\frac{\partial u}{\partial x}(x, t = 0)$ are specified. This is done using standard Fourier series analysis.

6.1.2 Free boundaries at both ends

A free end is subject to zero force, so $F = -AY\frac{\partial u}{\partial x} = 0$ and $\frac{\partial u}{\partial x} = 0$. Consider a bar with both ends free, i.e., $\frac{\partial u}{\partial x}(0, t) = \frac{\partial u}{\partial x}(L, t) = 0$, then,

$$- ikA + ikB = 0,$$
$$- ikAe^{-ikL} + ikBe^{ikL} = 0. \tag{6.20}$$

The general complex solution is then,

$$u(x, t) = \sum_{n=1}^{\infty} 2A_n e^{i\omega_n t} \cos k_n x, \tag{6.21}$$

where,

$$\omega_n = \frac{n\pi c}{L},$$
$$k_n = \frac{\omega_n}{c}, \tag{6.22}$$

and the complete real solution is,

$$u(x, t) = \sum_{n=1}^{\infty} u_n(x, t) = \sum_{n=1}^{\infty} (\alpha_n \cos \omega_n t + \beta_n \sin \omega_n t) \cos k_n x, \tag{6.23}$$

where α_n and β_n are real numbers.

6.1.3 Free boundary at one end and fixed boundary at the other end

If one end is free, say $x = 0$, and the other end fixed then $\frac{\partial u}{\partial x}(0, t) = u(L, t) = 0$. In this case, the general real solution becomes,

$$u(x, t) = \sum_{n=1}^{\infty} u_n(x, t) = \sum_{n=1}^{\infty} (\alpha_n \cos \omega_n t + \beta_n \sin \omega_n t) \cos k_n x, \tag{6.24}$$

where,

$$\omega_n = \frac{(2n - 1)\pi c}{2L},$$
$$k_n = \frac{\omega_n}{c}. \tag{6.25}$$

Note that the frequency of the fundamental vibration, $f_1 = \frac{\omega_1}{2\pi} = \frac{\pi c}{4L}$ is half that of the bar with rigid boundaries at both ends or free boundaries at both ends.

6.1.4 Free boundary at one end and mass-loaded boundary at the other end

Often, a bar is subject to ends that are neither rigid nor free. A typical case is when one end is connected to a mass, say at $x = L$. If the bar is free at $x = 0$ then,

$$u(x, t) = \sum_{n=1}^{\infty} u_n(x, t) = \sum_{n=1}^{\infty} (\alpha_n \cos \omega_n t + \beta_n \sin \omega_n t) \cos k_n x, \qquad (6.26)$$

but the possible angular frequencies and wavevectors ω_n, k_n are determined from Newton's second law applied to the mass-loaded end,

$$-AY \frac{\partial u}{\partial x} \bigg|_{x=L} = m \frac{\partial^2 u}{\partial t^2} \bigg|_{x=L}, \qquad (6.27)$$

where m is the mass connected to the bar at $x = L$. Use of equation (6.31) in equation (6.32) yields,

$$k_n AY \sin k_n L = -m \omega_n^2 \cos k_n L, \qquad (6.28)$$

or,

$$\tan k_n L = -\frac{m \omega_n c}{AY}. \qquad (6.29)$$

Since $Y = \rho c^2$ and using that the mass of the bar is $m_b = \rho AL$ the latter expression can be rewritten as

$$\tan k_n L = -\frac{m}{m_b} k_n L. \qquad (6.30)$$

This is a transcendental equation that can be easily solved numerically to obtain the discrete set of possible k_n values.

6.1.5 Fixed boundary at one end and mass-loaded boundary at the other end

If one end is rigid, say at $x = 0$, and the other end is connected to a mass, the general solution is,

$$u(x, t) = \sum_{n=1}^{\infty} u_n(x, t) = \sum_{n=1}^{\infty} (\alpha_n \cos \omega_n t + \beta_n \sin \omega_n t) \sin k_n x. \qquad (6.31)$$

Again, the allowed angular frequencies and wavevectors ω_n, k_n are determined from,

$$-AY \frac{\partial u}{\partial x} \bigg|_{x=L} = m \frac{\partial^2 u}{\partial t^2} \bigg|_{x=L}, \qquad (6.32)$$

which yields,

$$\cot k_n L = \frac{m}{m_b} k_n L. \qquad (6.33)$$

The generalization to cases where arbitrary impedances are connected to the ends of the bar can be handled using a similar procedure.

6.2 Transverse vibrations of bars

An important type of vibration of a bar is transverse vibrations, see figure 6.2. For a bar of length L and uniform cross-sectional area A performing vibrations in the x–z plane, consider an element of the bar of length dx subject to bending characterized by its radius of curvature R. It follows from figure 6.2 that the elongation δx of a strip, located a distance z from the undeformed neutral axis, is,

$$\delta x = \frac{\partial u}{\partial x} dx, \tag{6.34}$$

where $\frac{\partial u}{\partial x}$ is the longitudinal strain of the strip. The longitudinal force acting on the strip, dF is given by,

$$d\mathbf{F} = YdA\frac{\partial u}{\partial x}\mathbf{e}_x = YdA\frac{\delta x}{dx}\mathbf{e}_x, \tag{6.35}$$

where dA is the cross-sectional area of the strip.

A geometrical argument shows that,

$$\frac{z}{R} = \frac{\delta x}{dx}, \tag{6.36}$$

thus,

$$d\mathbf{F} = \frac{Y}{R}zdA\mathbf{e}_x. \tag{6.37}$$

The differential bending moment associated with $d\mathbf{F}$ is,

$$d\mathbf{M} = -z\mathbf{e}_z \times d\mathbf{F} = -\mathbf{e}_y Y\frac{z^2}{R}dA. \tag{6.38}$$

The above derivation of $d\mathbf{M}$ is valid for the strips in the lower half of the bar. The expression for strips in the upper half of the bar is similar except that both $d\mathbf{F}$ and the arm change sign whereby the result for $d\mathbf{M}$ is the same. The net bending moment M of the element given by,

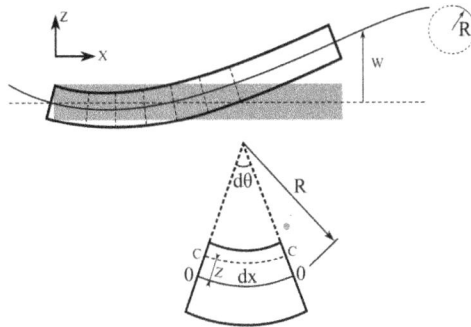

Figure 6.2. Deformations and forces on an element of a bar subject to transverse vibrations.

$$\mathbf{M} = \int d\mathbf{M} = -\mathbf{e}_y \frac{Y}{R} \int z^2 dA = -\mathbf{e}_y \frac{Y}{R} I, \tag{6.39}$$

where I is the second moment of area. Note also that since the bar is subject to transverse vibrations only, the total longitudinal force $\int dF = 0$. This result is a consequence of negative forces above the neutral line being canceled by positive forces below it. By introducing the radius of gyration, κ,

$$\kappa^2 = \frac{1}{A} \int z^2 dA, \tag{6.40}$$

the bending moment becomes,

$$\mathbf{M} = -\mathbf{e}_y A \kappa^2 \frac{Y}{R}. \tag{6.41}$$

Since the radius of curvature, for small transverse vibrations, is,

$$R \approx \frac{1}{\dfrac{\partial^2 w}{\partial x^2}}, \tag{6.42}$$

where w is the deformation of the bar, the bending moment can be written as,

$$\mathbf{M} = -\mathbf{e}_y Y A \kappa^2 \frac{\partial^2 w}{\partial x^2}. \tag{6.43}$$

If a positive bending moment corresponds to $\mathbf{M}$ oriented as $\mathbf{e}_y$,

$$M = -Y A \kappa^2 \frac{\partial^2 w}{\partial x^2}. \tag{6.44}$$

6.2.1 Energy of a bent bar

The energy density $\tilde{F}$ associated with elastic vibrations is given by,

$$\tilde{F} = \frac{1}{2} T_{ik} S_{ik}. \tag{6.45}$$

For a thin bar vibrating in a vacuum, the boundary conditions are,

$$T_{ik} n_k = 0, \tag{6.46}$$

where $\mathbf{n}$ is the normal vector at the boundary of the bar. In our case, the normal vector at the bar boundary is $\pm \mathbf{e}_y$ or $\pm \mathbf{e}_z$ depending on which of the four sides is considered. For the two boundary sides with normal vectors parallel to the y direction, it follows that,

$$T_{xy} = T_{yy} = T_{zy} = 0. \tag{6.47}$$

Since the bar is thin and equation (6.47) holds at opposite boundaries, the stress components $T_{xy} = T_{yy} = T_{zy}$ must be small everywhere inside the bar. To a first approximation, it is reasonable to consider equation (6.47) holds everywhere inside the bar. Applying the same argument to the two remaining boundary sides with normal vector parallel to $\mathbf{e}_z$, it also follows that,

$$T_{xz} = T_{yz} = T_{zz} = 0. \tag{6.48}$$

Thus, the only nonzero stress component is T_{xx}, and, from equations (6.36)–(6.37), the energy density reduces to,

$$\tilde{F} = \frac{1}{2}T_{xx}S_{xx} = \frac{1}{2}\frac{dF}{dA}\frac{\delta x}{dx} = \frac{1}{2}Y\left(\frac{z}{R}\right)^2, \tag{6.49}$$

where dF is the longitudinal force acting on the strip of area dA, so the total elastic potential energy $\mathcal{F}$ of the bar becomes,

$$
\begin{aligned}
\mathcal{F} &= \int_V \tilde{F}\,dV = \int_V \tilde{F}\,dA\,dx = \int_V \frac{1}{2}Y\left(\frac{z}{R}\right)^2 dA\,dx = \int_L \frac{1}{2}YA\frac{\kappa^2}{R^2}\,dx \\
&= \int_L \frac{1}{2}YA\kappa^2\left(\frac{\partial^2 w}{\partial x^2}\right)^2 dx,
\end{aligned}
\tag{6.50}
$$

where equation (6.40) was used in obtaining the fourth equality.

6.2.2 Lagrangian and transverse wave equation

With the above prerequisites, the Lagrangian $\mathcal{L}$ can now be written as,

$$
\begin{aligned}
\mathcal{L} = \mathcal{T} - \mathcal{F} &= \int_L \frac{1}{2}\left(\frac{\partial w}{\partial t}\right)^2 \rho A\,dx - \int_L \frac{1}{2}YA\kappa^2\left(\frac{\partial^2 w}{\partial x^2}\right)^2 dx \\
&= \int_L \frac{1}{2}A\left[\left(\frac{\partial w}{\partial t}\right)^2 \rho - Y\kappa^2\left(\frac{\partial^2 w}{\partial x^2}\right)^2\right]dx,
\end{aligned}
\tag{6.51}
$$

where $\mathcal{T}$ is the kinetic energy of the bar. Minimizing the action,

$$S = \int \mathcal{L}\,dt, \tag{6.52}$$

yields the Euler–Lagrange equation,

$$\frac{\partial^2 w}{\partial t^2} + (\kappa c)^2 \frac{\partial^4 w}{\partial x^4} = 0, \tag{6.53}$$

where $c^2 = Y/\rho$ is used. Equation (6.53) is the transverse wave equation of bars.

6.2.3 Exercise: Euler–Lagrange equation for a single function of two variables

Consider a single function f of two variables x_1, x_2. Then, the Euler–Lagrange equation is [1],

$$\frac{\partial \tilde{L}}{\partial f} - \frac{\partial}{\partial x_1}\left(\frac{\partial \tilde{L}}{\partial\left(\frac{\partial f}{\partial x_1}\right)}\right) - \frac{\partial}{\partial x_2}\left(\frac{\partial \tilde{L}}{\partial\left(\frac{\partial f}{\partial x_2}\right)}\right) +$$

$$\frac{\partial^2}{\partial x_1^2}\left(\frac{\partial \tilde{L}}{\partial\left(\frac{\partial^2 f}{\partial x_1^2}\right)}\right) + \frac{\partial^2}{\partial x_1 \partial x_2}\left(\frac{\partial \tilde{L}}{\partial\left(\frac{\partial^2 f}{\partial x_1 \partial x_2}\right)}\right) + \frac{\partial^2}{\partial x_2^2}\left(\frac{\partial \tilde{L}}{\partial\left(\frac{\partial^2 f}{\partial x_2^2}\right)}\right) = 0,$$

(6.54)

where $\tilde{L}$ is the Lagrangian density defined by,

$$\mathcal{L} = \int_L \tilde{L} dx. \tag{6.55}$$

Use this equation to derive equation (6.53) from equation (6.51).

6.2.4 Mono-frequency solutions to the transverse wave equation

Considering mono-frequency solutions to equation (6.53), the ansatz,

$$w(x, t) = \Psi(x)e^{i\omega t}, \tag{6.56}$$

is made, and a fourth-order ordinary differential equation for Ψ is obtained,

$$\frac{d^4\Psi}{dx^4} = \left(\frac{\omega}{v}\right)^4 \Psi, \tag{6.57}$$

$$v = \sqrt{\omega \kappa c}.$$

The general solution for Ψ is,

$$\Psi(x) = A'e^{gx} + B'e^{-gx} + C'e^{igx} + D'e^{-igx},$$

$$g = \frac{\omega}{v}, \tag{6.58}$$

where A', B', C', D' are complex coefficients.

The general solution $w(x, t)$ can be written in terms of real coefficients A, B, C, D and a real phase φ,

$$w(x, t) = (A\cosh(gx) + B\sinh(gx) + C\cos(gx) + D\sin(gx))\cos(\omega t + \phi). \tag{6.59}$$

The boundary and initial conditions determine the coefficients A, B, C, D, ϕ.

6.2.5 Boundary conditions

It is necessary to impose two boundary conditions at each end of the transverse vibrating bar to determine the vibrational solution. Three types of boundary conditions are frequently used,

(a) Clamped end

If an end of a bar is clamped both the displacement and the slope must be zero at the end for all times, that is,

$$w = 0,$$
$$\frac{\partial w}{\partial x} = 0. \tag{6.60}$$

(b) Free end

If an end of a bar is free, both the bending moment and shear force must be zero at the end for all times, that is,

$$\frac{\partial^2 w}{\partial x^2} = 0,$$
$$\frac{\partial^3 w}{\partial x^3} = 0. \tag{6.61}$$

(c) Simply supported end

If an end of a bar is simply supported, for example, if the bar end is constrained by two knife edges mounted perpendicular to the plane of the transverse motion, both the displacement and the bending moment at the end must vanish,

$$w = 0,$$
$$\frac{\partial^2 w}{\partial x^2} = 0. \tag{6.62}$$

6.2.6 Bar clamped at one end and free at the other end

Consider a bar of length L is rigidly clamped at $x = 0$ and free at $x = L$. Applying the boundary conditions $w = \frac{\partial w}{\partial x} = 0$ at $x = 0$ to equation (6.59) yields,

$$A + C = 0,$$
$$B + D = 0, \tag{6.63}$$

and the general solution becomes,

$$w(x, t) = [A(\cosh(gx) - \cos(gx)) + B(\sinh(gx) - \sin(gx))]\cos(\omega t + \phi). \tag{6.64}$$

Applying next the boundary conditions $\frac{\partial^2 w}{\partial x^2} = \frac{\partial^3 w}{\partial x^3} = 0$ at $x = L$ gives,

$$A(\cosh(gL) + \cos(gL)) = - B(\sinh(gL) + \sin(gL)),$$
$$A(\sinh(gL) - \sin(gL)) = - B(\cosh(gL) + \cos(gL)). \quad (6.65)$$

These conditions are satisfied simultaneously at some discrete frequencies. The allowed frequencies are determined by dividing one equation by the other and then cross-multiplying. Use of the identities,

$$\cos^2 x + \sin^2 x = 1,$$
$$\cosh^2 x - \sinh^2 x = 1, \quad (6.66)$$

yields,

$$\cosh(gL)\cos(gL) = -1. \quad (6.67)$$

Solutions are found numerically,

$$gL = \frac{\omega L}{v} = (1.194, 2.988, 5, 7, \ldots)\frac{\pi}{2}, \quad (6.68)$$

or, since $v = \sqrt{\omega \kappa c}$, the natural frequencies of a transversely vibrating clamped-free bar, are,

$$f = (1.194^2, 2.988^2, 5^2, 7^2, \ldots)\frac{\pi \kappa c}{8L^2}. \quad (6.69)$$

6.2.7 Bar free at both ends

If the bar is free at both ends, the solution method follows steps similar to the clamped-free bar case. The boundary conditions at $x = 0$ yield,

$$A = C,$$
$$B = D. \quad (6.70)$$

The same boundary conditions applied to $x = L$ lead to,

$$\cosh(gL)\cos(gL) = 1, \quad (6.71)$$

and the natural frequencies of a transversely vibrating free-free bar, are,

$$f = (3.011^2, 5^2, 7^2, 9^2, \ldots)\frac{\pi \kappa c}{8L^2}. \quad (6.72)$$

6.3 Torsional waves in a bar

A bar can vibrate torsionally as well as longitudinally and transversely. With reference to figure 6.3, the shearing stress T_s required to produce a shearing strain $rd\phi/dx$ is,

$$T_s = \mathcal{G}r\frac{d\phi}{dx}, \tag{6.73}$$

where $\mathcal{G}$ is the modulus of rigidity. The force dF is the stress multiplied by the area over which it acts, that is,

$$dF = \mathcal{G}(dw\,dr)r\frac{d\phi}{dx}, \tag{6.74}$$

and the torque $d\tau$ required to produce the strain in a circular shell of radius r is obtained as dF multiplied by its arm r and integrating over the circumference of the bar, $\int dw = 2\pi r$,

$$d\tau = \mathcal{G}(2\pi r^3)\frac{d\phi}{dx}dr. \tag{6.75}$$

The torque over the element of length dx is found by integrating over r from $r = 0$ to $r = a$,

$$\tau = \mathcal{G}\frac{1}{2}\pi a^4\frac{d\phi}{dx}. \tag{6.76}$$

The net torque on the element of length dx is the difference between torques $\tau(x + dx)$ and $\tau(x)$,

$$\tau(x + dx) - \tau(x) = \frac{d\tau}{dx}dx = \mathcal{G}\frac{1}{2}\pi a^4\frac{d^2\phi}{dx^2}dx. \tag{6.77}$$

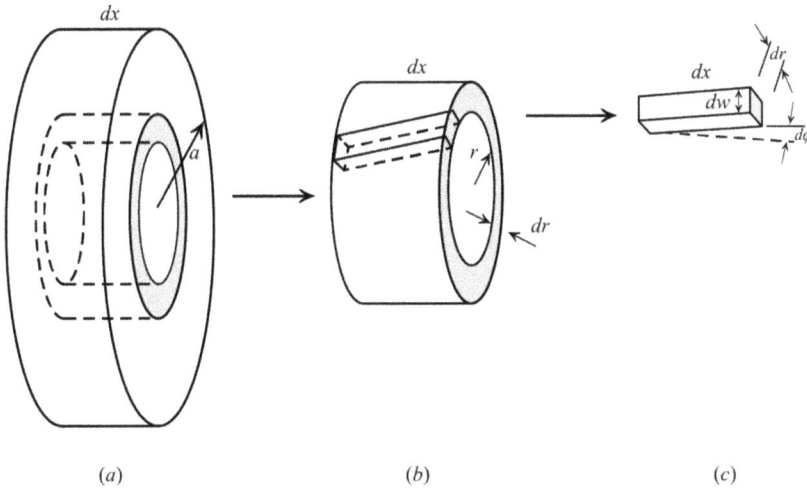

(a)   (b)   (c)

Figure 6.3. Torsional distortion of a bar. Left plot, an element of a bar with length dx and radius a. Middle plot, a cylindrical shell of radius r, thickness dr, and length dx. Right plot, a segment of width dw, thickness dr, and length dx strained by an angle $d\phi$.

This net torque must be equal to the moment of inertia of the cylinder, $\frac{a^2}{2}dm$, where $dm = \rho\pi a^2 dx$ is the mass of the cylinder, multiplied by its angular acceleration $\frac{\partial^2\phi}{\partial t^2}$. The torsional wave equation of a bar is then obtained,

$$\frac{\partial^2\phi}{\partial t^2} - c^2\frac{\partial^2\phi}{\partial x^2} = 0, \tag{6.78}$$

where $c = \sqrt{\frac{G}{\rho}}$ is the phase speed of torsional waves. Since the wave equation of torsional waves in a bar is similar to the wave equation of longitudinal vibrations of a bar or transverse vibrations of a string, the same techniques can be used to find solutions.

Reference

[1] Courant R and Hilbert D 1958 *Methods of Mathematical Physics* **vol I** first English edition (New York: Interscience)

Chapter 7

Vibrations of membranes

Membrane vibrations represent yet another complexity added above string vibrations in that the displacement is a function of two spatial coordinates. Membranes, however, carry the same simplicity as strings in that their thickness can be neglected for the study of membrane vibrations. In chapter 7, the theory of membrane vibrations is presented. Eigenfrequencies for freely vibrating and forced membranes are solved. Both rectangular and circular membrane cases are presented.

7.1 Theory of vibrating membranes

Consider a membrane stretched uniformly in all directions (figure 7.1). The surface mass density is ρ_S (units kg m^{-2}) and the tension per unit length of the membrane is $\mathcal{T}$ (units N m^{-1}). This implies that opposite sides of a membrane element of length dl are pulled apart with a force $\mathcal{T}dl$. If the membrane plane is the x–z plane, the net force in the vertical direction y from the two sides parallel to the z axis is,

$$\mathcal{T}dz\left[\left(\frac{\partial y}{\partial x}\right)_{x+dx} - \left(\frac{\partial y}{\partial x}\right)_x\right] = \mathcal{T}\frac{\partial^2 y}{\partial x^2}dxdz. \tag{7.1}$$

Similarly, the two other opposing sides of the membrane parallel to the x axis contribute a net force in the vertical direction,

$$\mathcal{T}dx\left[\left(\frac{\partial y}{\partial z}\right)_{z+dz} - \left(\frac{\partial y}{\partial z}\right)_z\right] = \mathcal{T}\frac{\partial^2 y}{\partial z^2}dxdz. \tag{7.2}$$

Adding the two force contributions gives the total force in the vertical direction, which by use of Newton's second equation gives,

$$\frac{\partial^2 y}{\partial x^2} + \frac{\partial^2 y}{\partial z^2} = \frac{1}{c^2}\frac{\partial^2 y}{\partial t^2}, \tag{7.3}$$

where $c = \sqrt{\frac{\mathcal{T}}{\rho_S}}$ is the membrane wave speed.

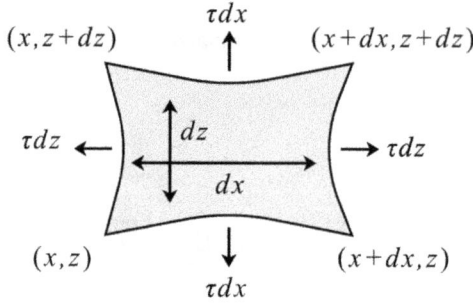

Figure 7.1. Infinitesimal area element of a membrane subject to transverse vibrations.

Rewriting in vector notation the differential equation becomes,

$$\nabla^2 y - \frac{1}{c^2}\frac{\partial^2 y}{\partial t^2} = 0, \tag{7.4}$$

where ∇ is the two-dimensional Laplacian. General solutions can be found in the form,

$$y(x, z, t) = \Psi(x, z)e^{j\omega t}. \tag{7.5}$$

Setting $k = \omega/c$, the two-dimensional Helmholtz equation is obtained,

$$\nabla^2 \Psi + k^2 \Psi = 0. \tag{7.6}$$

7.1.1 Exercise: membrane wave equation

Prove equation (7.3) using Newton's second law.

7.2 Rectangular membrane with fixed edges

If the membrane is rectangular-shaped and the edges are fixed (figure 7.2), the following boundary conditions apply,

$$y(x = 0, z, t) = y(x = L_x, z, t) = y(x, z = 0, t) = y(x, z = L_z, t), \tag{7.7}$$

where L_x (L_z) are the membrane side lengths parallel to the x (z) directions. Solutions can be found by separation-of-variables,

$$\Psi(x, z) = X(x)Z(z), \tag{7.8}$$

where

$$\frac{d^2 X}{dx^2} + k_x^2 X = 0,$$
$$\frac{d^2 Z}{dz^2} + k_z^2 Z = 0, \tag{7.9}$$

$$(0, l_z) \qquad\qquad\qquad (l_x, l_z)$$

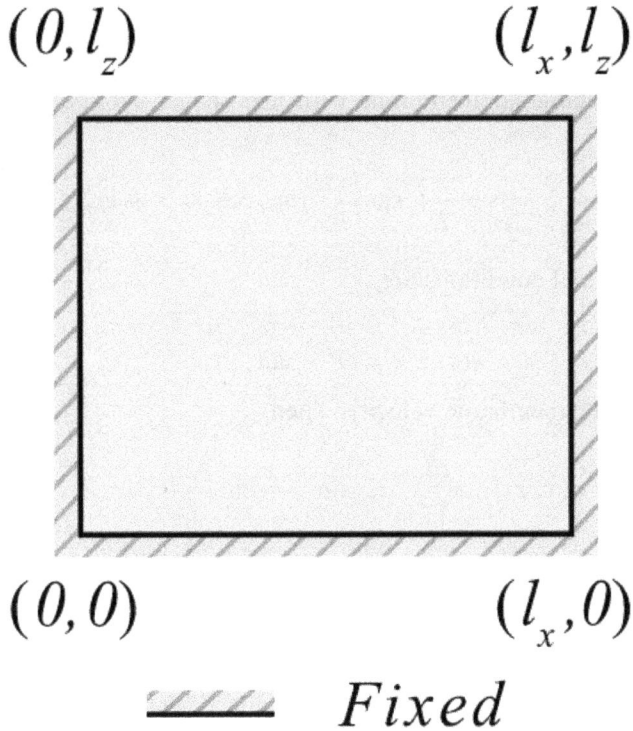

$$(0, 0) \qquad\qquad\qquad (l_x, 0)$$

$$\overset{\text{\tiny ////}}{=\!=\!=} \quad \textit{Fixed}$$

Figure 7.2. Rectangular membrane with fixed boundaries subject to transverse vibrations.

and $k_x^2 + k_z^2 = k^2$.

Solutions that satisfy the boundary conditions are,

$$y(x, z, t) = A\sin(k_x x)\sin(k_z z)e^{j\omega t},$$
$$k_x = \frac{n\pi}{L_x},$$
$$k_z = \frac{m\pi}{L_z}, \tag{7.10}$$

where n and m are positive integers. The general complex membrane displacement can now be written as,

$$y(x, z, t) = \sum_{nm} y_{nm}(x, z, t) = \sum_{nm} A_{nm}\sin(k_{xn}x)\sin(k_{zm}z)e^{j\omega_{nm}t}. \tag{7.11}$$

The allowed membrane frequencies are,

$$f_{nm} = \frac{\omega_{nm}}{2\pi} = \frac{c}{2}\sqrt{\left(\frac{n}{L_x}\right)^2 + \left(\frac{m}{L_z}\right)^2}, \tag{7.12}$$

and the associated complex amplitudes A_{nm} are determined from the initial conditions.

7.2.1 Rectangular membrane vibrations. Fixed edges and initial conditions

It follows from the solution, equation (7.11), that the general solution to the vibration of a rectangular membrane with fixed edges subject to arbitrary initial conditions is,

$$y(x, z, t) = \sum_{n=1, m=1}^{\infty} \sin \frac{n\pi x}{L_x} \sin \frac{m\pi z}{L_z} (a_{nm} \cos \omega_{nm} t + b_{nm} \sin \omega_{nm} t). \tag{7.13}$$

Suppose the initial conditions are,

$$\begin{aligned} y(x, z, t = 0) &= \alpha(x, z), \\ u(x, z, t = 0) &= \beta(x, z), \end{aligned} \tag{7.14}$$

where $u = \frac{\partial y}{\partial t}$ is the membrane velocity. Then,

$$\begin{aligned} \alpha(x, z) &= \sum_{n=1, m=1}^{\infty} a_{nm} \sin \frac{n\pi x}{L_x} \sin \frac{m\pi z}{L_z}, \\ \beta(x, z) &= \sum_{n=1, m=1}^{\infty} b_{nm} \omega_{nm} \sin \frac{n\pi x}{L_x} \sin \frac{m\pi z}{L_z}, \end{aligned} \tag{7.15}$$

and orthonormality of the normal modes gives,

$$\begin{aligned} a_{nm} &= \frac{4}{L_x L_z} \int_0^{L_x} \int_0^{L_z} \alpha(x, z) \sin \frac{n\pi x}{L_x} \sin \frac{m\pi z}{L_z} dx \, dz, \\ b_{nm} &= \frac{4}{L_x L_z \omega_{nm}} \int_0^{L_x} \int_0^{L_z} \beta(x, z) \sin \frac{n\pi x}{L_x} \sin \frac{m\pi z}{L_z} dx \, dz. \end{aligned} \tag{7.16}$$

Finally, insertion of the latter expressions for the a_{nm}, b_{nm} coefficients in equation (7.13) yields the complete time-dependent solution $y(x, z, t)$.

7.2.2 Circular membrane with fixed rim

The circular membrane equation is the two-dimensional Helmholtz equation written in circular coordinates r, θ,

$$\frac{\partial^2 \Psi}{\partial r^2} + \frac{1}{r} \frac{\partial \Psi}{\partial r} + \frac{1}{r^2} \frac{\partial^2 \Psi}{\partial \theta^2} + k^2 \Psi = 0. \tag{7.17}$$

If the membrane is fixed at $r = a$ (figure 7.3), solutions can be sought by separation-of-variables,

$$\begin{aligned} \Psi &= R(r) \Theta(\theta), \\ R(a) &= 0. \end{aligned} \tag{7.18}$$

The general solution is,

$$\begin{aligned} R(r) &= A J_m(kr) + B Y_m(kr), \\ \Theta(\theta) &= \cos(m\theta + \gamma_m), \end{aligned} \tag{7.19}$$

Figure 7.3. Circular membrane with a fixed boundary subject to transverse vibrations.

where m is an integer and γ_m a phase factor determined by the initial conditions. J_m and Y_m are Bessel and Neumann functions, respectively. Note that if the center, $r = 0$, is part of the membrane, B must be zero since the Neumann function diverges at $r = 0$. For an annulus-shaped membrane, however, $B \neq 0$.

For a circular membrane (where the center $r = 0$ is part of the membrane), the complete complex solution is a sum of all modes,

$$y(r,\, \theta,\, t) = \sum_{mn} y_{mn}(r,\, \theta,\, t) = \sum_{mn} A_{mn} J_m(k_{mn} r) \cos(m\theta + \gamma_{mn}) e^{j\omega_{mn} t},$$
$$k_{mn} a = j_{mn},$$

$$(7.20)$$

where j_{mn} is the nth zero of J_m, and A_{mn}, γ_{mn} are determined by the initial conditions. The natural membrane frequencies are,

$$f_{mn} = \frac{j_{mn} c}{2\pi a}.$$

$$(7.21)$$

7.2.3 Exercise: general solution to the ring-shaped membrane with clamped boundaries

Consider a circular membrane spanning the $r - \theta$ domain: $R_i \leqslant r \leqslant R_o$ and $0 \leqslant \theta \leqslant 2\pi$ corresponding to a ring-shaped membrane with inner radius R_i and outer radius R_o (figure 7.4). Determine the general solution in the case with clamped boundaries.

7.2.4 Exercise: general solution to the half-cut ring-shaped membrane with clamped boundaries

Consider a circular half-cut membrane spanning the $r - \theta$ domain: $R_i \leqslant r \leqslant R_o$ and $0 \leqslant \theta \leqslant \pi$ corresponding to an inner radius R_i and outer radius R_o (figure 7.5). Determine the general solution in the case with clamped boundaries.

7.2.5 Exercise: general solution to the ring-shaped membrane with free boundaries

Determine the general solution to the ring-shaped membrane with free boundaries at $r = R_i$ and $r = R_o$.

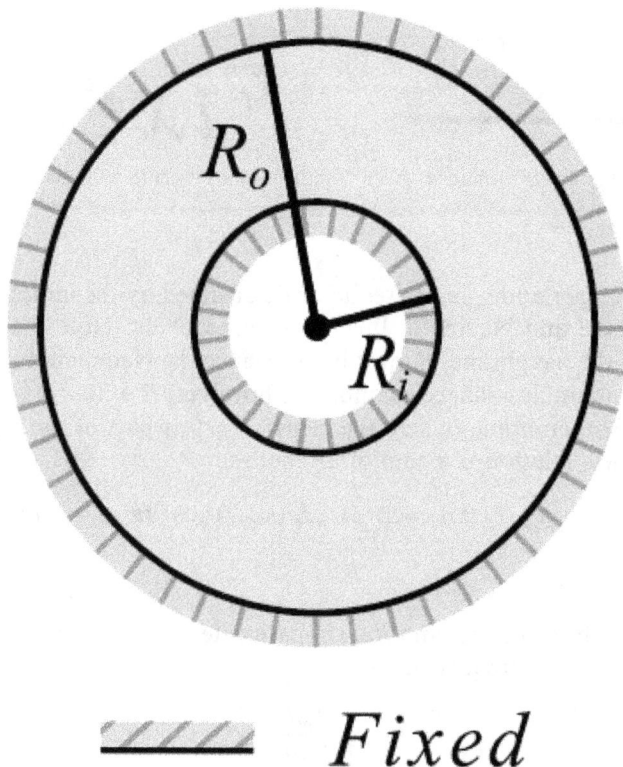

Figure 7.4. Circular ring membrane with fixed boundaries subject to transverse vibrations.

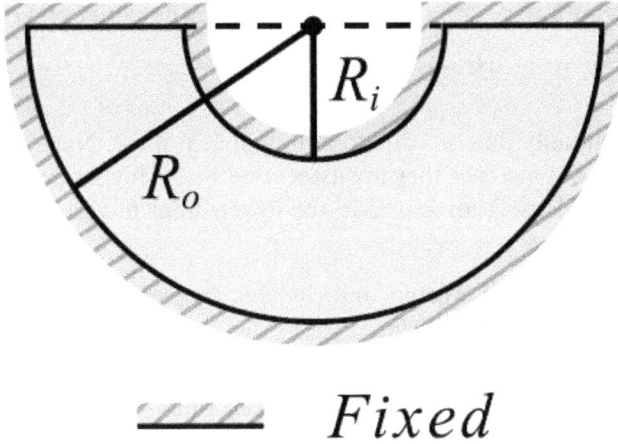

Figure 7.5. Circular half-cut ring membrane with fixed boundaries subject to transverse vibrations.

7.2.6 Exercise: general solution to the ring-shaped membrane with a fixed boundary at $r = R_i$ and a free boundary at $r = R_o$

Determine the general solution to the ring-shaped membrane with a fixed boundary at $r = R_i$ and a free boundary at $r = R_o$.

7.2.7 The damped, freely vibrating membrane

In practice, damping phenomena diminish vibrations. Hence, we phenomenologically introduce damping in the membrane equation,

$$\frac{\partial^2 y}{\partial t^2} + 2\beta \frac{\partial y}{\partial t} - c^2 \nabla^2 y = 0. \tag{7.22}$$

Assuming mono-frequency solutions,

$$y(\mathbf{r}, t) = \Psi(\mathbf{r}) e^{j\omega t}, \tag{7.23}$$

the Helmholtz equation is obtained,

$$\nabla^2 \Psi + k^2 \Psi = 0,$$
$$k^2 = \left(\frac{\omega}{c}\right)^2 - 2j\frac{\beta\omega}{c^2}. \tag{7.24}$$

Apparently, k is complex, however, if the membrane is forced to vanish at its edges, k must be real! A solution to equation (7.24) is

$$k = \frac{|\omega|}{c},$$
$$\omega = \omega_d + j\beta,$$
$$|\omega|^2 = \omega_d^2 + \beta^2, \tag{7.25}$$

where ω_d, β are both real, and the complete solution to equation (7.22) is

$$y(\mathbf{r}, t) = \sum_{mn} \Psi_{mn} e^{-\beta_{mn}t} e^{j\omega_{d,mn}t}. \tag{7.26}$$

The solution eventually dies out due to damping as it must. Note that higher-order modes usually die out faster as they are associated with a higher degree of membrane flexing (and flexing losses) compared to the lower-order modes.

7.2.8 Exercise: allowed wavevectors and complex frequencies for a damped circular membrane with a clamped rim

Specify a condition to determine the allowed wavevectors and complex frequencies for a damped circular membrane with a clamped rim.

7.2.9 Exercise: allowed wavevectors and complex frequencies for a damped ring-shaped membrane with clamped boundaries at $r = R_i$ and $r = R_o$

Specify a condition to determine the allowed wavevectors and complex frequencies for a damped ring-shaped membrane with boundaries at $r = R_i$ and $r = R_o$. In the case of a membrane with tension $T = 20\,000$ N m^{-1}, surface mass density $\rho_S = 1.0$ kg m^{-2}, damping coefficient $\beta = 20$ s^{-1}, inner radius $R_i = 0.20$ m, and outer radius $R_i = 0.50$ m, write a Matlab program to determine the numerical values of the allowed wavevectors and complex frequencies.

7.2.10 Forced membrane

In the case of a forced membrane, the membrane differential equation becomes,

$$\frac{\partial^2 y}{\partial t^2} + 2\beta \frac{\partial y}{\partial t} - c^2 \nabla^2 y = \frac{P}{\rho_S} f(t). \tag{7.27}$$

where P is a pressure and $f(t)$ is a dimensionless function of time. If $f = e^{j\omega t}$ then the steady-state solution must take the form,

$$y = \Psi e^{j\omega t}, \tag{7.28}$$

and Ψ obeys,

$$(-\omega^2 + 2j\beta\omega - c^2 \nabla^2)\Psi = \frac{P}{\rho_S}. \tag{7.29}$$

A general solution is a particular solution Ψ_p to the inhomogeneous differential equation added to a general homogeneous solution Ψ_h. The latter obeys,

$$\nabla^2 \Psi_h + k^2 \Psi_h = 0, \tag{7.30}$$

where

$$k = k_R - jk_I,$$

$$k_R = \frac{\omega}{c}\sqrt{\left(1 + \left(\frac{\beta}{\omega}\right)^2\right)},$$

$$\frac{k_I}{k_R} = \frac{\beta}{\omega}\frac{1}{\left(1 + \left(\frac{\beta}{\omega}\right)^2\right)} \approx \frac{\beta}{\omega}, \qquad (7.31)$$

where the latter approximation applies to first order in the damping coefficient β. A particular solution can be easily found since P does not depend on $\mathbf{r}$,

$$\Psi_p = -\frac{P}{\rho_S}\frac{1}{k^2 c^2}. \qquad (7.32)$$

Assuming a circular membrane with fixed edges the solution is,

$$\Psi = \frac{P}{\rho_S c^2 k^2}\left(\frac{J_0(kr)}{J_0(ka)} - 1\right). \qquad (7.33)$$

It follows that the amplitude of the membrane displacement is proportional to the pressure P. Note that the k^2 term in the prefactor shows that the displacement decreases substantially as the frequency ω of the force increases. Note also that large amplitudes of the membrane displacement occur whenever the frequency is such that $J_0(ka)$ is close to zero since then the denominator inside the brackets approaches zero.

7.2.11 Write a Matlab program to compute the membrane displacement versus radius and time for the frequencies $f = \frac{\omega}{2\pi} = 0.5$ Hz, 1 Hz, 5 Hz

7.2.12 Theory of a rectangular membrane with fixed rim struck at (x_0, z_0) at time $t = 0$ with a speed $v(x, z, 0) = \mathcal{V}\delta(\mathbf{r} - \mathbf{r}_0)$

For a fixed rim, the normal modes are,

$$\psi_{nm}(x, y) = \sin(k_{x,n}x)\sin(k_{z,m}z), \qquad (7.34)$$

and,

$$\frac{\omega_{nm}^2}{c^2} = k_{x,n}^2 + k_{z,m}^2, \qquad (7.35)$$

$$\int_0^{L_x}\int_0^{L_z}\psi_{nm}(x, z)\psi_{n'm'}(x, z)dxdz = \frac{L_x L_z}{4}\delta_{nn'}\delta_{mm'}. \qquad (7.36)$$

If the membrane is struck at a position (x_0, z_0) with an impulse at $t = 0$, then the initial speed is,

$$v(x, y, t = 0) = \mathcal{V}\delta(x - x_0)\delta(z - z_0), \tag{7.37}$$

where $\mathcal{V}$ is a constant characterizing the strength of the struck. The real displacement can be written as,

$$y(x, z, t) = \sum_{n,m}\alpha_{nm} \sin(k_{x,n}x) \sin(k_{z,m}z) \sin(\omega_{nm}t + \phi_{nm}), \tag{7.38}$$

where α_{nm} is the excitation amplitude of mode (n, m). Demanding,

$$\frac{\partial y(x, z, t)}{\partial t}\Big|_{t=0} = v(x, z, t = 0), \tag{7.39}$$

yields,

$$y(x, z, t) = \frac{4\mathcal{V}}{L_x L_z}\sum_{n,m}\frac{1}{\omega_{nm}} \sin(k_{x,n}x_0) \sin(k_{z,m}z_0) \sin(k_{x,n}x) \sin(k_{z,m}z) \sin(\omega_{nm}t). \tag{7.40}$$

In practical evaluations, the series in n and m are truncated at some large values N and M, respectively. Note also that we neglected damping which will lead to damping of the excited vibration over time. This can be done phenomenologically by introducing a damping term $\exp(-\beta_{nm}t)$ in the expression for the displacement and an appropriate renormalization of the frequency in accordance with section 7.2.7 on damped membrane vibrations.

7.2.13 Write a Matlab program to compute the membrane displacement as a function of position (x, z) and time t for the rectangular membrane case struck at (x_0, z_0) at time $t = 0$ with a speed $v(x, z, 0) = \mathcal{V}\delta(\mathbf{r} - \mathbf{r}_0)$

Use the program to compute a video of the rectangular membrane displacement versus position and time

7.2.14 Theory of a circular membrane with fixed rim struck at (r_0, θ_0) at time $t = 0$ with a speed $v(r, \theta, 0) = \mathcal{V}\delta(\mathbf{r} - \mathbf{r}_0)$

The problem follows the structure outlined in subsection 7.2.12. Firstly, notice that the general displacement can be written as,

$$y(r, \theta, t) = \sum_{m,n}\alpha_{mn}J_m(k_{mn}r) \cos(m\theta + \gamma_{mn}) \sin(\omega_{mn}t + \phi_{mn}), \tag{7.41}$$

where $k_{mn} = \frac{\omega_{mn}}{c}$. If the coordinate axes are chosen such that the angle of the point where the strike hits the membrane is 0 then, for all modes, the largest modal displacement corresponds to $\theta = 0$ and $\gamma_{mn} = 0$. Further, the membrane displacement must be zero at $t = 0$ so $\phi_{mn} = 0$ for all modes. Then, demanding that the speed of the membrane at $t = 0$ is $\mathcal{V}\delta(\mathbf{r} - \mathbf{r}_0)$ yields,

$$\sum_{m,n}\alpha_{mn}\omega_{mn}J_m(k_{mn}r) \cos(m\theta) = \mathcal{V}\frac{1}{r}\delta(r - r_0)\delta(\theta). \tag{7.42}$$

From the orthonormality conditions,

$$\int_0^R \int_0^{2\pi} (J_m(k_{mn}r)\cos(m\theta))^2\, rdrd\theta = \begin{cases} \pi R^2 \big(J_m'(k_{mn}R)\big)^2, & m = 0, \\ \dfrac{\pi R^2}{2}\big(J_m'(k_{mn}R)\big)^2, & m > 0, \end{cases} \tag{7.43}$$

where $k_{mn}R = j_{mn}$ and j_{mn} is the nth zero point of J_m, the membrane vibration due to the struck becomes,

$$y(r,\,\theta,\,t) = \frac{\mathcal{V}}{\pi R^2}\sum_{m,n}\frac{\epsilon_m}{\omega_{mn}}\frac{J_m(k_{mn}r_0)}{\big(J_m'(k_{mn}R)\big)^2}J_m(k_{mn}r)\cos(m\theta)\sin(\omega_{mn}t), \tag{7.44}$$

where $\epsilon_m = 1$ if $m = 0$ and $\epsilon_m = 2$ for all other m values. Again, for practical evaluations, the series in m and n are truncated at some large values M and N, respectively. Note also that we neglected damping which will lead to damping of the excited vibration over time. This can be done phenomenologically by introducing a damping term $\exp(-\beta_{mn}t)$ in the expression for the displacement and an appropriate renormalization of the frequency in accordance with section 7.2.7 on damped membrane vibrations.

7.2.15 Write a Matlab program to compute the membrane displacement as a function of position $(r,\,\theta)$ and time t for a circular membrane with fixed rim struck at $(r_0,\,\theta_0)$ at time $t = 0$ with a speed $v(r,\,\theta,\,0) = \mathcal{V}\delta(\mathbf{r} - \mathbf{r}_0)$

Use the program to compute a video of the circular membrane displacement versus position and time

7.2.16 Forced vibrations of a circular membrane

Suppose a circular membrane is subject to a harmonically varying load perpendicular to the membrane plane and acting within a circle of radius R at the origin. The pressure load takes the form,

$$p(r,\,t) = \begin{cases} p_0 e^{-i\omega t}, & \text{if } r < R, \\ 0, & \text{if } r > R. \end{cases}$$

This will lead to a total force $\pi R^2 p_0 e^{-i\omega t}$ acting on the inner circle $r < R$. To balance this force component the membrane experiences a restoring force in the vertical direction from the part of the membrane outside $r = R$,

$$\pi R^2 p_0 e^{-i\omega t} + 2\pi RT\frac{\partial y}{\partial r}\,|_{r=R} = 0, \tag{7.45}$$

where T is the restoring force per length along the edge $r = R$. Writing $y(r,\,t) = Y(r)e^{-i\omega t}$, the boundary equation becomes,

$$\frac{dY}{dr}\,|_{r=R} = -\frac{p_0 R}{2T}. \tag{7.46}$$

The membrane equation is,

$$\frac{d^2Y}{dr^2} + \frac{1}{r}\frac{dY}{dr} + k^2Y = 0, \tag{7.47}$$

and the solution, written in terms of the Hankel functions, becomes,

$$Y(r) = AH_0^{(1)}(kr) + BH_0^{(2)}(kr), \tag{7.48}$$

where $k = \omega/c$. Since the membrane vibration is generated by the external load acting within $r = R$, the radiation condition is that the Hankel function $H_0^{(2)}$ is not excited, i.e., $B = 0$. Imposing the boundary condition, equation (7.46), fixes the magnitude of A, and the membrane solution becomes,

$$y(r,\, t) = \frac{p_0 R}{2kT} \frac{H_0^{(1)}(kr)}{H_1^{(1)}(kR)} e^{-i\omega t}, \tag{7.49}$$

where the mathematical identity,

$$\frac{dH_0^{(1)}(kx)}{dx} = -kH_1^{(1)}(kx), \tag{7.50}$$

was used. From the asymptotic behavior of the Hankel function, it follows that the membrane vibration at large r values, far above R, obeys,

$$y(r,\, t) \sim \frac{p_0 R}{\sqrt{2\pi}\, k^{3/2} T H_1^{(1)}(kR)} \frac{1}{\sqrt{r}} e^{i(kr - \omega t - \pi/4)}, \tag{7.51}$$

so the amplitude of the vibration dies off as $r^{-1/2}$ at large distances from the membrane center.

Chapter 8

Cylindrical rod vibrations

Cylindrical rods are important structures in many applications in physics and engineering. The understanding of cylindrical rod vibrations requires a complete three-dimensional analysis, including three-dimensional boundary and initial conditions. Chapter 8 presents in detail the theory of cylindrical rod vibrations. The first part of this chapter is devoted to deriving equations for the variables Φ and H using Helmholtz decomposition. The important cases of longitudinal, torsional, and flexural rod vibrations are addressed and solved for eigenfrequencies and dispersion relations, including examples of Matlab program codes.

8.1 Wave equations of cylindrical rods

Consider a cylindrical rod composed of an isotropic solid material (figure 8.1). We shall derive the vibrational modes [1]. Using the result from section 3.3 a Helmholtz decomposition can be made for the displacement [2],

$$\mathbf{u} = \nabla\Phi + \nabla \times \mathbf{H}, \tag{8.1}$$

whereby the scalar components of **u** read,

$$u_r = \frac{\partial\Phi}{\partial r} + \frac{1}{r}\frac{\partial H_z}{\partial\theta} - \frac{\partial H_\theta}{\partial z}, \tag{8.2}$$

$$u_\theta = \frac{1}{r}\frac{\partial\Phi}{\partial\theta} + \frac{\partial H_r}{\partial z} - \frac{\partial H_z}{\partial r}, \tag{8.3}$$

$$u_z = \frac{\partial\Phi}{\partial z} + \frac{1}{r}\frac{\partial}{\partial r}(rH_\theta) - \frac{1}{r}\frac{\partial H_r}{\partial\theta}. \tag{8.4}$$

From equations (3.37) and (3.38), the potentials Φ and **H** satisfy the scalar and vector wave equations,

$$\nabla^2\Phi = \frac{1}{c_L^2}\frac{\partial^2\Phi}{\partial t^2}, \tag{8.5}$$

doi:10.1088/978-0-7503-5557-5ch8

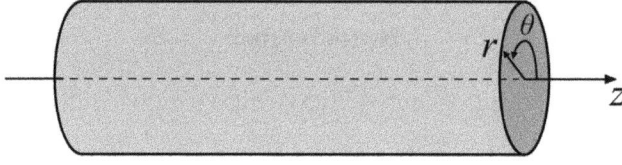

Figure 8.1. Sketch of a cylindrical solid rod subject to deformation.

$$\nabla^2 \mathbf{H} = \frac{1}{c_T^2} \frac{\partial^2 \mathbf{H}}{\partial t^2}, \tag{8.6}$$

where,

$$c_L^2 = \frac{\lambda + 2\mu}{\rho}, \tag{8.7}$$

$$c_T^2 = \frac{\mu}{\rho}. \tag{8.8}$$

Here, λ, μ denote the isotropic elastic constants (Lamé constants) and ρ is the solid material's mass density.

In cylindrical coordinates, the following expressions hold,

$$\nabla^2 \Phi = \frac{\partial^2 \Phi}{\partial r^2} + \frac{1}{r} \frac{\partial \Phi}{\partial r} + \frac{1}{r^2} \frac{\partial^2 \Phi}{\partial \theta^2} + \frac{\partial^2 \Phi}{\partial z^2}, \tag{8.9}$$

$$\nabla^2 \mathbf{H} = \left(\nabla^2 H_r - \frac{H_r}{r^2} - \frac{2}{r^2} \frac{\partial H_\theta}{\partial \theta} \right) \mathbf{e}_r + \left(\nabla^2 H_\theta - \frac{H_\theta}{r^2} + \frac{2}{r^2} \frac{\partial H_r}{\partial \theta} \right) \mathbf{e}_\theta + \nabla^2 H_z \mathbf{e}_z. \tag{8.10}$$

Searching for harmonic wave propagation in a cylinder, the following ansatz is made,

$$\Phi = f(r)\Theta(\theta)e^{i(kz-\omega t)}, \tag{8.11}$$

$$H_r = h_r(r)\Theta_r(\theta)e^{i(kz-\omega t)}, \tag{8.12}$$

$$H_\theta = h_\theta(r)\Theta_\theta(\theta)e^{i(kz-\omega t)}, \tag{8.13}$$

$$H_z = h_z(r)\Theta_z(\theta)e^{i(kz-\omega t)}. \tag{8.14}$$

It follows immediately by inspection of equations (8.5) and (8.6), (8.9) and (8.10) that the angular dependence of the Φ and $\mathbf{H}$ fields must be in either sine or cosine form, i.e.,

$$\Phi = f(r)\cos(n\theta)e^{i(kz-\omega t)}, \tag{8.15}$$

$$H_r = h_r(r)\sin(n\theta)e^{i(kz-\omega t)}, \tag{8.16}$$

$$H_\theta = h_\theta(r)\cos(n\theta)e^{i(kz-\omega t)}, \tag{8.17}$$

$$H_z = h_z(r)\sin(n\theta)e^{i(kz-\omega t)}. \tag{8.18}$$

Let us now proceed to determine the general solutions for the radial part of the fields. Starting with Φ, it follows from equation (8.5),

$$\frac{d^2f}{dr^2} + \frac{1}{r}\frac{df}{dr} + \left(\alpha^2 - \frac{n^2}{r^2}\right)f = 0, \tag{8.19}$$

where $\alpha^2 = \omega^2/c_L^2 - k^2$. This is a Bessel's equation of order n having the general solution,

$$f(r) = AJ_n(\alpha r), \tag{8.20}$$

since the second solution $Y_n(\alpha r)$ diverges at the origin and therefore can be discarded.

For h_z the resulting equation is of the same form as Φ, i.e.,

$$h_z(r) = B_3 J_n(\beta r), \tag{8.21}$$

where $\beta^2 = \omega^2/c_T^2 - k^2$.

The remaining two equations in h_r and h_θ are generally coupled, as can be easily verified by inspection of the vector Helmholtz equation, equation (8.6), and equation (8.10). The equations are,

$$\frac{d^2h_r}{dr^2} + \frac{1}{r}\frac{dh_r}{dr} + \frac{1}{r^2}(-n^2h_r + 2nh_\theta - h_r) - k^2h_r + \frac{\omega^2}{c_T^2}h_r = 0, \tag{8.22}$$

$$\frac{d^2h_\theta}{dr^2} + \frac{1}{r}\frac{dh_\theta}{dr} + \frac{1}{r^2}(-n^2h_\theta + 2nh_r - h_\theta) - k^2h_\theta + \frac{\omega^2}{c_T^2}h_\theta = 0. \tag{8.23}$$

The solutions are obtained by subtracting the second equation from the first,

$$\left(\frac{d^2}{dr^2} + \frac{1}{r}\frac{d}{dr} + \left(\beta^2 - \frac{(n+1)^2}{r^2}\right)\right)(h_r - h_\theta) = 0, \tag{8.24}$$

thus,

$$h_r - h_\theta = 2B_2 J_{n+1}(\beta r). \tag{8.25}$$

Adding the two equations gives a second solution,

$$\left(\frac{d^2}{dr^2} + \frac{1}{r}\frac{d}{dr} + \left(\beta^2 - \frac{(n-1)^2}{r^2}\right)\right)(h_r + h_\theta) = 0, \tag{8.26}$$

i.e.,

$$h_r + h_\theta = 2B_1 J_{n-1}(\beta r). \tag{8.27}$$

Therefore,

$$h_r = B_1 J_{n-1}(\beta r) + B_2 J_{n+1}(\beta r), \tag{8.28}$$

$$h_\theta = B_1 J_{n-1}(\beta r) - B_2 J_{n+1}(\beta r). \tag{8.29}$$

Gauge invariance implies freedom in the choice of the vector potential $\mathbf{H}$ that allows us to fix $B_1 = 0$. Thus,

$$h_r(r) = -h_\theta(r) = B_2 J_{n+1}(\beta r). \tag{8.30}$$

Using equations (8.15)–(8.18), the resulting displacement components become,

$$u_r = \left(f' + \frac{n}{r}h_z + ikh_r\right)\cos(n\theta)e^{i(kz-\omega t)}, \tag{8.31}$$

$$u_\theta = \left(-\frac{n}{r}f + ikh_r - h_z'\right)\sin(n\theta)e^{i(kz-\omega t)}, \tag{8.32}$$

$$u_z = \left(ikf - h_r' - \frac{(n+1)h_r}{r}\right)\cos(n\theta)e^{i(kz-\omega t)}. \tag{8.33}$$

8.1.1 Exercise

Prove equation (8.10).

Hint: Use the following expressions,

$$
\begin{aligned}
\mathbf{H} &= H_r\mathbf{e}_r + H_\theta\mathbf{e}_\theta + H_z\mathbf{e}_z, \\
\nabla^2 &= \frac{\partial^2}{\partial r^2} + \frac{1}{r}\frac{\partial}{\partial r} + \frac{1}{r^2}\frac{\partial^2}{\partial\theta^2} + \frac{\partial^2}{\partial z^2}, \\
\mathbf{e}_r &= \cos\theta\mathbf{e}_x + \sin\theta\mathbf{e}_y, \\
\mathbf{e}_\theta &= -\sin\theta\mathbf{e}_x + \cos\theta\mathbf{e}_y.
\end{aligned} \tag{8.34}
$$

8.1.2 Strain and stress components

In cylindrical coordinates, the strain components S_{ij} are,

$$
\begin{aligned}
S_{rr} &= \frac{\partial u_r}{\partial r}, \\
S_{\theta\theta} &= \frac{1}{r}\frac{\partial u_\theta}{\partial\theta} + \frac{u_r}{r}, \\
S_{zz} &= \frac{\partial u_z}{\partial z}, \\
S_{r\theta} &= \frac{1}{2}\left(\frac{1}{r}\frac{\partial u_r}{\partial\theta} + \frac{\partial u_\theta}{\partial r} - \frac{u_\theta}{r}\right), \\
S_{rz} &= \frac{1}{2}\left(\frac{\partial u_z}{\partial r} + \frac{\partial u_r}{\partial z}\right), \\
S_{\theta z} &= \frac{1}{2}\left(\frac{\partial u_\theta}{\partial z} + \frac{1}{r}\frac{\partial u_z}{\partial\theta}\right),
\end{aligned} \tag{8.35}
$$

and the stress components T_{ij} of an isotropic solid read,

$$
\begin{aligned}
T_{ij} &= \lambda(S_{rr} + S_{\theta\theta} + S_{zz})\delta_{ij} + 2\mu S_{ij} \\
&= \lambda\left(\frac{\partial u_r}{\partial r} + \frac{u_r}{r} + \frac{1}{r}\frac{\partial u_\theta}{\partial \theta} + \frac{\partial u_z}{\partial z}\right)\delta_{ij} + 2\mu S_{ij},
\end{aligned}
\tag{8.36}
$$

where δ_{ij} is a Kronecker delta.

8.2 Longitudinal vibrations

In the case of longitudinal vibrations (figure 8.2), there is no θ dependence nor a u_θ displacement component ($n = 0$). Then,

$$
h_r = B_2 J_1(\beta r),
\tag{8.37}
$$

so,

$$
u_r = (-\alpha A J_1(\alpha r) + ik B_2 J_1(\beta r))e^{i(kz-\omega t)},
\tag{8.38}
$$

$$
u_z = \left(ik A J_0(\alpha r) - B_2\frac{d}{dr}(J_1(\beta r)) - \frac{B_2}{r}J_1(\beta r)\right)e^{i(kz-\omega t)}.
\tag{8.39}
$$

Using the Bessel function identities,

$$
\frac{2n}{x}J_n(x) = J_{n-1}(x) + J_{n+1}(x),
\tag{8.40}
$$

$$
\frac{dJ_n(x)}{dx} = \frac{1}{2}(J_{n-1}(x) - J_{n+1}(x)),
\tag{8.41}
$$

for $n = 0$ allows us to write,

$$
u_r = (-\alpha A J_1(\alpha r) + ik B_2 J_1(\beta r))e^{i(kz-\omega t)},
\tag{8.42}
$$

$$
u_z = (ik A J_0(\alpha r) - B_2\beta J_0(\beta r))e^{i(kz-\omega t)}.
\tag{8.43}
$$

Next, expressions for two stress components in cylindrical coordinates applicable to an isotropic solid are used,

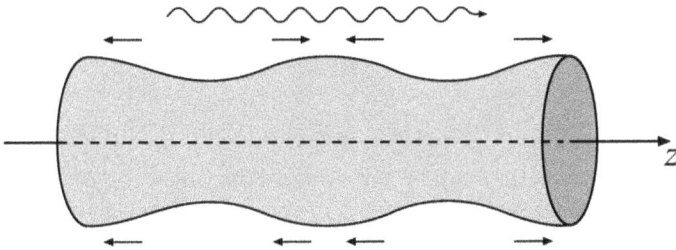

Figure 8.2. Sketch of a cylindrical solid rod subject to longitudinal deformation.

$$T_{rr} = 2\mu \frac{\partial u_r}{\partial r} + \lambda \left(\frac{u_r}{r} + \frac{\partial u_r}{\partial r} + \frac{\partial u_z}{\partial z} \right), \tag{8.44}$$

$$T_{rz} = \mu \left(\frac{\partial u_r}{\partial z} + \frac{\partial u_z}{\partial r} \right). \tag{8.45}$$

The boundary conditions for a cylindrical rod are,

$$T_{rr}(r = a) = 0, \tag{8.46}$$

$$T_{rz}(r = a) = 0, \tag{8.47}$$

where a is the radius of the rod.

Substituting equations (8.42) and (8.43) in equations (8.46) and (8.47), two linear equations in the unknowns A and B_2 are obtained and the determinantal equation becomes,

$$\frac{2\alpha}{a}(\beta^2 + k^2)J_1(\alpha a)J_1(\beta a) - (\beta^2 - k^2)^2 J_0(\alpha a)J_1(\beta a) - 4k^2\alpha\beta J_1(\alpha a)J_0(\beta a) = 0. \tag{8.48}$$

The latter equation is known as the Pochhammer frequency equation for longitudinal mode vibrations of a solid rod. Solving the above expression leads to discrete solutions $k = k_i$ as a function of ω where i labels the solution.

In figure 8.3 a program is provided for computing dispersion curves from the Pochhammer frequency equation. In figure 8.4 a numerical solution of the Pochhammer frequency equation is presented.

Equation (8.47) gives a relation between the two coefficients A and B_2,

$$\frac{\partial u_r}{\partial z} + \frac{\partial u_z}{\partial r} = -ikA\alpha J_1(\alpha a) - k^2 B_2 J_1(\beta a) - ikA\alpha J_1(\alpha a) + \beta^2 B_2 J_1(\beta a) = 0, \tag{8.49}$$

i.e.,

$$\frac{A}{B_2} = \frac{\beta^2 - k^2}{2ik\alpha} \frac{J_1(\beta a)}{J_1(\alpha a)}. \tag{8.50}$$

8.2.1 Exercise

Derive the Pochhammer frequency equation from the stress-free boundary conditions.

8.2.2 Exercise

Show that the latter equation can be rewritten as the following relation between the phase velocity $c_p = \omega/k$ and fd where f is the frequency and $d = 2a$ is the rod diameter,

```matlab
close all
clear all

aa = 1.3e0;                            % radius

numb_ii = 30;                          % number of frequency values
numb_jj = 100000;                      % number of k values
numb_bands = 10;                       % solving for max 'numb_bands' bands

freq_vec = zeros(numb_ii,1);           % initialization of frequency
                                       % vector to solve for
k_vec = zeros(numb_jj,1);              % initialization of k vector
cc_sol = zeros(numb_bands,numb_ii);    % initialization of k solution vector
                                       % (one k vector per band)

determ_real_vec=zeros(numb_jj,1);
determ_imag_vec=zeros(numb_jj,1);
determ_abs_vec=zeros(numb_jj,1);

isol_vec = zeros(numb_ii,1);
mat = zeros(3,3);                      % initialization of determinant matrix
max_freq = 5e4;                        % maximum frequency value (k*a)
max_a_vs_lambda = 0.8e0;               % maximum a/lambda value to search for

EE = 1e11;
rho = 2e3;
nu = 0.29e0;

lambda = EE*nu/(1e0 + nu)/(1e0-2e0*nu);
mu = EE/2e0/(1e0 + nu);

c_L2 = (lambda + 2e0*mu)/rho;
c_T2 = mu/rho;

c0 = sqrt(EE/rho);
for ii=1:numb_ii
  a_vs_lambda = 0e0 + (ii-1)/(numb_ii-1)*max_a_vs_lambda;
  a_vs_lambda_vec(ii) = a_vs_lambda;
  kk = 2e0*pi*a_vs_lambda/aa;
  isol = 0;
  for jj=1:numb_jj
    freq = 1e0*kk*sqrt(c_T2)/2e0/pi + ...
        (jj-1)/(numb_jj-1)*(1.2e0*kk*sqrt(c_T2)/2e0/pi);
    omega = 2e0*pi*freq;
    c_vec(jj) = 2e0*pi*freq/kk/c0;
    alpha = sqrt(omega^2/c_L2 - kk^2);
    beta = sqrt(omega^2/c_T2 - kk^2);
    determ = 2e0*alpha/aa*(beta^2 + ...
        kk^2)*besselj(1,alpha*aa)*besselj(1,beta*aa) - ...
        (beta^2-kk^2)^2*besselj(0,alpha*aa)*besselj(1,beta*aa) - ...
        4e0*kk^2*alpha*beta*besselj(1,alpha*aa)*besselj(0,beta*aa);

    determ_real_vec(jj) = real(determ);
```

Figure 8.3. Program for computing dispersion curves by solving the Pochhammer frequency equation.

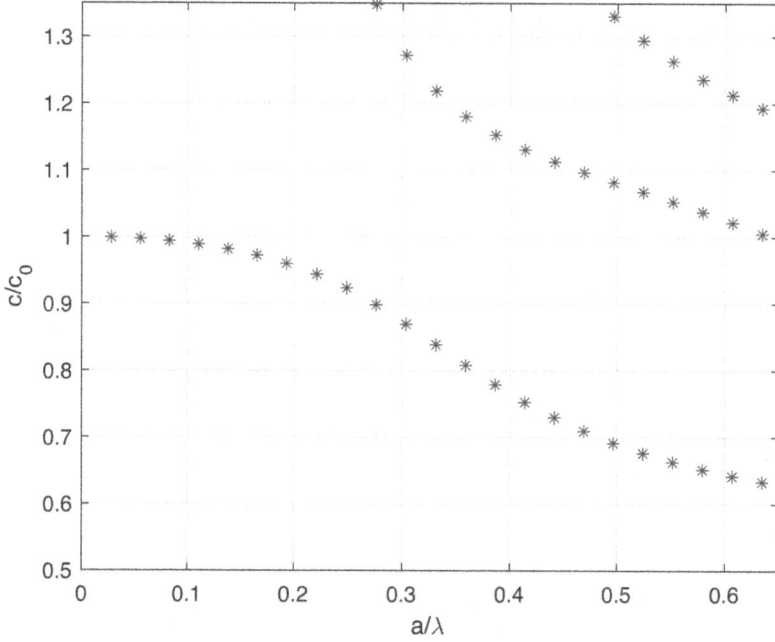

Figure 8.4. Numerical solution of the Pochhammer frequency equation. Parameters are radius $a = 1$ m, E modulus $E = 1 \times 10^{11}$ Pa, mass density $\rho = 2000$ kg m^{-3}, and Poisson ratio $\nu = 0.29$.

$$\frac{2}{c_T^2}\sqrt{\frac{c_p^2}{c_L^2} - 1}\, J_1\!\left(\frac{\pi}{c_p}(fd)\sqrt{c_p^2/c_L^2 - 1}\right) J_1\!\left(\frac{\pi}{c_p}(fd)\sqrt{c_p^2/c_T^2 - 1}\right)$$

$$-\frac{\pi(fd)}{c_p^3}\left(c_p^2/c_T^2 - 2\right)^2 J_0\!\left(\frac{\pi}{c_p}(fd)\sqrt{c_p^2/c_L^2 - 1}\right) J_1\!\left(\frac{\pi}{c_p}(fd)\sqrt{c_p^2/c_T^2 - 1}\right) \qquad (8.51)$$

$$-\frac{4\pi(fd)}{c_p^3}\sqrt{c_p^2/c_L^2 - 1}\,\sqrt{c_p^2/c_T^2 - 1}\, J_1\!\left(\frac{\pi}{c_p}(fd)\sqrt{c_p^2/c_L^2 - 1}\right) J_0\!\left(\frac{\pi}{c_p}(fd)\sqrt{c_p^2/c_T^2 - 1}\right) = 0.$$

8.2.3 Exercise

Plot the phase velocity c_p versus fd for an aluminum beam ($c_L = 6300$ m s^{-1}, and $c_T = 3100$ m s^{-1}) in the range $fd = 0 - 3 \times 10^4$ Hz m.

8.3 Elasticity equations

In the absence of body forces, the equation-of-motion for an isotropic elastic solid is the Navier equations, from equation (3.28),

$$(\lambda + \mu)\boldsymbol{\nabla}\boldsymbol{\nabla} \cdot \mathbf{u} + \mu\boldsymbol{\nabla}^2\mathbf{u} = \rho\frac{\partial^2\mathbf{u}}{\partial t^2}, \qquad (8.52)$$

which, using the identity,

$$\nabla^2 \mathbf{u} = \nabla \nabla \cdot \mathbf{u} - \nabla \times \nabla \times \mathbf{u}, \tag{8.53}$$

can be recast as,

$$(\lambda + 2\mu)\nabla \nabla \cdot \mathbf{u} - \mu \nabla \times \nabla \times \mathbf{u} = \rho \frac{\partial^2 \mathbf{u}}{\partial t^2}. \tag{8.54}$$

Another useful form of the Navier equations follows from equation (8.54),

$$(\lambda + 2\mu)\frac{\partial \phi}{\partial r} - \frac{2\mu}{r}\frac{\partial \omega_z}{\partial \theta} + 2\mu\frac{\partial \omega_\theta}{\partial z} = \rho \frac{\partial^2 u_r}{\partial t^2}, \tag{8.55}$$

$$(\lambda + 2\mu)\frac{1}{r}\frac{\partial \phi}{\partial \theta} - 2\mu\frac{\partial \omega_r}{\partial z} + 2\mu\frac{\partial \omega_z}{\partial r} = \rho \frac{\partial^2 u_\theta}{\partial t^2}, \tag{8.56}$$

$$(\lambda + 2\mu)\frac{\partial \phi}{\partial z} - \frac{2\mu}{r}\frac{\partial}{\partial r}(r\omega_\theta) + \frac{2\mu}{r}\frac{\partial \omega_r}{\partial \theta} = \rho \frac{\partial^2 u_z}{\partial t^2}, \tag{8.57}$$

where,

$$\phi = \nabla \cdot \mathbf{u} = \frac{1}{r}\frac{\partial(r u_r)}{\partial r} + \frac{1}{r}\frac{\partial u_\theta}{\partial \theta} + \frac{\partial u_z}{\partial z}, \tag{8.58}$$

$$2\boldsymbol{\omega} = (2\omega_r, 2\omega_\theta, 2\omega_z) = \nabla \times \mathbf{u}, \tag{8.59}$$

i.e.,

$$2\omega_r = \frac{1}{r}\frac{\partial u_z}{\partial \theta} - \frac{\partial u_\theta}{\partial z}, \tag{8.60}$$

$$2\omega_\theta = \frac{\partial u_r}{\partial z} - \frac{\partial u_z}{\partial r}, \tag{8.61}$$

$$2\omega_z = \frac{1}{r}\left(\frac{\partial(r u_\theta)}{\partial r} - \frac{\partial u_r}{\partial \theta}\right). \tag{8.62}$$

8.4 Torsional waves

Torsional waves result when u_r and u_z vanish (figure 8.5). In this case, $n = 0$ and equation (8.56) becomes,

$$\frac{\partial^2 u_\theta}{\partial r^2} + \frac{1}{r}\frac{\partial u_\theta}{\partial r} - \frac{u_\theta}{r^2} + \frac{\partial^2 u_\theta}{\partial z^2} = \frac{1}{c_T^2}\frac{\partial^2 u_\theta}{\partial t^2}. \tag{8.63}$$

Solutions are assumed in the form of plane waves,

$$u_\theta = V(r)e^{i(kz - \omega t)}, \tag{8.64}$$

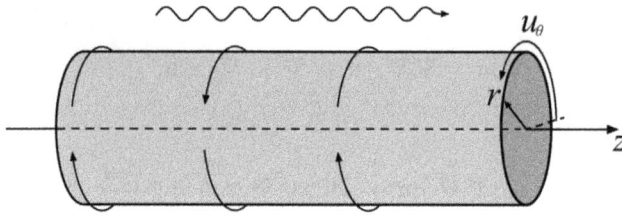

Figure 8.5. Sketch of a cylindrical solid rod subject to torsional deformation.

i.e.,

$$\frac{d^2V}{dr^2} + \frac{1}{r}\frac{dV}{dr} - \frac{V}{r^2} - k^2V = -\frac{\omega^2}{c_T^2}V. \tag{8.65}$$

Thus,

$$u_\theta = \frac{B}{\beta}J_1(\beta r)e^{i(kz-\omega t)}, \tag{8.66}$$

where B is arbitrary and determined by initial conditions. Imposing stress-free conditions at the rod surface implies,

$$\left[\frac{\beta}{r}\left(J_0(\beta r) - \frac{J_1(\beta r)}{\beta r}\right) - \frac{J_1(\beta r)}{r^2}\right]\Bigg|_{r=a} = 0 \implies \tag{8.67}$$

$$\frac{\beta}{a}J_0(\beta a) - \frac{J_1(\beta a)}{a^2} - \frac{J_1(\beta a)}{a^2} = 0 \implies \tag{8.68}$$

$$\beta a J_0(\beta a) = 2J_1(\beta a), \tag{8.69}$$

where the Bessel function identities, equations (8.40) and (8.41), and,

$$T_{r\theta}(r=a) = \mu\left[\frac{1}{r}\frac{\partial u_r}{\partial \theta} + r\frac{\partial}{\partial r}\left(\frac{u_\theta}{r}\right)\right]\Bigg|_{r=a} = 0 \implies \tag{8.70}$$

$$\frac{\partial}{\partial r}\left(\frac{u_\theta}{r}\right)\Bigg|_{r=a} = 0 \implies \tag{8.71}$$

$$\frac{d}{dr}\left(\frac{J_1(\beta r)}{r}\right)\Bigg|_{r=a} = 0 \implies \tag{8.72}$$

$$\left(\frac{\beta}{r}\frac{dJ_1(\beta r)}{d(\beta r)} - \frac{J_1(\beta r)}{r^2}\right)\Bigg|_{r=a} = 0, \tag{8.73}$$

were used.

From equation (8.69) the first three roots are,

$$\beta_0 = 0,$$
$$\beta_1 a = 5.126,$$
$$\beta_2 a = 8.417.$$

Consider the lowest torsional mode $\beta_0 = 0$. Using the asymptotic relation,

$$J_1(x) \rightarrow \frac{1}{2}x \text{ as } x \rightarrow 0, \tag{8.74}$$

it follows from equation (8.66) that,

$$u_\theta = \frac{1}{2}Bre^{i(kz-\omega t)}. \tag{8.75}$$

Evidently, the motion is a rotation of each cross section as a whole about its center. As $\beta_0 = 0$, the phase velocity $c_p = \omega/k$ is equal to c_T independent of frequency. Hence, the lowest mode is not dispersive yet all higher modes are.

8.5 Flexural waves

The flexural modes are solutions to the elastic equations that depend on the azimuthal coordinate θ (figure 8.6). The most important flexural waves are those corresponding to $n = 1$ for which the general displacement components are,

$$u_r = U(r)\cos\theta e^{i(kz-\omega t)}, \tag{8.76}$$

$$u_\theta = V(r)\sin\theta e^{i(kz-\omega t)}, \tag{8.77}$$

$$u_z = W(r)\cos\theta e^{i(kz-\omega t)}. \tag{8.78}$$

According to references [3, 4],

$$U(r) = A\frac{dJ_1(\alpha r)}{dr} + Bk\frac{dJ_1(\beta r)}{dr} + \frac{C}{r}J_1(\beta r), \tag{8.79}$$

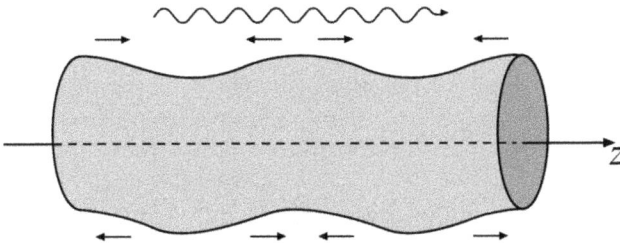

Figure 8.6. Sketch of a cylindrical solid rod subject to flexural deformations.

$$V(r) = -\frac{A}{r}J_1(\alpha r) - \frac{Bk}{r}J_1(\beta r) - C\frac{dJ_1(\beta r)}{dr}, \tag{8.80}$$

$$W(r) = ikAJ_1(\alpha r) - iB\beta^2 J_1(\beta r). \tag{8.81}$$

8.5.1 Proof of equations (8.79)–(8.81)

Let us now derive equations (8.79)–(8.81) for flexural waves using equations (8.31)–(8.33) and the general expressions of the radial functions f, h_z, h_r.

From equations (8.31)–(8.33) it follows that,

$$U = f' + \frac{n}{r}h_z + ikh_r,$$

$$V = -\frac{n}{r}f + ikh_r - h_z', \tag{8.82}$$

$$W = ikf - h_r' - \frac{n+1}{r}h_r,$$

and,

$$f = A'J_n(\alpha r),$$

$$h_z = B'J_n(\beta r), \tag{8.83}$$

$$h_r = C'J_{n+1}(\beta r),$$

where A', B', C' are constants. Hence, for flexural waves ($n = 1$),

$$U(r) = A'\frac{d}{dr}(J_1(\alpha r)) + \frac{B'}{r}J_1(\beta r) + ikC'J_2(\beta r). \tag{8.84}$$

Since,

$$J_2(x) = \frac{J_1(x)}{x} - \frac{dJ_1(x)}{dx}, \tag{8.85}$$

the latter expression can be written as,

$$U(r) = A'\frac{d}{dr}(J_1(\alpha r)) + \frac{B'}{r}J_1(\beta r) + \frac{ikC'}{\beta r}J_1(\beta r) - ikC'\frac{dJ_1(\beta r)}{d(\beta r)}. \tag{8.86}$$

Setting,

$$A = A',$$

$$B = -i\frac{C'}{\beta}, \tag{8.87}$$

$$C = B' + \frac{ikC'}{\beta},$$

the expression for U becomes,

$$U(r) = A\frac{dJ_1(\alpha r)}{dr} + Bk\frac{dJ_1(\beta r)}{dr} + \frac{C}{r}J_1(\beta r). \tag{8.88}$$

Similar manipulations applied to V yield, upon using the Bessel function identities equations (8.40)–(8.41),

$$\begin{aligned}
V(r) &= -\frac{n}{r}f + ikh_r - h'_z = -A'\frac{J_1(\alpha r)}{r} + ikC'J_2(\beta r) - \beta B'\frac{dJ_1(\beta r)}{d(\beta r)} \\
&= -A'\frac{J_1(\alpha r)}{r} + ikC'\frac{J_1(\beta r)}{\beta r} - ikC'\frac{dJ_1(\beta r)}{d(\beta r)} - \beta B'\frac{dJ_1(\beta r)}{d(\beta r)} \\
&= -A'\frac{J_1(\alpha r)}{r} + ikC'\frac{J_1(\beta r)}{\beta r} - (B'\beta + ikC')\frac{dJ_1(\beta r)}{d(\beta r)} \\
&= -A\frac{J_1(\alpha r)}{r} - Bk\beta\frac{J_1(\beta r)}{\beta r} - \beta C\frac{dJ_1(\beta r)}{d(\beta r)} \\
&= -\frac{A}{r}J_1(\alpha r) - \frac{Bk}{r}J_1(\beta r) - C\frac{dJ_1(\beta r)}{dr}.
\end{aligned} \tag{8.89}$$

For W, the result is,

$$\begin{aligned}
W(r) &= ikf - h'_r - \frac{2h_r}{r} = ikA'J_1(\alpha r) - C'\frac{dJ_2(\beta r)}{dr} - 2C'\frac{J_2(\beta r)}{r} \\
&= ikA'J_1(\alpha r) - \beta C'J_1(\beta r) = ikAJ_1(\alpha r) - iB\beta^2 J_1(\beta r),
\end{aligned} \tag{8.90}$$

where the identity,

$$J_1(x) = \frac{dJ_2(x)}{dx} + \frac{2}{x}J_2(x), \tag{8.91}$$

is used in obtaining the third equality. The three expressions, equations (8.88)–(8.90), agree with equations (8.79)–(8.81) as we sought to demonstrate.

8.5.2 Dispersion equation for flexural waves ($n = 1$)

The dispersion equation is found by setting the stress components T_{rr}, T_{rz}, $T_{r\theta}$ to zero at $r = a$. This yields three equations in the coefficients A, B, C and the determinantal equation gives the frequency condition. The determinantal equation becomes [4], using equations (8.79)–(8.81) for the displacement,

$$\begin{vmatrix} (a_1 - a_2)J_1(\alpha a) + a_2 J_3(\alpha a) & a_3(J_3(\beta a) - 3J_1(\beta a)) & a_4(J_0(\beta a) - J_2(\beta a)) - a_5 J_1(\beta a) \\ a_5 J_1(\alpha a) - a_6(J_0(\alpha a) - J_2(\alpha a)) & a_4(J_2(\beta a) - J_0(\beta a)) + a_5 J_1(\beta a) & \left(a_3 - \frac{a_5}{2}\right)J_1(\beta a) + \frac{a_3}{2}(J_1(\beta a) - J_3(\beta a)) + \frac{a_4}{2}(J_0(\beta a) - J_2(\beta a)) \\ a_7(J_0(\alpha a) - J_2(\alpha a)) & a_8(J_0(\beta a) - J_2(\beta a)) & a_9 J_1(\beta a) \end{vmatrix} = 0,$$

where the following dimension-less parameters are used, introducing the bulk modulus E, Poisson ratio ν, and mass density ρ,

$$c = -\omega/k,$$

$$c_0 = \sqrt{\frac{E}{\rho}},$$

$$a_1 = \frac{1}{1+\nu} - \left(\frac{c}{c_0}\right)^2\left(\frac{1}{1-\nu}\right),$$

$$a_2 = \frac{1}{2}\left(\frac{c}{c_0}\right)^2\left(\frac{1-2\nu}{1-\nu}\right) - \frac{1}{2(1+\nu)},$$

$$a_3 = \left(\frac{c}{c_0}\right)^2 - \frac{1}{2(1+\nu)},$$

$$a_4 = \frac{1}{(1+\nu)ka}\left[2\left(\frac{c}{c_0}\right)^2(1+\nu) - 1\right]^{1/2},$$

$$a_5 = \frac{1}{(1+\nu)\frac{1}{2}(ka)^2},$$

$$a_6 = \frac{1}{(1+\nu)ka}\left[\left(\frac{c}{c_0}\right)^2\frac{(1+\nu)(1-2\nu)}{1-\nu} - 1\right]^{1/2},$$

$$a_7 = (ka)a_6,$$

$$a_8 = ka\left[1 - \left(\frac{c}{c_0}\right)^2(1+\nu)\right]a_4,$$

$$a_9 = \frac{1}{2}(ka)a_5,$$

$$\alpha a = ka(1+\nu)a_7,$$

$$\beta a = (ka)^2(1+\nu)a_4.$$

(8.92)

8.6 General three-dimensional dispersion equation for infinite cylindrical rods

In this section, the complete set of governing equations to determine the general three-dimensional displacement vector of infinite cylindrical rods and the corresponding dispersion equations are presented. With the functional dependence on the azimuthal (θ) and axial (z) coordinates already established, equations (8.15)–(8.18), the boundary conditions ($T_{rr}(a) = T_{r\theta}(a) = T_{rz}(a) = 0$) for infinite cylindrical rods eventually reduce to conditions on the radial functions f, $h_r = -h_\theta$, h_z only.

The starting point is,

$$u_r = \left(f' + \frac{n}{r}h_z + ikh_r\right)\cos(n\theta)\exp(i(kz - \omega t)),$$

$$u_\theta = \left(-\frac{n}{r}f + ikh_r - h_z'\right)\sin(n\theta)\exp(i(kz - \omega t)), \tag{8.93}$$

$$u_z = \left(ikf - h_r' - \frac{(n + 1)h_r}{r}\right)\cos(n\theta)\exp(i(kz - \omega t)).$$

Earlier, it was obtained that (arbitrary coefficients A, B, C),

$$f = AJ_n(\alpha r),$$
$$h_z = BJ_n(\beta r), \tag{8.94}$$
$$h_r = CJ_{n+1}(\beta r).$$

The relevant stress components are,

$$T_{rr} = \lambda\left(\frac{\partial u_r}{\partial r} + \frac{u_r}{r} + \frac{1}{r}\frac{\partial u_\theta}{\partial \theta} + \frac{\partial u_z}{\partial z}\right) + 2\mu\frac{\partial u_r}{\partial r}, \tag{8.95}$$

$$T_{r\theta} = \mu\left(\frac{1}{r}\frac{\partial u_r}{\partial \theta} + \frac{\partial u_\theta}{\partial r} - \frac{u_\theta}{r}\right), \tag{8.96}$$

$$T_{rz} = \mu\left(\frac{\partial u_r}{\partial z} + \frac{\partial u_z}{\partial r}\right). \tag{8.97}$$

Expressions are necessary for the following terms,

$$\frac{\partial u_r}{\partial r} = \left(A\alpha^2 J_n''(\alpha r) + \frac{n}{r}B\beta J_n'(\beta r) - \frac{n}{r^2}BJ_n(\beta r) + ikC\beta J_{n+1}'(\beta r)\right)\cos(n\theta)\exp(i(kz - \omega t)), \tag{8.98}$$

$$\frac{u_r}{r} = \left(A\alpha\frac{J_n'(\alpha r)}{r} + \frac{n}{r^2}BJ_n(\beta r) + ikC\frac{J_{n+1}(\beta r)}{r}\right)\cos(n\theta)\exp(i(kz - \omega t)), \tag{8.99}$$

$$\frac{1}{r}\frac{u_\theta}{\partial \theta} = \left(-\frac{n^2}{r^2}AJ_n(\alpha r) - \frac{\beta n}{r}BJ_n'(\beta r) + iknC\frac{J_{n+1}(\beta r)}{r}\right)\cos(n\theta)\exp(i(kz - \omega t)), \tag{8.100}$$

$$\frac{\partial u_z}{\partial z} = \left(-k^2 AJ_n(\alpha r) - ikC\beta J_{n+1}'(\beta r) - \frac{ik(n + 1)CJ_{n+1}(\beta r)}{r}\right)\cos(n\theta)\exp(i(kz - \omega t)), \tag{8.101}$$

$$\frac{1}{r}\frac{\partial u_r}{\partial \theta} = \left(-nA\alpha\frac{J_n'(\alpha r)}{r} - \frac{n^2}{r^2}BJ_n(\beta r) - iknC\frac{J_{n+1}(\beta r)}{r}\right)\sin(n\theta)\exp(i(kz - \omega t)), \tag{8.102}$$

$$\frac{\partial u_\theta}{\partial r} = \left(\frac{n}{r^2}AJ_n(\alpha r) - \frac{n}{r}A\alpha J_n'(\alpha r) - B\beta^2 J_n''(\beta r) + ikC\beta J_{n+1}'(\beta r)\right)\sin(n\theta)\exp(i(kz - \omega t)), \tag{8.103}$$

$$\frac{u_\theta}{r} = \left(-A\frac{n}{r^2}J_n(\alpha r) - \frac{B\beta J_n'(\beta r)}{r} + ikC\frac{J_{n+1}(\beta r)}{r}\right)\sin(n\theta)\exp(i(kz - \omega t)), \quad (8.104)$$

$$\frac{\partial u_r}{\partial z} = \left(ikA\alpha J_n'(\alpha r) + \frac{iknB}{r}J_n(\beta r) - k^2CJ_{n+1}(\beta r)\right)\cos(n\theta)\exp(i(kz - \omega t)), \quad (8.105)$$

$$\frac{\partial u_z}{\partial r} = \left(ik\alpha A J_n'(\alpha r) - C\beta^2 J_{n+1}''(\beta r) + \frac{(n+1)}{r^2}CJ_{n+1}(\beta r) - \frac{(n+1)}{r}\beta C J_{n+1}'(\beta r)\right)\cos(n\theta)\exp(i(kz - \omega t)). \quad (8.106)$$

Three linear equations can now be written down in the coefficients A, B, C by using the boundary conditions. Setting $T_{rr}(a) = 0$,

$$\left[(\lambda + 2\mu)\alpha^2 J_n''(\alpha a) + \lambda\alpha\frac{J_n'(\alpha a)}{a} - \lambda\frac{n^2}{a^2}J_n(\alpha a) - \lambda k^2 J_n(\alpha a)\right]A$$

$$+ \left[(\lambda + 2\mu)\left(\frac{n}{a}\beta J_n'(\beta a) - \frac{n}{a^2}J_n(\beta a)\right) + \lambda\frac{n}{a^2}J_n(\beta a) - \lambda\frac{n\beta}{a}J_n'(\beta a)\right]B$$

$$+ \left[\begin{array}{c}(\lambda + 2\mu)ik\beta J_{n+1}'(\beta a) + ik\lambda\dfrac{J_{n+1}(\beta a)}{a} + ink\lambda\dfrac{J_{n+1}(\beta a)}{a} \\ + \lambda\left(-ik\beta J_{n+1}'(\beta a) - ik\dfrac{(n+1)J_{n+1}(\beta a)}{a}\right)\end{array}\right]C = 0. \quad (8.107)$$

Setting $T_{r\theta}(a) = 0$,

$$\left[-n\alpha\frac{J_n'(\alpha a)}{a} + 2\frac{n}{a^2}J_n(\alpha a) - \frac{n}{a}\alpha J_n'(\alpha a)\right]A$$

$$+ \left[-\frac{n^2}{a^2}J_n(\beta a) - \beta^2 J_n''(\beta a) + \beta\frac{J_n'(\beta a)}{a}\right]B \quad (8.108)$$

$$+ \left[-ikn\frac{J_{n+1}(\beta a)}{a} + ik\beta J_{n+1}'(\beta a) - ik\frac{J_{n+1}(\beta a)}{a}\right]C = 0.$$

Setting $T_{rz}(a) = 0$,

$$\left[2ik\alpha J_n'(\alpha a)\right]A$$

$$+ \left[\frac{ikn}{a}J_n(\beta a)\right]B \quad (8.109)$$

$$+ \left[-k^2 J_{n+1}(\beta a) - \beta^2 J_{n+1}''(\beta a) + \frac{n+1}{a^2}J_{n+1}(\beta a) - \frac{n+1}{a}\beta J_{n+1}'(\beta a)\right]C = 0.$$

Solving the determinantal equation in the coefficients A, B, C, the dispersion equation $(\omega_i(k))$ is obtained where i is the band index.

8.7 Program for computing three-dimensional dispersion curves of infinite cylindrical rods

In this section, some relations between parameters of relevance for computing three-dimensional dispersion curves of infinite cylindrical rods are given. It is convenient to fix the Poisson ratio at some value, say, $\nu = 0.29$. The following set of relations is useful,

$$\lambda = \frac{E\nu}{(1 + \nu)(1 - 2\nu)},$$

$$\mu = \frac{E}{2(1 + \nu)} \Longrightarrow$$

$$\frac{\lambda}{\mu} = \frac{2\nu}{1 - 2\nu},$$

$$c_L = \sqrt{\frac{\lambda + 2\mu}{\rho}},$$

$$c_T = \sqrt{\frac{\mu}{\rho}} \Longrightarrow$$

$$\frac{c_L^2}{c_T^2} = \frac{\lambda + 2\mu}{\mu} = \frac{2\nu}{1 - 2\nu} + 2,$$

$$k_L = \frac{\omega}{c_L},$$

$$k_T = \frac{\omega}{c_T},$$

$$\alpha^2 = \frac{\omega^2}{c_L^2} - k^2 = k_L^2 - k^2,$$

$$\beta^2 = \frac{\omega^2}{c_T^2} - k^2 = k_T^2 - k^2 = \frac{c_L^2}{c_T^2} k_L^2 - k^2.$$

$$(8.110)$$

The following exact expressions for certain first- and second-order derivatives of Bessel functions will also be used,

$$J_n''(x) = \frac{1}{4} J_{n+2}(x) - \frac{1}{2} J_n(x) + \frac{1}{4} J_{n-2}(x),$$

$$J_{n+1}''(x) = \frac{1}{4} J_{n+3}(x) - \frac{1}{2} J_{n+1}(x) + \frac{1}{4} J_{n-1}(x),$$

$$J_n'(x) = \frac{1}{2} J_{n-1}(x) - \frac{1}{2} J_{n+1}(x),$$

$$J_{n+1}'(x) = \frac{1}{2} J_n(x) - \frac{1}{2} J_{n+2}(x),$$

$$(8.111)$$

to evaluate Bessel functions at $x = \alpha a, \beta a$.

In figures 8.7–8.9, a Matlab code for calculating dispersion curves of infinite cylindrical rods for an arbitrary azimuthal index n is shown.

```
%%%%%%%%%%%%%%%%%%%%%%%%%%%%%%%%%%%%%%%%%%%%%%%%%%%%%%%%%%%%%%%%%%%%%%%%%%%%
%                                                                        %
% Program for calculating dispersion curves of infinite cylindrical rods %
% Arbitrary azimuthal index n can be chosen                              %
%                                                                        %
%%%%%%%%%%%%%%%%%%%%%%%%%%%%%%%%%%%%%%%%%%%%%%%%%%%%%%%%%%%%%%%%%%%%%%%%%%%%

close all
clear all

format long

nu = 0.29e0;                            % Poisson constant
lam_o_mu = 2e0*nu/(1e0 - 2e0*nu);       % lambda_over_mu ratio
cL_o_cT_sq = lam_o_mu + 2e0;            % cL_o_cT squared: (c_L/c_T)^2
                                        % = (lambda + 2*mu)/mu

nn = 0;                                 % azimuthal integer (n=1 for flexural waves)

aa = 0.1e0;                             % radius

numb_ii = 100;                          % number of k values
numb_jj = 10000;                        % number of k_L values
numb_bands = 10;                        % solving for max 'numb_bands' bands

kka_vec = zeros(numb_ii,1);             % initialization of k vector to solve for
kla_vec = zeros(numb_jj,1);             % initialization of k_L vector
kla_sol = zeros(numb_bands,numb_ii);    % initialization of k_L solution vector
                                        % (one k_L vector per band)
isol_vec = zeros(numb_ii,1);
mat = zeros(3,3);                       % initialization of determinant matrix
max_kka = 0.2e0*2e0*pi;                 % maximum k*a value (k*a)
max_kla = 0.5e0*2e0*pi;                 % maximum k_L*a value (a*omega/c_L)

for ii=1:numb_ii
  kka = 0e0 + (ii-1)/(numb_ii-1)*max_kka;
  kka_vec(ii) = kka;
  isol = 0;
  determ_vec = zeros(numb_jj,1);
  for jj=1:numb_jj
    kla = 0e0 + (jj-1)/(numb_jj-1)*max_kla;
    kla_vec(jj) = kla;

    bet_a = sqrt(cL_o_cT_sq*kla^2 - kka^2);
    alp_a = sqrt(kla^2 - kka^2);

    Jn_pp_alp_a = 1e0/4e0*besselj(nn+2,alp_a) ...
        - 1e0/2e0*besselj(nn,alp_a) + 1e0/4e0*besselj(nn-2,alp_a);
    Jn_pp_bet_a = 1e0/4e0*besselj(nn+2,bet_a) ...
        - 1e0/2e0*besselj(nn,bet_a) + 1e0/4e0*besselj(nn-2,bet_a);
    Jnp1_pp_bet_a = 1e0/4e0*besselj(nn+3,bet_a) ...
        - 1e0/2e0*besselj(nn+1,bet_a) + 1e0/4e0*besselj(nn-1,bet_a);
```

Figure 8.7. Program for calculating dispersion curves of infinite cylindrical rods for an arbitrary azimuthal index n (page 1).

```
Jn_p_alp_a = 1e0/2e0*besselj(nn-1,alp_a) ...
        - 1e0/2e0*besselj(nn+1,alp_a);
Jn_p_bet_a = 1e0/2e0*besselj(nn-1,bet_a) ...
        - 1e0/2e0*besselj(nn+1,bet_a);
Jnp1_p_bet_a = 1e0/2e0*besselj(nn,bet_a) ...
        - 1e0/2e0*besselj(nn+2,bet_a);

Jn_alp_a = besselj(nn,alp_a);
Jn_bet_a = besselj(nn,bet_a);
Jnp1_alp_a = besselj(nn+1,alp_a);
Jnp1_bet_a = besselj(nn+1,bet_a);

% matrix entries

mat(1,1) = cL_o_cT_sq*alp_a^2*Jn_pp_alp_a + alp_a*Jn_p_alp_a ...
        - nn^2*Jn_alp_a - kka^2*Jn_alp_a;
mat(1,2) = cL_o_cT_sq*( nn*bet_a*Jn_p_bet_a - nn*Jn_bet_a ) ...
        + nn*Jn_bet_a - nn*bet_a*Jn_p_bet_a;
mat(1,3) = cL_o_cT_sq*i*kka*bet_a*Jnp1_p_bet_a + i*kka*Jnp1_bet_a ...
        + i*nn*kka*Jnp1_bet_a ...
        - i*kka*bet_a*Jnp1_p_bet_a - (nn + 1)*i*kka*Jnp1_bet_a;

mat(2,1) = - nn*alp_a*Jn_p_alp_a + ...
        2e0*nn*Jn_alp_a - nn*alp_a*Jn_p_alp_a;
mat(2,2) = - nn^2*Jn_bet_a - bet_a^2*Jn_pp_bet_a ...
        + bet_a*Jn_p_bet_a;
mat(2,3) = - i*kka*nn*Jnp1_bet_a + i*kka*bet_a*Jnp1_p_bet_a ...
        - i*kka*Jnp1_bet_a;

mat(3,1) = 2e0*i*kka*alp_a*Jn_p_alp_a;
mat(3,2) = i*kka*nn*Jn_bet_a;
mat(3,3) = -kka^2*Jnp1_bet_a - bet_a^2*Jnp1_pp_bet_a ...
        + (nn+1)*Jnp1_bet_a - (nn+1)*bet_a*Jnp1_p_bet_a;

if (jj>1)
  determ_old = abs(determ);
end

determ = det(mat);
determ_real_vec(jj) = real(determ);
determ_imag_vec(jj) = imag(determ);
determ_abs_vec(jj) = abs(determ);

if (jj>2 && determ_abs_vec(jj-1)<determ_abs_vec(jj-2) ...
        && determ_abs_vec(jj-1)<determ_abs_vec(jj))
    isol = isol + 1;
    if (isol<numb_bands)
      kla_sol(isol,ii) = kla_vec(jj-1);
    end
end

determ_vec(jj)=abs(determ);
determ_new = abs(determ);
```

Figure 8.8. Program for calculating dispersion curves of infinite cylindrical rods for an arbitrary azimuthal index n (page 2).

```
    end
    isol_vec(ii) = isol;
end

subplot(3,1,1),plot(kla_vec,determ_real_vec)
grid
subplot(3,1,2),plot(kla_vec,determ_imag_vec,'r')
grid
subplot(3,1,3),plot(kla_vec,determ_abs_vec,'g')
grid

figure

plot(kka_vec,kla_sol(1,:),'b*')
hold
plot(kka_vec,kla_sol(2,:),'b*')
plot(kka_vec,kla_sol(3,:),'b*')
plot(kka_vec,kla_sol(4,:),'b*')
grid
xlabel('k*a')
ylabel('k_L*a')
```

Figure 8.9. Program for calculating dispersion curves of infinite cylindrical rods for an arbitrary azimuthal index n (page 3).

8.8 Circumferential waves in a hollow elastic cylinder

A slightly more complicated geometry is the hollow cylinder (figure 8.10). Let us consider propagation of circumferential elastic waves. This problem was addressed in references [5–7]. Assume propagation of circumferential waves for which $u_z = 0$ and there is no dependence on the cylinder axis coordinate z. In this case, solutions in the form,

$$
\begin{aligned}
u_r &= u_r(r, \theta), \\
u_\theta &= u_\theta(r, \theta), \\
u_z &= 0,
\end{aligned}
\tag{8.112}
$$

will be sought where, according to equations (8.2) and (8.3) with $\phi = \Phi$ and $\psi = H_z$,

$$
u_r = \frac{\partial \phi}{\partial r} + \frac{1}{r}\frac{\partial \psi}{\partial \theta},
\tag{8.113}
$$

$$
u_\theta = \frac{1}{r}\frac{\partial \phi}{\partial \theta} - \frac{\partial \psi}{\partial r},
\tag{8.114}
$$

and the functions ϕ, ψ obey the wave equations, see equations (8.5) and (8.6),

$$
\left(\frac{\partial^2}{\partial r^2} + \frac{1}{r}\frac{\partial}{\partial r} + \frac{1}{r^2}\frac{\partial^2}{\partial \theta^2} \right)\phi + \frac{\omega^2}{c_L^2}\phi = 0,
\tag{8.115}
$$

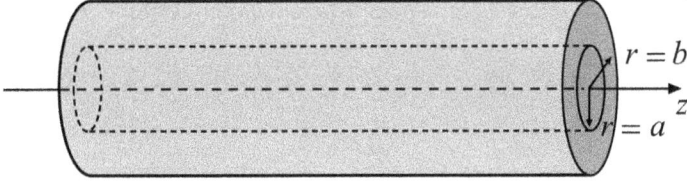

Figure 8.10. Sketch of a cylindrical hollow solid rod.

$$\left(\frac{\partial^2}{\partial r^2} + \frac{1}{r}\frac{\partial}{\partial r} + \frac{1}{r^2}\frac{\partial^2}{\partial \theta^2}\right)\psi + \frac{\omega^2}{c_T^2}\psi = 0. \tag{8.116}$$

At the cylinder surface ($r = b$ and $r = a$), the stress components T_{rr}, $T_{r\theta}$, T_{rz} vanish. The strain and stress components follow from equations (8.95) and (8.96), however, only expressions for the stress components T_{rr}, $T_{r\theta}$ are needed since T_{rz} vanishes identically,

$$T_{rr} = \lambda\left(\frac{\partial u_r}{\partial r} + \frac{u_r}{r} + \frac{1}{r}\frac{\partial u_\theta}{\partial \theta}\right) + 2\mu\frac{\partial u_r}{\partial r}, \tag{8.117}$$

$$T_{r\theta} = \mu\left(\frac{\partial u_\theta}{\partial r} - \frac{u_\theta}{r} + \frac{1}{r}\frac{\partial u_r}{\partial \theta}\right). \tag{8.118}$$

It is easy to guess the form of separable solutions,

$$\phi = \Phi(r)e^{i(kb\theta - \omega t)},$$
$$\psi = \Psi(r)e^{i(kb\theta - \omega t)}.$$

Substitution into equations (8.115) and (8.116) yields,

$$\Phi'' + \frac{1}{r}\Phi' + \left[\left(\frac{\omega}{c_L}\right)^2 - \left(\frac{kb}{r}\right)^2\right]\Phi = 0, \tag{8.119}$$

$$\Psi'' + \frac{1}{r}\Psi' + \left[\left(\frac{\omega}{c_T}\right)^2 - \left(\frac{kb}{r}\right)^2\right]\Psi = 0, \tag{8.120}$$

for which the general solution is,

$$\Phi(r) = A_1 J_{kb}\left(\frac{\omega r}{c_L}\right) + A_2 Y_{kb}\left(\frac{\omega r}{c_L}\right), \tag{8.121}$$

$$\Psi(r) = B_1 J_{kb}\left(\frac{\omega r}{c_T}\right) + B_2 Y_{kb}\left(\frac{\omega r}{c_T}\right). \tag{8.122}$$

Note that Y_{kb} must be included in the general solution since the Neumann function does not diverge within the hollow cylinder region where $r \geqslant a > 0$.

Substitution of the latter expressions into the stress components gives,

$$T_{rr}(r, \theta) = \frac{\mu e^{i(kb\theta - \omega t)}}{r^2}[\chi^2 r^2 \Phi'' + (\chi^2 - 2)r\Phi' - (\chi^2 - 2)k^2 b^2 \Phi + 2ikb(r\Psi' - \Psi)], \quad (8.123)$$

$$T_{r\theta}(r, \theta) = \frac{\mu e^{i(kb\theta - \omega t)}}{r^2}[-r^2 \Psi'' + r\Psi' - k^2 b^2 \Psi + 2ikb(r\Phi' - \Phi)], \quad (8.124)$$

where $\chi = c_L/c_T$.

8.8.1 Exercise

Derive equations (8.123)–(8.124) by noting that,

$$\lambda + 2\mu = \mu\chi^2,$$
$$\lambda = \mu(\chi^2 - 2).$$

In the usual case of traction-free boundaries, the four unknown constants A_1, A_2, B_1, B_2 are determined by setting the stress components T_{rr}, $T_{r\theta}$ to zero at $r = b$ and $r = a$. This yields four linear equations in four unknowns and non-trivial solutions are found by setting the determinant of the system of equations to zero. The following dispersion equation is then obtained,

$$|D_{nm}| = 0, \quad (8.125)$$

where Bessel function identities for the first- and second-order derivatives are used to get,

$$D_{11} = \chi^2 b^2 \frac{\omega^2}{c_L^2}\left(\frac{1}{4}J_{M+2}\left(\frac{\omega b}{c_L}\right) - \frac{1}{2}J_M\left(\frac{\omega b}{c_L}\right) + \frac{1}{4}J_{M-2}\left(\frac{\omega b}{c_L}\right)\right)$$
$$+ (\chi^2 - 2)b\frac{\omega}{c_L}\left(\frac{1}{2}J_{M-1}\left(\frac{\omega b}{c_L}\right) - \frac{1}{2}J_{M+1}\left(\frac{\omega b}{c_L}\right)\right) - (\chi^2 - 2)k^2 b^2 J_M\left(\frac{\omega b}{c_L}\right), \quad (8.126)$$

$$D_{12} = \chi^2 b^2 \frac{\omega^2}{c_L^2}\left(\frac{1}{4}Y_{M+2}\left(\frac{\omega b}{c_L}\right) - \frac{1}{2}Y_M\left(\frac{\omega b}{c_L}\right) + \frac{1}{4}Y_{M-2}\left(\frac{\omega b}{c_L}\right)\right)$$
$$+ (\chi^2 - 2)b\frac{\omega}{c_L}\left(\frac{1}{2}Y_{M-1}\left(\frac{\omega b}{c_L}\right) - \frac{1}{2}Y_{M+1}\left(\frac{\omega b}{c_L}\right)\right) - (\chi^2 - 2)k^2 b^2 Y_M\left(\frac{\omega b}{c_L}\right), \quad (8.127)$$

$$D_{13} = 2ikb\left(b\frac{\omega}{c_T}\left(\frac{1}{2}J_{M-1}\left(\frac{\omega b}{c_T}\right) - \frac{1}{2}J_{M+1}\left(\frac{\omega b}{c_T}\right)\right) - J_M\left(\frac{\omega b}{c_T}\right)\right), \quad (8.128)$$

$$D_{14} = 2ikb\left(b\frac{\omega}{c_T}\left(\frac{1}{2}Y_{M-1}\left(\frac{\omega b}{c_T}\right) - \frac{1}{2}Y_{M+1}\left(\frac{\omega b}{c_T}\right)\right) - Y_M\left(\frac{\omega b}{c_T}\right)\right), \quad (8.129)$$

$$D_{21} = \chi^2 a^2 \frac{\omega^2}{c_L^2}\left(\frac{1}{4}J_{M+2}\left(\frac{\omega a}{c_L}\right) - \frac{1}{2}J_M\left(\frac{\omega a}{c_L}\right) + \frac{1}{4}J_{M-2}\left(\frac{\omega a}{c_L}\right)\right)$$
$$+ (\chi^2 - 2)a\frac{\omega}{c_L}\left(\frac{1}{2}J_{M-1}\left(\frac{\omega a}{c_L}\right) - \frac{1}{2}J_{M+1}\left(\frac{\omega a}{c_L}\right)\right) - (\chi^2 - 2)k^2 a^2 J_M\left(\frac{\omega a}{c_L}\right), \tag{8.130}$$

$$D_{22} = \chi^2 a^2 \frac{\omega^2}{c_L^2}\left(\frac{1}{4}Y_{M+2}\left(\frac{\omega a}{c_L}\right) - \frac{1}{2}Y_M\left(\frac{\omega a}{c_L}\right) + \frac{1}{4}Y_{M-2}\left(\frac{\omega a}{c_L}\right)\right)$$
$$+ (\chi^2 - 2)a\frac{\omega}{c_L}\left(\frac{1}{2}Y_{M-1}\left(\frac{\omega a}{c_L}\right) - \frac{1}{2}Y_{M+1}\left(\frac{\omega a}{c_L}\right)\right) - (\chi^2 - 2)k^2 a^2 Y_M\left(\frac{\omega a}{c_L}\right), \tag{8.131}$$

$$D_{23} = 2ikb\left(a\frac{\omega}{c_T}\left(\frac{1}{2}J_{M-1}\left(\frac{\omega a}{c_T}\right) - \frac{1}{2}J_{M+1}\left(\frac{\omega a}{c_T}\right)\right) - J_M\left(\frac{\omega a}{c_T}\right)\right), \tag{8.132}$$

$$D_{24} = 2ikb\left(a\frac{\omega}{c_T}\left(\frac{1}{2}Y_{M-1}\left(\frac{\omega a}{c_T}\right) - \frac{1}{2}Y_{M+1}\left(\frac{\omega a}{c_T}\right)\right) - Y_M\left(\frac{\omega a}{c_T}\right)\right), \tag{8.133}$$

$$D_{31} = 2ikb\left(b\frac{\omega}{c_L}\left(\frac{1}{2}J_{M-1}\left(\frac{\omega b}{c_L}\right) - \frac{1}{2}J_{M+1}\left(\frac{\omega b}{c_L}\right)\right) - J_M\left(\frac{\omega b}{c_L}\right)\right), \tag{8.134}$$

$$D_{32} = 2ikb\left(b\frac{\omega}{c_L}\left(\frac{1}{2}Y_{M-1}\left(\frac{\omega b}{c_L}\right) - \frac{1}{2}Y_{M+1}\left(\frac{\omega b}{c_L}\right)\right) - Y_M\left(\frac{\omega b}{c_L}\right)\right), \tag{8.135}$$

$$D_{33} = -b^2 \frac{\omega^2}{c_T^2}\left(\frac{1}{4}J_{M+2}\left(\frac{\omega b}{c_T}\right) - \frac{1}{2}J_M\left(\frac{\omega b}{c_T}\right) + \frac{1}{4}J_{M-2}\left(\frac{\omega b}{c_T}\right)\right)$$
$$+ b\frac{\omega}{c_T}\left(\frac{1}{2}J_{M-1}\left(\frac{\omega b}{c_T}\right) - \frac{1}{2}J_{M+1}\left(\frac{\omega b}{c_T}\right)\right) - k^2 b^2 J_M\left(\frac{\omega b}{c_T}\right), \tag{8.136}$$

$$D_{34} = -b^2 \frac{\omega^2}{c_T^2}\left(\frac{1}{4}Y_{M+2}\left(\frac{\omega b}{c_T}\right) - \frac{1}{2}Y_M\left(\frac{\omega b}{c_T}\right) + \frac{1}{4}Y_{M-2}\left(\frac{\omega b}{c_T}\right)\right)$$
$$+ b\frac{\omega}{c_T}\left(\frac{1}{2}Y_{M-1}\left(\frac{\omega b}{c_T}\right) - \frac{1}{2}Y_{M+1}\left(\frac{\omega b}{c_T}\right)\right) - k^2 b^2 Y_M\left(\frac{\omega b}{c_T}\right), \tag{8.137}$$

$$D_{41} = 2ika\left(a\frac{\omega}{c_L}\left(\frac{1}{2}J_{M-1}\left(\frac{\omega a}{c_L}\right) - \frac{1}{2}J_{M+1}\left(\frac{\omega a}{c_L}\right)\right) - J_M\left(\frac{\omega a}{c_L}\right)\right), \tag{8.138}$$

$$D_{42} = 2ika\left(a\frac{\omega}{c_L}\left(\frac{1}{2}Y_{M-1}\left(\frac{\omega a}{c_L}\right) - \frac{1}{2}Y_{M+1}\left(\frac{\omega a}{c_L}\right)\right) - Y_M\left(\frac{\omega a}{c_L}\right)\right), \qquad (8.139)$$

$$\begin{aligned}D_{43} = &-a^2\frac{\omega^2}{c_T^2}\left(\frac{1}{4}J_{M+2}\left(\frac{\omega a}{c_T}\right) - \frac{1}{2}J_M\left(\frac{\omega a}{c_T}\right) + \frac{1}{4}J_{M-2}\left(\frac{\omega a}{c_T}\right)\right) \\ &+a\frac{\omega}{c_T}\left(\frac{1}{2}J_{M-1}\left(\frac{\omega a}{c_T}\right) - \frac{1}{2}J_{M+1}\left(\frac{\omega a}{c_T}\right)\right) - k^2a^2J_M\left(\frac{\omega a}{c_T}\right),\end{aligned} \qquad (8.140)$$

$$\begin{aligned}D_{44} = &-a^2\frac{\omega^2}{c_T^2}\left(\frac{1}{4}Y_{M+2}\left(\frac{\omega a}{c_T}\right) - \frac{1}{2}Y_M\left(\frac{\omega a}{c_T}\right) + \frac{1}{4}Y_{M-2}\left(\frac{\omega a}{c_T}\right)\right) \\ &+a\frac{\omega}{c_T}\left(\frac{1}{2}Y_{M-1}\left(\frac{\omega a}{c_T}\right) - \frac{1}{2}Y_{M+1}\left(\frac{\omega a}{c_T}\right)\right) - k^2a^2Y_M\left(\frac{\omega a}{c_T}\right),\end{aligned} \qquad (8.141)$$

Note that due to the geometrical constraint, $\Phi(r, \theta) = \Phi(r, \theta + 2\pi)$, $\Psi(r, \theta) = \Psi(r, \theta + 2\pi)$, the product kb must be an integer which is denoted M in equations (8.126)–(8.141). From the functional dependence $e^{i(kb\theta - \omega t)}$ it follows that the angular phase velocity α of circumferential waves is,

$$\alpha = \frac{\omega}{kb} = \frac{\omega}{M}, \qquad (8.142)$$

and the phase velocity c_p becomes,

$$c_p = r\alpha = (\omega/k) \cdot (r/b) = \frac{\omega r}{M}. \qquad (8.143)$$

References

[1] Graff K F 1991 *Wave Motion in Elastic Solids* (New York: Dover)

[2] Morse P M and Feshbach H 1953 *Methods of Theoretical Physics* **vol 1** 1st edn (Minneapolis, MN: Feshbach Publishing)

[3] Love A E H 1906 *A Treatise on the Mathematical Theory of Elasticity* 2nd edn (Cambridge: Cambridge University Press)

[4] Norman Abramson H 1957 Flexural waves in elastic beams of circular cross section *J. Acoust. Soc. Am.* **29** 42–6

[5] Viktorov I A 1967 *Rayleigh and Lamb Waves–Physical Theory and Applications* (New York: Plenum)

[6] Qu J, Berthelot Y and Li Z 1996 Dispersion of guided circumferential waves in a circular annulus *Review of Progress in Quantitative Nondestructive Evaluation* ed D O Thompson and D E Chimenti (New York: Plenum) vol 15

[7] Rose J L 2004 *Ultrasonic Waves in Solid Media* (Cambridge: Cambridge University Press)

Part III

Piezoelectricity and applications

IOP Publishing

Piezoelectricity in Classical and Modern Systems

Morten Willatzen

Chapter 9

A piezoelectric toy model

In the first chapter in the second part of the book, a conceptually simple but instructive toy model of piezoelectricity is described using four mechanically connected point charges. The model system is originally due to Auld [1]. The first toy model demonstrates that if the system of point charges satisfies inversion symmetry, then piezoelectricity is absent. The second toy model shows that when the point charges are placed such that inversion symmetry is broken, piezoelectricity is induced. Both toy models are characterized by a total charge equal to zero.

9.1 Non-piezoelectric system

Consider the one-dimensional system shown in figure 9.1. Four charged particles are connected by mechanical forces and electrostatic forces. The outermost charges ($+q$ both) can move while the inner two charges ($-q$ both) cannot move due to the rigid connection between them. The same type of spring (with spring constant K) is connected between the outermost particles and their immediate neighbor. The unstrained length of the springs is l_0 and the distance from the center point between the two inner charges and the outer charge to the left (right) is l_a (l_b), respectively.

The mechanical force on particle a is,

$$f_a = K(l_a - l - l_0), \qquad (9.1)$$

whereas the mechanical force on particle b is,

$$f_b = -K(l_b - l - l_0). \qquad (9.2)$$

Evidently, the total force on particle a is,

$$(F_a)_x = K(l_a - l - l_0) + \frac{q^2}{4\pi\epsilon}\left(\frac{1}{(l_a - l)^2} + \frac{1}{(l_a + l)^2} - \frac{1}{(l_a + l_b)^2}\right), \qquad (9.3)$$

and, similarly, the total force on particle b is,

doi:10.1088/978-0-7503-5557-5ch9

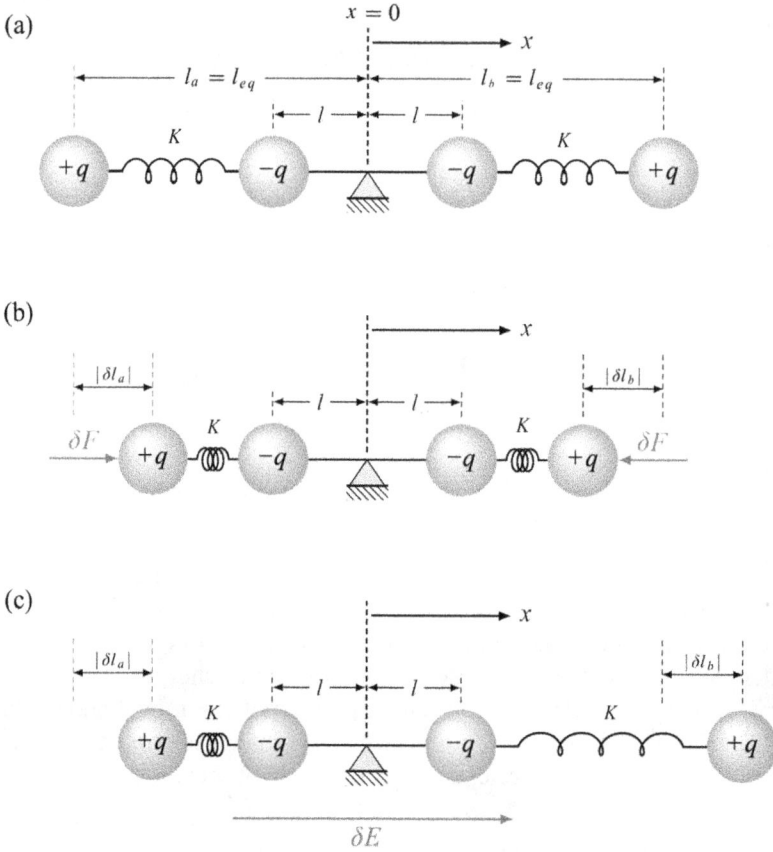

Figure 9.1. Non-piezoelectric system. Adapted from reference [1] with permission from Krieger Publishing Company. All rights reserved.

$$(F_b)_x = -K(l_b - l - l_0) - \frac{q^2}{4\pi\epsilon}\left(\frac{1}{(l_b - l)^2} + \frac{1}{(l_b + l)^2} - \frac{1}{(l_a + l_b)^2}\right), \quad (9.4)$$

where ϵ is the permittivity of the system. For a *stable* equilibrium,

$$\left(\frac{\partial[(F_a)_x]_{l_a = l_b}}{\partial l_a}\right)_{l_a = l_{equil}} > 0. \quad (9.5)$$

It is clear that,

$$(P_x)_{equil} = -ql_{equil} + ql - ql + ql_{equil} = 0, \quad (9.6)$$

so there is no *spontaneous electrical polarization*.

9.1.1 Mechanical force

Consider now the application of a mechanical force. Because the applied forces are symmetric, it is clear that points *a* and *b* are displaced inward by equal amounts,

$$|\delta l_a| = |\delta l_b|, \tag{9.7}$$

and the mechanical response is,

$$\delta L = -2|\delta l_a|. \tag{9.8}$$

The electrical response is,

$$\delta P_x = q|\delta l_a| - q|\delta l_b| = 0. \tag{9.9}$$

Hence, an applied mechanical force does not lead to an electrical response.

9.1.2 Electrical force

Consider now the application of an electric field. A new equilibrium situation is established for which,

$$(F_a)_x = (F_a)_{x_{equil}} + \left(\frac{\partial[(F_a)_x]}{\partial l_a}\right)_{equil} \delta l_a + \left(\frac{\partial[(F_a)_x]}{\partial l_b}\right)_{equil} \delta l_b, \tag{9.10}$$

$$(F_b)_x = (F_b)_{x_{equil}} + \left(\frac{\partial[(F_b)_x]}{\partial l_a}\right)_{equil} \delta l_a + \left(\frac{\partial[(F_b)_x]}{\partial l_b}\right)_{equil} \delta l_b, \tag{9.11}$$

and the equilibrium forces are zero. It follows immediately from inspection of equations (9.3) and (9.4) that, at equilibrium ($l_a = l_b$),

$$\left(\frac{\partial[(F_a)_x]}{\partial l_a}\right)_{equil} = -\left(\frac{\partial[(F_b)_x]}{\partial l_b}\right)_{equil} \equiv A, \tag{9.12}$$

$$\left(\frac{\partial[(F_a)_x]}{\partial l_b}\right)_{equil} = -\left(\frac{\partial[(F_b)_x]}{\partial l_a}\right)_{equil} \equiv B. \tag{9.13}$$

Upon application of the electrical field the total force on each charge is balanced, i.e.,

$$\begin{aligned} (F_a)_x + q\delta E_x &= 0, \\ (F_b)_x + q\delta E_x &= 0, \end{aligned} \tag{9.14}$$

and equations (9.10)–(9.13) lead to,

$$A\delta l_a + B\delta l_b = -q\delta E_x, \tag{9.15}$$

$$-B\delta l_a - A\delta l_b = -q\delta E_x, \tag{9.16}$$

so,

$$\delta l_b = -\delta l_a = \frac{q\delta E_x}{A - B}. \tag{9.17}$$

Hence, the electrical response is non-zero,

$$\delta P_x = -q\delta l_a + q\delta l_b = \frac{2q^2 \delta E_x}{A - B}, \tag{9.18}$$

but the mechanical response vanishes,

$$\delta L = \delta l_a + \delta l_b = 0. \tag{9.19}$$

9.2 Piezoelectric system

Consider next the one-dimensional system shown in figure 9.2. Four charged particles are connected by mechanical forces and electrostatic forces. The outermost charges ($+Rq$ to the left and $-Rq$ to the right) can move while the inner two charges ($-q$ to the left and $+q$ to the right) cannot due to the rigid connection between them.

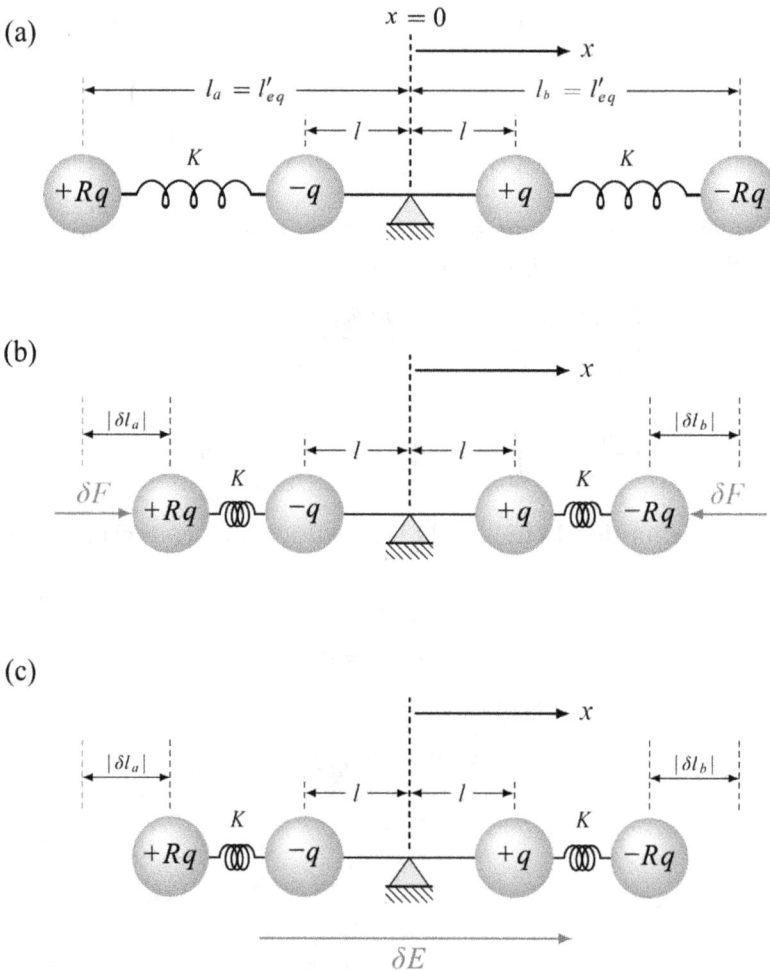

Figure 9.2. Piezoelectric system. Adapted from reference [1] with permission from Krieger Publishing Company. All rights reserved.

Again, the same type of spring (with spring constant K) is connected between the outermost charges and their immediate neighbor particle. The distance from the center point, located exactly at the middle point between the two inner charges, and the outer charge to the left (right) is l_a (l_b), respectively.

The two springs will be symmetrically compressed by the electrostatic charges. In equilibrium,

$$l_a = l_b = l'_{\text{equil}}. \tag{9.20}$$

The electrical polarization is,

$$P_{x\text{equil}} = -Rql'_{\text{equil}} + ql + ql - Rql'_{\text{equil}} = 2q\left(l - Rl'_{\text{equil}}\right). \tag{9.21}$$

Note that this value may be either zero or non-zero depending on the value of R, i.e., the charges on particles a and b! Hence, *a spontaneous polarization exists in the system* unless, by accidence, $R = l/l'_{\text{equil}}$!

From figure 9.2, it is clear that the mechanical and electrical responses to a mechanical or electrical force are both symmetric and non-zero,

$$\begin{aligned}
|\delta l_a| &= |\delta l_b|, \\
\delta L &= 2\delta l_a, \\
\delta P_x &= -Rq\delta l_a - Rq\delta l_b = -2Rq\delta l_a.
\end{aligned} \tag{9.22}$$

Compiling the results above, it was shown that,

$$\delta P_x = \chi \delta E_x + d\delta F_x, \tag{9.23}$$

$$\delta L = \hat{d}\delta E_x + s\delta F_x, \tag{9.24}$$

For a non-piezoelectric structure the coefficients d, $\hat{d}$ are zero! The latter expressions are known as the piezoelectric constitutive equations.

For a three-dimensional solid, the piezoelectric constitutive equations read,

$$\delta P_i = \chi_{ij}\delta E_j + d_{ij}\delta F_j, \tag{9.25}$$

$$\delta L_i = \hat{d}_{ij}\delta E_j + s_{ij}\delta F_j, \tag{9.26}$$

where i, j take the values 1, 2, 3 and, as always, summation over repeated indices applies.

9.2.1 Exercise

- Find explicit expressions for $(F_a)_x$ and $(F_b)_x$ in the piezoelectric system.
- Argue that upon application of an electric field, the following expressions apply in equilibrium,

$$\begin{aligned}
(F_a)_x + Rq\delta E_x &= 0, \\
(F_b)_x - Rq\delta E_x &= 0.
\end{aligned} \tag{9.27}$$

- Show that $\delta L = \delta l_a + \delta l_b = -\frac{2Rq}{A+B}\delta E_x$ and $\delta P_x = \frac{2R^2q^2}{A+B}\delta E_x$.

Hence,

$$\chi = \frac{2R^2q^2}{A+B},$$
$$\hat{d} = -\frac{2Rq}{A+B}.$$

(9.28)

9.2.2 Corollary on piezoelectricity and inversion asymmetry

As an important corollary, observe that the piezoelectric system shown in figure 9.2 displays inversion asymmetry around the centerline $x = 0$. Indeed, upon reflection in the plane $x = 0$, the charge $+Rq$ on the left takes the place of the charge $-Rq$ on the right and vice versa. Similar for the charges connected by the rigid rod: $-q$ on the left takes the place of $+q$ on the right and vice versa. Hence the system is not inversion-symmetric.

On the contrary, the non-piezoelectric system shown in figure 9.1 obeys inversion symmetry: $+q$ on the far left takes the place of the charge $+q$ on the far right upon reflection in the $x = 0$ plane, etc. Hence the non-piezoelectric system displays inversion symmetry.

It appears from the above considerations that piezoelectricity requires the system to be inversion-asymmetric!

Reference

[1] Auld B A 1990 *Acoustic Fields and Waves in Solids* **vol I** 2nd edn (Malabar, FL: Krieger)

IOP Publishing

Piezoelectricity in Classical and Modern Systems

Morten Willatzen

Chapter 10

Piezoelectricity in solid crystals

Having studied the basic ingredients for the existence of piezoelectricity in non-centrosymmetric structures in the previous chapter, chapter 10 is concerned with the piezoelectric terms allowed by symmetry in the general expression for the free energy of solid crystals. A systematic derivation of the piezoelectric tensor for zinc blende cubic, and wurtzite hexagonal structures is given.

10.1 Piezoelectricity in cubic (diamond and zincblende) structures

Since the diamond structure (point group O_h ($m\bar{3}m$)) is centrosymmetric, this structure is not piezoelectric. Another important cubic structure, zincblende (point group T_d ($\bar{4}3m$)), is non-centrosymmetric, and therefore materials with a T_d ($\bar{4}3m$) crystal structure can be piezoelectric. First, observe that the following transformations are symmetries of the T_d ($\bar{4}3m$) point group,

$$
\begin{aligned}
x &\to -x, & y &\to -y, & z &\to z, \\
x &\to -x, & y &\to y, & z &\to -z, \\
x &\to x, & y &\to -y, & z &\to -z \\
x &\to y, & y &\to -x, & z &\to -z \\
y &\to z, & z &\to -y, & x &\to -x \\
x &\to z, & z &\to -x, & y &\to -y,
\end{aligned}
\tag{10.1}
$$

Then, the possible piezoelectric contributions to the free energy are,

$$
F_{\text{piezo}} = -2e_{xyz}S_{xy}E_z - 2e_{xzy}S_{xz}E_y - 2e_{yzx}S_{yz}E_x.
\tag{10.2}
$$

The three terms obey the symmetries in equation (10.1) if,

$$
e_{xyz} = e_{xzy} = e_{yzx},
\tag{10.3}
$$

and,

$$
F_{\text{piezo}} = -2e_{xyz}(S_{xy}E_z + S_{xz}E_y + S_{yz}E_x) = -e_{xyz}(S_6E_z + S_5E_y + S_4E_x).
\tag{10.4}
$$

The piezoelectric terms in the stress–strain relations can now be obtained,

$$T_1 = \frac{\partial F_{\text{piezo}}}{\partial S_1} = 0,$$

$$T_2 = \frac{\partial F_{\text{piezo}}}{\partial S_2} = 0,$$

$$T_3 = \frac{\partial F_{\text{piezo}}}{\partial S_3} = 0,$$

$$T_4 = \frac{\partial F_{\text{piezo}}}{\partial S_4} = -e_{xyz} E_x,$$

$$T_5 = \frac{\partial F_{\text{piezo}}}{\partial S_5} = -e_{xyz} E_y,$$

$$T_6 = \frac{\partial F_{\text{piezo}}}{\partial S_6} = -e_{xyz} E_z.$$

(10.5)

Writing the piezoelectric part of the stress–strain relations in Voigt matrix form yields,

$$
\begin{bmatrix} T_1 \\ T_2 \\ T_3 \\ T_4 \\ T_5 \\ T_6 \end{bmatrix} = -
\begin{bmatrix}
0 & 0 & 0 \\
0 & 0 & 0 \\
0 & 0 & 0 \\
e_{xyz} & 0 & 0 \\
0 & e_{xyz} & 0 \\
0 & 0 & e_{xyz}
\end{bmatrix}
\begin{bmatrix} E_x \\ E_y \\ E_z \end{bmatrix}.
$$

(10.6)

Hence, there is only one piezoelectric coefficient for the zincblende structure (T_d ($\bar{4}3m$)).

10.2 Piezoelectricity in hexagonal structures

In the following, let us derive the form of the piezoelectric tensor in two hexagonal non-centrosymmetric structures: C_{6v} ($6mm$) and D_{3h} ($\bar{6}m2$).

10.2.1 C_{6v} ($6mm$)

Piezoelectricity requires lack of inversion symmetry. An important hexagonal structure that lacks inversion symmetry is the C_{6v} (or $6mm$) structure. In this case, the most general form of the free energy terms related to piezoelectricity is,

$$F_{\text{piezo}} = -2e_{\xi\eta z} S_{\xi\eta} E_z - 2e_{\xi z\eta} S_{\xi z} E_\eta - 2e_{\eta z\xi} S_{\eta z} E_\xi - e_{zzz} S_{zz} E_z,$$

(10.7)

due to the requirement that the same number of ξ's and η's must appear in the symmetry-allowed free energy terms Note that if inversion was a symmetry, then all coefficients e_{ijk} in equation (10.7) vanish since all the terms on the right-hand side change sign under inversion. We also used that a reflection in the x–y plane ($z \to -z$) is *not* a symmetry of C_{6v} ($6mm$) which allows for piezoelectric free energy contributions from e_{ijk} terms involving an odd number of z indices.

Replacing the ξ, η variables by x, y in equation (10.7) yields,

$$
\begin{aligned}
F_{\text{piezo}} =& - 2e_{\xi\eta z}(S_{xx} + S_{yy})E_z - 2e_{\xi z\eta}(S_{xz} + iS_{yz})(E_x - iE_y) \\
& - 2e_{\eta z\xi}(S_{xz} - iS_{yz})(E_x + iE_y) - e_{zzz}S_{zz}E_z \\
=& - 2e_{\xi\eta z}E_zS_{xx} - 2e_{\xi\eta z}E_zS_{yy} - (2e_{\xi z\eta}(E_x - iE_y) + 2e_{\eta z\xi}(E_x + iE_y))S_{xz} \\
& - i(2e_{\xi z\eta}(E_x - iE_y) - 2e_{\eta z\xi}(E_x + iE_y))S_{yz} - e_{zzz}E_zS_{zz}.
\end{aligned}
\tag{10.8}
$$

Rewriting the latter expression using Voigt notation gives,

$$
\begin{aligned}
F_{\text{piezo}} =& - 2e_{\xi\eta z}E_zS_1 - 2e_{\xi\eta z}E_zS_2 - \frac{1}{2}(2e_{\xi z\eta}(E_x - iE_y) + 2e_{\eta z\xi}(E_x + iE_y))S_5 \\
& - \frac{1}{2}i(2e_{\xi z\eta}(E_x - iE_y) - 2e_{\eta z\xi}(E_x + iE_y))S_4 - e_{zzz}E_zS_3.
\end{aligned}
\tag{10.9}
$$

The piezoelectric terms in the stress–strain relations can now be obtained,

$$
\begin{aligned}
T_1 =& \frac{\partial F_{\text{piezo}}}{\partial S_1} = -2e_{\xi\eta z}E_z, \\[4pt]
T_2 =& \frac{\partial F_{\text{piezo}}}{\partial S_2} = -2e_{\xi\eta z}E_z, \\[4pt]
T_3 =& \frac{\partial F_{\text{piezo}}}{\partial S_3} = -e_{zzz}E_z, \\[4pt]
T_4 =& \frac{\partial F_{\text{piezo}}}{\partial S_4} = -\frac{1}{2}i(2e_{\xi z\eta}(E_x - iE_y) - 2e_{\eta z\xi}(E_x + iE_y)) \\
=& - i(e_{\xi z\eta} - e_{\eta z\xi})E_x - (e_{\xi z\eta} + e_{\eta z\xi})E_y, \\[4pt]
T_5 =& \frac{\partial F_{\text{piezo}}}{\partial S_5} = -\frac{1}{2}(2e_{\xi z\eta}(E_x - iE_y) + 2e_{\eta z\xi}(E_x + iE_y)) \\
=& - (e_{\xi z\eta} + e_{\eta z\xi})E_x + i(e_{\xi z\eta} - e_{\eta z\xi})E_y, \\[4pt]
T_6 =& \frac{\partial F_{\text{piezo}}}{\partial S_6} = 0.
\end{aligned}
\tag{10.10}
$$

The latter expression can be further simplified by noting that a reflection in the plane perpendicular to the y axis is a symmetry of C_{6v},

$$
\begin{aligned}
x &\rightarrow x, \\
y &\rightarrow -y, \\
z &\rightarrow z,
\end{aligned}
$$

i.e.,

$$
\begin{aligned}
\xi &\rightarrow \eta, \\
\eta &\rightarrow \xi,
\end{aligned}
$$

so,

$$
e_{\xi z\eta} = e_{\eta z\xi}.
\tag{10.11}
$$

The expressions for T_4, T_5 can then be simplified,

$$T_4 = -2e_{\xi z \eta} E_y,$$
$$T_5 = -2e_{\xi z \eta} E_x. \tag{10.12}$$

Writing the piezoelectric part of the stress–strain relations in Voigt matrix form yields,

$$
\begin{bmatrix} T_1 \\ T_2 \\ T_3 \\ T_4 \\ T_5 \\ T_6 \end{bmatrix} = -
\begin{bmatrix}
0 & 0 & 2e_{\xi \eta z} \\
0 & 0 & 2e_{\xi \eta z} \\
0 & 0 & e_{zzz} \\
0 & 2e_{\xi z \eta} & 0 \\
2e_{\xi z \eta} & 0 & 0 \\
0 & 0 & 0
\end{bmatrix}
\begin{bmatrix} E_x \\ E_y \\ E_z \end{bmatrix}. \tag{10.13}
$$

Substituting,

$$e_{31} = 2e_{\xi \eta z},$$
$$e_{33} = e_{zzz},$$
$$e_{15} = 2e_{\xi z \eta},$$

gives,

$$
\begin{bmatrix} T_1 \\ T_2 \\ T_3 \\ T_4 \\ T_5 \\ T_6 \end{bmatrix} = -
\begin{bmatrix}
0 & 0 & e_{31} \\
0 & 0 & e_{31} \\
0 & 0 & e_{33} \\
0 & e_{15} & 0 \\
e_{15} & 0 & 0 \\
0 & 0 & 0
\end{bmatrix}
\begin{bmatrix} E_x \\ E_y \\ E_z \end{bmatrix}. \tag{10.14}
$$

Hence, only 3 independent piezoelectric coefficients exist in C_{6v} (6mm) structures.

10.2.2 Exercise

Use the symmetry properties of the variables ξ and η in subsection 3.2.1 to prove that the most general form of the piezoelectric tensor is given by F_{piezo} in equation (10.7).

10.2.3 D_{3h} ($\bar{6}m2$)

For the D_{3h} (or $\bar{6}m2$) hexagonal structure, the most general form of the free energy terms related to piezoelectricity is,

$$F_{\text{piezo}} = -e_{\xi \xi \xi} S_{\xi \xi} E_\xi - e_{\eta \eta \eta} S_{\eta \eta} E_\eta, \tag{10.15}$$

where it was used that a rotation of $2\pi/3$ about the z axis and a reflection in the x–y plane ($z \rightarrow -z$) are both symmetries of D_{3h} ($\bar{6}m2$).

A rotation of $2\pi/3$ about the z axis implies that,

$$S_{\xi\xi} E_\xi \rightarrow S_{\xi\xi} E_\xi,$$
$$S_{\eta\eta} E_\eta \rightarrow S_{\eta\eta} E_\eta,$$

(10.16)

since,

$$S_{\xi\xi} \rightarrow e^{i4\pi/3} S_{\xi\xi},$$
$$E_\xi \rightarrow e^{i2\pi/3} E_\xi$$
$$S_{\eta\eta} \rightarrow e^{-i4\pi/3} S_{\eta\eta},$$
$$E_\eta \rightarrow e^{-i2\pi/3} E_\eta.$$

(10.17)

Writing out equation (10.15) yields,

$$
\begin{aligned}
F_{\text{piezo}} = &-e_{\xi\xi\xi}(S_{xx} - S_{yy} + 2iS_{xy})(E_x + iE_y) - e_{\eta\eta\eta}(S_{xx} - S_{yy} - 2iS_{xy})(E_x - iE_y)\\
= &-(e_{\xi\xi\xi}E_x + ie_{\xi\xi\xi}E_y + e_{\eta\eta\eta}E_x - ie_{\eta\eta\eta}E_y)S_1\\
&-(-e_{\xi\xi\xi}E_x - ie_{\xi\xi\xi}E_y - e_{\eta\eta\eta}E_x + ie_{\eta\eta\eta}E_y)S_2\\
&-(ie_{\xi\xi\xi}E_x - e_{\xi\xi\xi}E_y - ie_{\eta\eta\eta}E_x - e_{\eta\eta\eta}E_y)S_6.
\end{aligned}
$$

(10.18)

The stress–strain terms involving piezoelectric contributions are therefore,

$$T_1 = \frac{\partial F_{\text{piezo}}}{\partial S_1} = -(e_{\xi\xi\xi}E_x + ie_{\xi\xi\xi}E_y + e_{\eta\eta\eta}E_x - ie_{\eta\eta\eta}E_y),$$

$$T_2 = \frac{\partial F_{\text{piezo}}}{\partial S_2} = -(-e_{\xi\xi\xi}E_x - ie_{\xi\xi\xi}E_y - e_{\eta\eta\eta}E_x + ie_{\eta\eta\eta}E_y),$$

(10.19)

$$T_6 = \frac{\partial F_{\text{piezo}}}{\partial S_6} = -(ie_{\xi\xi\xi}E_x - e_{\xi\xi\xi}E_y - ie_{\eta\eta\eta}E_x - e_{\eta\eta\eta}E_y).$$

Since the free energy is a real quantity, we also have $e_{\xi\xi\xi} = e_{\eta\eta\eta}$, and the latter relations simplify to,

$$T_1 = -2e_{\xi\xi\xi}E_x,$$
$$T_2 = 2e_{\xi\xi\xi}E_x,$$
$$T_6 = 2e_{\xi\xi\xi}E_y.$$

(10.20)

In matrix form, writing $e \equiv 2e_{\xi\xi\xi}$, we have established the piezoelectric constitutive equations of D_{3h} ($\bar{6}m2$),

$$
\begin{bmatrix} T_1 \\ T_2 \\ T_3 \\ T_4 \\ T_5 \\ T_6 \end{bmatrix}
= -
\begin{bmatrix}
e & 0 & 0 \\
-e & 0 & 0 \\
0 & 0 & 0 \\
0 & 0 & 0 \\
0 & 0 & 0 \\
0 & -e & 0
\end{bmatrix}
\begin{bmatrix} E_x \\ E_y \\ E_z \end{bmatrix}.
$$

(10.21)

10.2.4 Exercise

Use the symmetry properties of the variables ξ and η in subsection 3.2.3 to prove that the most general form of the piezoelectric tensor is given by F_{piezo} in equation (10.15).

IOP Publishing

Piezoelectricity in Classical and Modern Systems

Morten Willatzen

Chapter 11

Group theory, transformation properties, and application to material properties

In chapter 11, the transformation properties of the stiffness tensor, the piezoelectric tensor, and the permittivity tensor derived in chapter 2 are used to derive the general form of these quantities allowed by symmetry. This method serves as an alternative formal recipe to the method used in the preceding chapters and sections. In the first part of the chapter, the 32 three-dimensional point groups are classified according to whether they are piezoelectric, non-piezoelectric, polar (pyroelectric), non-polar, ferroelectric, or non-ferroelectric.

11.1 Group tables and material parameters

A point group is a group of symmetry operations that leave at least one point fixed (unmoved). Thus, all operations containing translations are excluded. Point groups can be subdivided into crystallographic and non-crystallographic point groups. A crystallographic point group is a point group that maps a point lattice onto itself. Consequently, rotations and roto-inversions are restricted to the well-known crystallographic cases 1, 2, 3, 4, 6 and $\bar{1}, \bar{2}, \bar{3}, \bar{4}, \bar{6}$; a result known as the crystallographic restriction theorem. There are 32 crystallographic point groups in three dimensions. There are no such restrictions on the non-crystallographic point groups. Thus, there are an infinite number of three-dimensional non-crystallographic point groups.

In figure 11.1, the different crystal classes and three-dimensional point groups are listed. The point groups are classified into centrosymmetric and non-centrosymmetric point groups. The non-centrosymmetric point groups can be subdivided into piezoelectric and polar point groups. Out of the 32 crystallographic point groups, 21 are non-centrosymmetric. 20 non-centrosymmetric point groups are piezoelectric. 10 of the piezoelectric point groups are polar with a non-zero spontaneous polarization. Ferroelectric materials display spontaneous polarization that can be inverted in the presence of a sufficiently strong electric field. Thus, ferroelectric materials all belong to the 10 polar point groups (tables 11.1–11.3).

doi:10.1088/978-0-7503-5557-5ch11

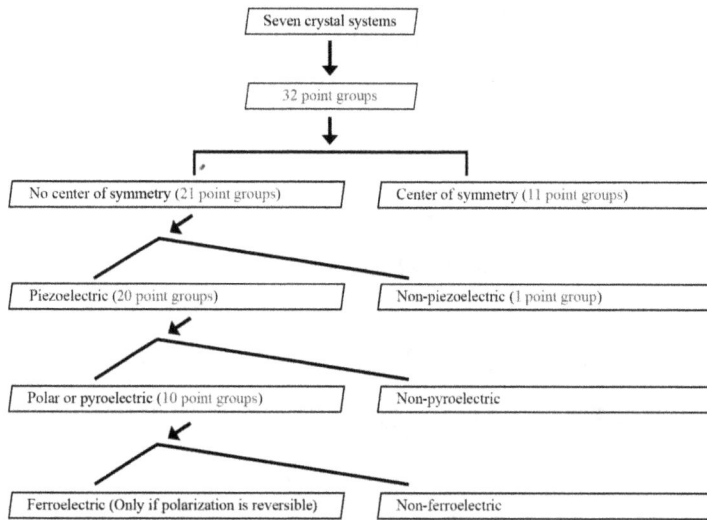

Figure 11.1. Schematic illustration of the different classes of crystal systems and their properties.

Table 11.1. The 32 three-dimensional crystallographic point groups. Hermann–Mauguin (left) and Schoenflies (right) symbols.

7 Crystal systems and 32 three-dimensional point groups													
Triclinic		Monoclinic		Orthorhombic		Tetragonal		Trigonal		Hexagonal		Cubic	
1	C_1	2	C_2	222	D_2	4	C_4	3	C_3	6	C_6	23	T
$\bar{1}$	C_i	m	C_s (C_{1h})	$2mm$	C_{2v}	$\bar{4}$	S_4	$\bar{3}$	C_{3i}	$\bar{6}$	C_{3h}	$\frac{2}{m}\bar{3}$	T_h
		$\frac{2}{m}$	C_{2h}	$\frac{2\,2\,2}{m\,m\,m}$	D_{2h}	$\frac{4}{m}$	C_{4h}	32	D_3	$\frac{6}{m}$	C_{6h}	432	O
						422	D_4	$3m$	C_{3v}	622	D_6	$\bar{4}3m$	T_d
						$4mm$	C_{4v}	$\bar{3}\frac{2}{m}$	D_{3d}	$6mm$	C_{6v}	$\frac{4}{m}\bar{3}\frac{2}{m}$	O_h
						$\bar{4}2m$	D_{2d}			$\bar{6}m2$	D_{3h}		
						$\frac{4\,2\,2}{m\,m\,m}$	D_{4h}			$\frac{6\,2\,2}{m\,m\,m}$	D_{6h}		

Table 11.2. The 20 three-dimensional piezoelectric crystallographic point groups. Hermann–Mauguin (left) and Schoenflies (right) symbols.

7 Crystal systems and 20 three-dimensional piezoelectric point groups													
Triclinic		Monoclinic		Orthorhombic		Tetragonal		Trigonal		Hexagonal		Cubic	
1	C_1	2	C_2	222	D_2	4	C_4	3	C_3	6	C_6	23	T
		m	C_s (C_{1h})	$2mm$	C_{2v}	$\bar{4}$	S_4	32	D_3	$\bar{6}$	C_{3h}	$\bar{4}3m$	T_d
						422	D_4	$3m$	C_{3v}	622	D_6		
						$4mm$	C_{4v}			$6mm$	C_{6v}		
						$\bar{4}2m$	D_{2d}			$\bar{6}m2$	D_{3h}		

Table 11.3. The 10 three-dimensional polar (pyroelectric) crystallographic point groups. Hermann–Mauguin (left) and Schoenflies (right) symbols. The polar point groups display spontaneous polarization.

7 Crystal systems and 10 three-dimensional polar (pyroelectric) point groups													
Triclinic		Monoclinic		Orthorhombic		Tetragonal		Trigonal		Hexagonal		Cubic	
1	C_1	2	C_2	2mm	C_{2v}	4	C_4	3	C_3	6	C_6		
		m	C_s (C_{1h})			4mm	C_{4v}	3m	C_{3v}	6mm	C_{6v}		

11.2 Stiffness matrices

Matrix elements c_{IJ} refer to the case where crystal axes X, Y, Z coincide with the coordinate axes x, y, z.

Triclinic system
21 constants

$$\begin{bmatrix} c_{11} & c_{12} & c_{13} & c_{14} & c_{15} & c_{16} \\ c_{12} & c_{22} & c_{23} & c_{24} & c_{25} & c_{26} \\ c_{13} & c_{23} & c_{33} & c_{34} & c_{35} & c_{36} \\ c_{14} & c_{24} & c_{34} & c_{44} & c_{45} & c_{46} \\ c_{15} & c_{25} & c_{35} & c_{45} & c_{55} & c_{56} \\ c_{16} & c_{26} & c_{36} & c_{46} & c_{56} & c_{66} \end{bmatrix}$$

Monoclinic system
13 constants

$$\begin{bmatrix} c_{11} & c_{12} & c_{13} & 0 & c_{15} & 0 \\ c_{12} & c_{22} & c_{23} & 0 & c_{25} & 0 \\ c_{13} & c_{23} & c_{33} & 0 & c_{35} & 0 \\ 0 & 0 & 0 & c_{44} & 0 & c_{46} \\ c_{15} & c_{25} & c_{35} & 0 & c_{55} & 0 \\ 0 & 0 & 0 & c_{46} & 0 & c_{66} \end{bmatrix}$$

Orthorhombic system
9 constants

$$\begin{bmatrix} c_{11} & c_{12} & c_{13} & 0 & 0 & 0 \\ c_{12} & c_{22} & c_{23} & 0 & 0 & 0 \\ c_{13} & c_{23} & c_{33} & 0 & 0 & 0 \\ 0 & 0 & 0 & c_{44} & 0 & 0 \\ 0 & 0 & 0 & 0 & c_{55} & 0 \\ 0 & 0 & 0 & 0 & 0 & c_{66} \end{bmatrix}$$

Tetragonal system

Classes 4, $\bar{4}$, $\dfrac{4}{m}$

7 constants

$$\begin{bmatrix} c_{11} & c_{12} & c_{13} & 0 & 0 & c_{16} \\ c_{12} & c_{11} & c_{13} & 0 & 0 & -c_{16} \\ c_{13} & c_{13} & c_{33} & 0 & 0 & 0 \\ 0 & 0 & 0 & c_{44} & 0 & 0 \\ 0 & 0 & 0 & 0 & c_{44} & 0 \\ c_{16} & -c_{16} & 0 & 0 & 0 & c_{66} \end{bmatrix}$$

Tetragonal system

Classes 422, 4mm, $\bar{4}2m$, $\dfrac{4}{m}\dfrac{2}{m}\dfrac{2}{m}$

6 constants

$$\begin{bmatrix} c_{11} & c_{12} & c_{13} & 0 & 0 & 0 \\ c_{12} & c_{11} & c_{13} & 0 & 0 & 0 \\ c_{13} & c_{13} & c_{33} & 0 & 0 & 0 \\ 0 & 0 & 0 & c_{44} & 0 & 0 \\ 0 & 0 & 0 & 0 & c_{44} & 0 \\ 0 & 0 & 0 & 0 & 0 & c_{66} \end{bmatrix}$$

Trigonal system

Classes 3, $\bar{3}$

7 constants

$$\begin{bmatrix} c_{11} & c_{12} & c_{13} & c_{14} & -c_{25} & 0 \\ c_{12} & c_{11} & c_{13} & -c_{14} & c_{25} & 0 \\ c_{13} & c_{13} & c_{33} & 0 & 0 & 0 \\ c_{14} & -c_{14} & 0 & c_{44} & 0 & c_{25} \\ -c_{25} & c_{25} & 0 & 0 & c_{44} & c_{14} \\ 0 & 0 & 0 & c_{25} & c_{14} & \frac{1}{2}(c_{11}-c_{12}) \end{bmatrix}$$

Trigonal system

Classes 32, 3m, $\bar{3}\dfrac{2}{m}$

6 constants

$$\begin{bmatrix} c_{11} & c_{12} & c_{13} & c_{14} & 0 & 0 \\ c_{12} & c_{11} & c_{13} & -c_{14} & 0 & 0 \\ c_{13} & c_{13} & c_{33} & 0 & 0 & 0 \\ c_{14} & -c_{14} & 0 & c_{44} & 0 & 0 \\ 0 & 0 & 0 & 0 & c_{44} & c_{14} \\ 0 & 0 & 0 & 0 & c_{14} & \frac{1}{2}(c_{11}-c_{12}) \end{bmatrix}$$

Hexagonal system

5 constants

$$\begin{bmatrix} c_{11} & c_{12} & c_{13} & 0 & 0 & 0 \\ c_{12} & c_{11} & c_{13} & 0 & 0 & 0 \\ c_{13} & c_{13} & c_{33} & 0 & 0 & 0 \\ 0 & 0 & 0 & c_{44} & 0 & 0 \\ 0 & 0 & 0 & 0 & c_{44} & 0 \\ 0 & 0 & 0 & 0 & 0 & \frac{1}{2}(c_{11}-c_{12}) \end{bmatrix}$$

Cubic system

3 constants

$$\begin{bmatrix} c_{11} & c_{12} & c_{12} & 0 & 0 & 0 \\ c_{12} & c_{11} & c_{12} & 0 & 0 & 0 \\ c_{12} & c_{12} & c_{11} & 0 & 0 & 0 \\ 0 & 0 & 0 & c_{44} & 0 & 0 \\ 0 & 0 & 0 & 0 & c_{44} & 0 \\ 0 & 0 & 0 & 0 & 0 & c_{44} \end{bmatrix}$$

$$
\begin{array}{c}
\text{Isotropic} \\
\text{2 constants}
\end{array}
\begin{bmatrix}
c_{11} & c_{12} & c_{12} & 0 & 0 & 0 \\
c_{12} & c_{11} & c_{12} & 0 & 0 & 0 \\
c_{12} & c_{12} & c_{11} & 0 & 0 & 0 \\
0 & 0 & 0 & \frac{1}{2}(c_{11} - c_{12}) & 0 & 0 \\
0 & 0 & 0 & 0 & \frac{1}{2}(c_{11} - c_{12}) & 0 \\
0 & 0 & 0 & 0 & 0 & \frac{1}{2}(c_{11} - c_{12})
\end{bmatrix}
$$

11.3 Piezoelectric matrices

The d_{iJ} matrix elements correspond to the situation where the coordinate axes x, y, z coincide with the crystal axes X, Y, Z. In cases where differences occur, the relevant e_{iJ} matrix elements are provided to the right of the matrices. d_{iJ} matrices are given for the 20 piezoelectric point groups.

$$
\text{Triclinic 1}
\begin{bmatrix}
d_{x1} & d_{x2} & d_{x3} & d_{x4} & d_{x5} & d_{x6} \\
d_{y1} & d_{y2} & d_{y3} & d_{y4} & d_{y5} & d_{y6} \\
d_{z1} & d_{z2} & d_{z3} & d_{z4} & d_{z5} & d_{z6}
\end{bmatrix}
$$

$$
\text{Monoclinic 2}
\begin{bmatrix}
0 & 0 & 0 & d_{x4} & 0 & d_{x6} \\
d_{y1} & d_{y2} & d_{y3} & 0 & d_{y5} & 0 \\
0 & 0 & 0 & d_{z4} & 0 & d_{z6}
\end{bmatrix}
$$

$$
\text{Monoclinic } m
\begin{bmatrix}
d_{x1} & d_{x2} & d_{x3} & 0 & d_{x5} & 0 \\
0 & 0 & 0 & d_{y4} & 0 & d_{y6} \\
d_{z1} & d_{z2} & d_{z3} & 0 & d_{z5} & 0
\end{bmatrix}
$$

$$
\text{Orthorhombic 222}
\begin{bmatrix}
0 & 0 & 0 & d_{x4} & 0 & 0 \\
0 & 0 & 0 & 0 & d_{y5} & 0 \\
0 & 0 & 0 & 0 & 0 & d_{z6}
\end{bmatrix}
$$

$$
\text{Orthorhombic } 2mm
\begin{bmatrix}
0 & 0 & 0 & 0 & d_{x5} & 0 \\
0 & 0 & 0 & d_{y4} & 0 & 0 \\
d_{z1} & d_{z2} & d_{z3} & 0 & 0 & 0
\end{bmatrix}
$$

$$
\text{Tetragonal 4}
\begin{bmatrix}
0 & 0 & 0 & d_{x4} & d_{x5} & 0 \\
0 & 0 & 0 & d_{x5} & -d_{x4} & 0 \\
d_{z1} & d_{z1} & d_{z3} & 0 & 0 & 0
\end{bmatrix}
$$

Tetragonal $\bar{4}$
$$
\begin{bmatrix}
0 & 0 & 0 & d_{x4} & d_{x5} & 0 \\
0 & 0 & 0 & -d_{x5} & d_{x4} & 0 \\
d_{z1} & -d_{z1} & 0 & 0 & 0 & d_{z6}
\end{bmatrix}
$$

Tetragonal 422
$$
\begin{bmatrix}
0 & 0 & 0 & d_{x4} & 0 & 0 \\
0 & 0 & 0 & 0 & -d_{x4} & 0 \\
0 & 0 & 0 & 0 & 0 & 0
\end{bmatrix}
$$

Tetragonal 4mm
$$
\begin{bmatrix}
0 & 0 & 0 & 0 & d_{x5} & 0 \\
0 & 0 & 0 & d_{x5} & 0 & 0 \\
d_{z1} & d_{z1} & d_{z3} & 0 & 0 & 0
\end{bmatrix}
$$

Tetragonal $\bar{4}$2m
$$
\begin{bmatrix}
0 & 0 & 0 & d_{x4} & 0 & 0 \\
0 & 0 & 0 & 0 & d_{x4} & 0 \\
0 & 0 & 0 & 0 & 0 & d_{z6}
\end{bmatrix}
$$

Trigonal 3
$$
\begin{bmatrix}
d_{x1} & -d_{x1} & 0 & d_{x4} & d_{x5} & -2d_{y2} \\
-d_{y2} & d_{y2} & 0 & d_{x5} & -d_{x4} & -2d_{x1} \\
d_{z1} & d_{z1} & d_{z3} & 0 & 0 & 0
\end{bmatrix}
\quad
\begin{aligned}
e_{x6} &= -e_{y2} \\
e_{y6} &= -e_{x1}
\end{aligned}
$$

Trigonal 32
$$
\begin{bmatrix}
d_{x1} & -d_{x1} & 0 & d_{x4} & 0 & 0 \\
0 & 0 & 0 & 0 & -d_{x4} & -2d_{x1} \\
0 & 0 & 0 & 0 & 0 & 0
\end{bmatrix}
\quad e_{y6} = -e_{x1}
$$

Trigonal 3m
$$
\begin{bmatrix}
0 & 0 & 0 & 0 & d_{x5} & -2d_{y2} \\
-d_{y2} & d_{y2} & 0 & d_{x5} & 0 & 0 \\
d_{z1} & d_{z1} & d_{z3} & 0 & 0 & 0
\end{bmatrix}
\quad e_{x6} = -e_{y2}
$$

Hexagonal 6
$$
\begin{bmatrix}
0 & 0 & 0 & d_{x4} & d_{x5} & 0 \\
0 & 0 & 0 & d_{x5} & -d_{x4} & 0 \\
d_{z1} & d_{z1} & d_{z3} & 0 & 0 & 0
\end{bmatrix}
$$

Hexagonal $\bar{6}$
$$
\begin{bmatrix}
d_{x1} & -d_{x1} & 0 & 0 & 0 & -2d_{y2} \\
-d_{y2} & d_{y2} & 0 & 0 & 0 & -2d_{x1} \\
0 & 0 & 0 & 0 & 0 & 0
\end{bmatrix}
\quad
\begin{aligned}
e_{x6} &= -e_{y2} \\
e_{y6} &= -e_{x1}
\end{aligned}
$$

Hexagonal 622
$$
\begin{bmatrix}
0 & 0 & 0 & d_{x4} & 0 & 0 \\
0 & 0 & 0 & 0 & -d_{x4} & 0 \\
0 & 0 & 0 & 0 & 0 & 0
\end{bmatrix}
$$

Hexagonal 6mm
$$
\begin{bmatrix}
0 & 0 & 0 & 0 & d_{x5} & 0 \\
0 & 0 & 0 & d_{x5} & 0 & 0 \\
d_{z1} & d_{z1} & d_{z3} & 0 & 0 & 0
\end{bmatrix}
$$

$$\text{Hexagonal } \bar{6}m2 \begin{bmatrix} d_{x1} & -d_{x1} & 0 & 0 & 0 & 0 \\ 0 & 0 & 0 & 0 & 0 & -2d_{x1} \\ 0 & 0 & 0 & 0 & 0 & 0 \end{bmatrix} \quad e_{y6} = -e_{x1}$$

$$\text{Cubic 23 and } \bar{4}3m \begin{bmatrix} 0 & 0 & 0 & d_{x4} & 0 & 0 \\ 0 & 0 & 0 & 0 & d_{x4} & 0 \\ 0 & 0 & 0 & 0 & 0 & d_{x4} \end{bmatrix}$$

11.4 Permittivity matrices

The ϵ_{ij} matrix elements correspond to the situation where the coordinate axes x, y, z coincide with the crystal axes X, Y, Z.

$$\text{Triclinic} \begin{bmatrix} \epsilon_{xx} & \epsilon_{xy} & \epsilon_{xz} \\ \epsilon_{xy} & \epsilon_{yy} & \epsilon_{yz} \\ \epsilon_{xz} & \epsilon_{yz} & \epsilon_{zz} \end{bmatrix}$$

$$\text{Monoclinic} \begin{bmatrix} \epsilon_{xx} & 0 & \epsilon_{xz} \\ 0 & \epsilon_{yy} & 0 \\ \epsilon_{xz} & 0 & \epsilon_{zz} \end{bmatrix}$$

$$\text{Orthorhombic} \begin{bmatrix} \epsilon_{xx} & 0 & 0 \\ 0 & \epsilon_{yy} & 0 \\ 0 & 0 & \epsilon_{zz} \end{bmatrix}$$

$$\text{Hexagonal, Trigonal, Tetragonal} \begin{bmatrix} \epsilon_{xx} & 0 & 0 \\ 0 & \epsilon_{xx} & 0 \\ 0 & 0 & \epsilon_{zz} \end{bmatrix}$$

$$\text{Cubic, Isotropic} \begin{bmatrix} \epsilon_{xx} & 0 & 0 \\ 0 & \epsilon_{xx} & 0 \\ 0 & 0 & \epsilon_{xx} \end{bmatrix}$$

11.5 Transformation matrices for the symmetry group generators

Generator matrices for all the crystal classes (point groups) are listed below.

Triclinic

$$\text{Class 1} \quad \text{Gen. matr.} \quad 1 \rightarrow \begin{bmatrix} 1 & 0 & 0 \\ 0 & 1 & 0 \\ 0 & 0 & 1 \end{bmatrix}$$

Class $\bar{1}$ Gen. matr. $\bar{1} \rightarrow \begin{bmatrix} -1 & 0 & 0 \\ 0 & -1 & 0 \\ 0 & 0 & -1 \end{bmatrix}$

Monoclinic

Class 2   Gen. matr. $2 \rightarrow \begin{bmatrix} -1 & 0 & 0 \\ 0 & 1 & 0 \\ 0 & 0 & -1 \end{bmatrix}$

Class m Gen. matr. $m \rightarrow \begin{bmatrix} 1 & 0 & 0 \\ 0 & -1 & 0 \\ 0 & 0 & 1 \end{bmatrix}$

Class $\dfrac{2}{m}$ Gen. matr. $2 \rightarrow \begin{bmatrix} -1 & 0 & 0 \\ 0 & 1 & 0 \\ 0 & 0 & -1 \end{bmatrix}$, $m \rightarrow \begin{bmatrix} 1 & 0 & 0 \\ 0 & -1 & 0 \\ 0 & 0 & 1 \end{bmatrix}$

Orthorhombic

Class 222   Gen. matr. $2 \rightarrow \begin{bmatrix} -1 & 0 & 0 \\ 0 & -1 & 0 \\ 0 & 0 & 1 \end{bmatrix}$, $2 \rightarrow \begin{bmatrix} 1 & 0 & 0 \\ 0 & -1 & 0 \\ 0 & 0 & -1 \end{bmatrix}$

Class $2mm$ Gen. matr. $2 \rightarrow \begin{bmatrix} -1 & 0 & 0 \\ 0 & -1 & 0 \\ 0 & 0 & 1 \end{bmatrix}$, $m \rightarrow \begin{bmatrix} 1 & 0 & 0 \\ 0 & -1 & 0 \\ 0 & 0 & 1 \end{bmatrix}$

Class $\dfrac{2}{m}\dfrac{2}{m}\dfrac{2}{m}$ Gen. matr. $m \rightarrow \begin{bmatrix} -1 & 0 & 0 \\ 0 & 1 & 0 \\ 0 & 0 & 1 \end{bmatrix}$, $m \rightarrow \begin{bmatrix} 1 & 0 & 0 \\ 0 & -1 & 0 \\ 0 & 0 & 1 \end{bmatrix}$,

$m \rightarrow \begin{bmatrix} 1 & 0 & 0 \\ 0 & 1 & 0 \\ 0 & 0 & -1 \end{bmatrix}$

Tetragonal

Class 4   Gen. matr. $4 \rightarrow \begin{bmatrix} 0 & 1 & 0 \\ -1 & 0 & 0 \\ 0 & 0 & 1 \end{bmatrix}$

Class $\bar{4}$ Gen. matr. $\bar{4} \rightarrow \begin{bmatrix} 0 & -1 & 0 \\ 1 & 0 & 0 \\ 0 & 0 & -1 \end{bmatrix}$

Class $\dfrac{4}{m}$ Gen. matr. $4 \rightarrow \begin{bmatrix} 0 & 1 & 0 \\ -1 & 0 & 0 \\ 0 & 0 & 1 \end{bmatrix}, \; m \rightarrow \begin{bmatrix} 1 & 0 & 0 \\ 0 & 1 & 0 \\ 0 & 0 & -1 \end{bmatrix}$

Class 422   Gen. matr. $4 \rightarrow \begin{bmatrix} 0 & 1 & 0 \\ -1 & 0 & 0 \\ 0 & 0 & 1 \end{bmatrix}, \; 2 \rightarrow \begin{bmatrix} 1 & 0 & 0 \\ 0 & -1 & 0 \\ 0 & 0 & -1 \end{bmatrix}$

Class $4mm$ Gen. matr. $4 \rightarrow \begin{bmatrix} 0 & 1 & 0 \\ -1 & 0 & 0 \\ 0 & 0 & 1 \end{bmatrix}, \; m \rightarrow \begin{bmatrix} 1 & 0 & 0 \\ 0 & -1 & 0 \\ 0 & 0 & 1 \end{bmatrix}$

Class $\bar{4}2m$ Gen. matr. $\bar{4} \rightarrow \begin{bmatrix} 0 & -1 & 0 \\ 1 & 0 & 0 \\ 0 & 0 & -1 \end{bmatrix}, \; 2 \rightarrow \begin{bmatrix} 1 & 0 & 0 \\ 0 & -1 & 0 \\ 0 & 0 & -1 \end{bmatrix}$

Class $\dfrac{4}{m}\dfrac{2}{m}\dfrac{2}{m}$ Gen. matr. $4 \rightarrow \begin{bmatrix} 0 & 1 & 0 \\ -1 & 0 & 0 \\ 0 & 0 & 1 \end{bmatrix}, \; m \rightarrow \begin{bmatrix} 1 & 0 & 0 \\ 0 & 1 & 0 \\ 0 & 0 & -1 \end{bmatrix},$

$$m \rightarrow \begin{bmatrix} 1 & 0 & 0 \\ 0 & -1 & 0 \\ 0 & 0 & 1 \end{bmatrix}$$

Trigonal

Class 3   Gen. matr. $3 \rightarrow \begin{bmatrix} -\dfrac{1}{2} & \dfrac{\sqrt{3}}{2} & 0 \\ -\dfrac{\sqrt{3}}{2} & -\dfrac{1}{2} & 0 \\ 0 & 0 & 1 \end{bmatrix}$

Class $\bar{3}$ Gen. matr. $\bar{3} \rightarrow \begin{bmatrix} \dfrac{1}{2} & -\dfrac{\sqrt{3}}{2} & 0 \\ \dfrac{\sqrt{3}}{2} & \dfrac{1}{2} & 0 \\ 0 & 0 & -1 \end{bmatrix}$

Class 32   Gen. matr. $3 \rightarrow \begin{bmatrix} -\dfrac{1}{2} & \dfrac{\sqrt{3}}{2} & 0 \\ -\dfrac{\sqrt{3}}{2} & -\dfrac{1}{2} & 0 \\ 0 & 0 & 1 \end{bmatrix} \; 2 \rightarrow \begin{bmatrix} 1 & 0 & 0 \\ 0 & -1 & 0 \\ 0 & 0 & -1 \end{bmatrix}$

Class $3m$ Gen. matr. $3 \rightarrow \begin{bmatrix} -\dfrac{1}{2} & \dfrac{\sqrt{3}}{2} & 0 \\[2mm] -\dfrac{\sqrt{3}}{2} & -\dfrac{1}{2} & 0 \\[2mm] 0 & 0 & 1 \end{bmatrix}$ $m \rightarrow \begin{bmatrix} -1 & 0 & 0 \\ 0 & 1 & 0 \\ 0 & 0 & 1 \end{bmatrix}$

Class $\bar{3}\dfrac{2}{m}$ Gen. matr. $\bar{3} \rightarrow \begin{bmatrix} \dfrac{1}{2} & -\dfrac{\sqrt{3}}{2} & 0 \\[2mm] \dfrac{\sqrt{3}}{2} & \dfrac{1}{2} & 0 \\[2mm] 0 & 0 & -1 \end{bmatrix}$ $m \rightarrow \begin{bmatrix} -1 & 0 & 0 \\ 0 & 1 & 0 \\ 0 & 0 & 1 \end{bmatrix}$

Hexagonal

Class 6   Gen. matr. $6 \rightarrow \begin{bmatrix} \dfrac{1}{2} & \dfrac{\sqrt{3}}{2} & 0 \\[2mm] -\dfrac{\sqrt{3}}{2} & \dfrac{1}{2} & 0 \\[2mm] 0 & 0 & 1 \end{bmatrix}$

Class $\bar{6}$ Gen. matr. $\bar{6} \rightarrow \begin{bmatrix} -\dfrac{1}{2} & -\dfrac{\sqrt{3}}{2} & 0 \\[2mm] \dfrac{\sqrt{3}}{2} & -\dfrac{1}{2} & 0 \\[2mm] 0 & 0 & -1 \end{bmatrix}$

Class $\dfrac{6}{m}$ Gen. matr. $6 \rightarrow \begin{bmatrix} \dfrac{1}{2} & \dfrac{\sqrt{3}}{2} & 0 \\[2mm] -\dfrac{\sqrt{3}}{2} & \dfrac{1}{2} & 0 \\[2mm] 0 & 0 & 1 \end{bmatrix}$ $m \rightarrow \begin{bmatrix} 1 & 0 & 0 \\ 0 & 1 & 0 \\ 0 & 0 & -1 \end{bmatrix}$

Class 622   Gen. matr. $6 \rightarrow \begin{bmatrix} \dfrac{1}{2} & \dfrac{\sqrt{3}}{2} & 0 \\[2mm] -\dfrac{\sqrt{3}}{2} & \dfrac{1}{2} & 0 \\[2mm] 0 & 0 & 1 \end{bmatrix}$ $2 \rightarrow \begin{bmatrix} 1 & 0 & 0 \\ 0 & -1 & 0 \\ 0 & 0 & -1 \end{bmatrix}$

Class 6mm   Gen. matr. $6 \rightarrow$ $\begin{bmatrix} \dfrac{1}{2} & \dfrac{\sqrt{3}}{2} & 0 \\[2mm] -\dfrac{\sqrt{3}}{2} & \dfrac{1}{2} & 0 \\[2mm] 0 & 0 & 1 \end{bmatrix}$ $m \rightarrow$ $\begin{bmatrix} 1 & 0 & 0 \\ 0 & -1 & 0 \\ 0 & 0 & 1 \end{bmatrix}$

Class $\bar{6}m2$ Gen. matr. $\bar{6} \rightarrow$ $\begin{bmatrix} -\dfrac{1}{2} & -\dfrac{\sqrt{3}}{2} & 0 \\[2mm] \dfrac{\sqrt{3}}{2} & -\dfrac{1}{2} & 0 \\[2mm] 0 & 0 & -1 \end{bmatrix}$ $2 \rightarrow$ $\begin{bmatrix} 1 & 0 & 0 \\ 0 & -1 & 0 \\ 0 & 0 & -1 \end{bmatrix}$

Class $\dfrac{6}{m}\dfrac{2}{m}\dfrac{2}{m}$ Gen. matr. $6 \rightarrow$ $\begin{bmatrix} \dfrac{1}{2} & \dfrac{\sqrt{3}}{2} & 0 \\[2mm] -\dfrac{\sqrt{3}}{2} & \dfrac{1}{2} & 0 \\[2mm] 0 & 0 & 1 \end{bmatrix}$ $m \rightarrow$ $\begin{bmatrix} 1 & 0 & 0 \\ 0 & 1 & 0 \\ 0 & 0 & -1 \end{bmatrix}$,

$m \rightarrow$ $\begin{bmatrix} 1 & 0 & 0 \\ 0 & -1 & 0 \\ 0 & 0 & 1 \end{bmatrix}$

Cubic

Class 23   Gen. matr. $2 \rightarrow$ $\begin{bmatrix} -1 & 0 & 0 \\ 0 & -1 & 0 \\ 0 & 0 & 1 \end{bmatrix}$ $3 \rightarrow$ $\begin{bmatrix} 0 & 1 & 0 \\ 0 & 0 & 1 \\ 1 & 0 & 0 \end{bmatrix}$

Class $\dfrac{2}{m}\bar{3}$ Gen. matr. $2 \rightarrow$ $\begin{bmatrix} -1 & 0 & 0 \\ 0 & -1 & 0 \\ 0 & 0 & 1 \end{bmatrix}$ $3 \rightarrow$ $\begin{bmatrix} 0 & 1 & 0 \\ 0 & 0 & 1 \\ 1 & 0 & 0 \end{bmatrix}$,

$m \rightarrow$ $\begin{bmatrix} 1 & 0 & 0 \\ 0 & 1 & 0 \\ 0 & 0 & -1 \end{bmatrix}$

Class 432   Gen. matr. $4 \rightarrow$ $\begin{bmatrix} 0 & 1 & 0 \\ -1 & 0 & 0 \\ 0 & 0 & 1 \end{bmatrix}$ $2 \rightarrow$ $\begin{bmatrix} -1 & 0 & 0 \\ 0 & 0 & 1 \\ 0 & 1 & 0 \end{bmatrix}$

Class $\bar{4}3m$ Gen. matr. $\bar{4} \rightarrow$ $\begin{bmatrix} 0 & -1 & 0 \\ 1 & 0 & 0 \\ 0 & 0 & -1 \end{bmatrix}$ $m \rightarrow$ $\begin{bmatrix} 1 & 0 & 0 \\ 0 & 0 & 1 \\ 0 & 1 & 0 \end{bmatrix}$

$$\text{Class} \frac{4}{m}\bar{3}\frac{2}{m} \qquad \text{Gen. matr.} \quad 4 \rightarrow \begin{bmatrix} 0 & 1 & 0 \\ -1 & 0 & 0 \\ 0 & 0 & 1 \end{bmatrix} \quad 2 \rightarrow \begin{bmatrix} -1 & 0 & 0 \\ 0 & 0 & 1 \\ 0 & 1 & 0 \end{bmatrix},$$

$$m \rightarrow \begin{bmatrix} 1 & 0 & 0 \\ 0 & 1 & 0 \\ 0 & 0 & -1 \end{bmatrix}$$

11.6 Transformation properties of the stiffness tensor using point-group symmetry

Having established how the stiffness tensor transforms under transformations, it is possible to derive the general form of the stiffness tensor by use of symmetry for each crystal class (point group). From the transformation rule,

$$[c'] = [M][c][N]^{-1}, \tag{11.1}$$

where $[M]$ and $[N]$ are calculated from the coordinate transformation matrix $[a]$, it follows that for any symmetry operation $[a]$,

$$[c'] = [c], \tag{11.2}$$

thus,

$$[c][N] = [M][c]. \tag{11.3}$$

It immediately follows that equation (11.3) applies to any combination of symmetries (a) and (b), i.e.,

$$[c][N^{(ba)}] = [M^{(ba)}][c]. \tag{11.4}$$

where

$$M^{(ba)} = M^{(b)}M^{(a)}, \tag{11.5}$$

$$N^{(ba)} = N^{(b)}N^{(a)}. \tag{11.6}$$

It is therefore enough to only apply the generator matrices for each point group to determine the most general form of the stiffness tensor allowed by symmetry.

11.6.1 Application to the 432 cubic group

Using the transformation properties in section 11.5, the coordinate transformation matrices $[a]$ associated with the group generators are,

$$4 \rightarrow \begin{bmatrix} 0 & 1 & 0 \\ -1 & 0 & 0 \\ 0 & 0 & 1 \end{bmatrix} \quad 2 \rightarrow \begin{bmatrix} -1 & 0 & 0 \\ 0 & 0 & 1 \\ 0 & 1 & 0 \end{bmatrix}.$$

The transformation matrices $[M]$ and $[N]$ can now be written down,

$$[M^{(4)}] = [N^{(4)}] = \begin{bmatrix} 0 & 1 & 0 & 0 & 0 & 0 \\ 1 & 0 & 0 & 0 & 0 & 0 \\ 0 & 0 & 1 & 0 & 0 & 0 \\ 0 & 0 & 0 & 0 & -1 & 0 \\ 0 & 0 & 0 & 1 & 0 & 0 \\ 0 & 0 & 0 & 0 & 0 & -1 \end{bmatrix},$$

and

$$[M^{(2)}] = [N^{(2)}] = \begin{bmatrix} 1 & 0 & 0 & 0 & 0 & 0 \\ 0 & 0 & 1 & 0 & 0 & 0 \\ 0 & 1 & 0 & 0 & 0 & 0 \\ 0 & 0 & 0 & 1 & 0 & 0 \\ 0 & 0 & 0 & 0 & 0 & -1 \\ 0 & 0 & 0 & 0 & -1 & 0 \end{bmatrix}.$$

Starting from an arbitrary symmetric stiffness matrix and imposing the condition in equation (11.4) yields, for the generator operation 4,

$$\begin{bmatrix} c_{12} & c_{11} & c_{13} & c_{15} & -c_{14} & -c_{16} \\ c_{22} & c_{12} & c_{23} & c_{25} & -c_{24} & -c_{26} \\ c_{23} & c_{13} & c_{33} & c_{35} & -c_{34} & -c_{36} \\ c_{24} & c_{14} & c_{34} & c_{45} & -c_{44} & -c_{46} \\ c_{25} & c_{15} & c_{35} & c_{55} & -c_{45} & -c_{56} \\ c_{26} & c_{16} & c_{36} & c_{56} & -c_{46} & -c_{66} \end{bmatrix} = \begin{bmatrix} c_{12} & c_{22} & c_{23} & c_{24} & c_{25} & c_{26} \\ c_{11} & c_{12} & c_{13} & c_{14} & c_{15} & c_{16} \\ c_{13} & c_{23} & c_{33} & c_{34} & c_{35} & c_{36} \\ -c_{15} & -c_{25} & -c_{35} & -c_{45} & -c_{55} & -c_{56} \\ c_{14} & c_{24} & c_{34} & c_{44} & c_{45} & c_{46} \\ -c_{16} & -c_{26} & -c_{36} & -c_{46} & -c_{56} & -c_{66} \end{bmatrix}.$$

Comparing the two matrices forces several stiffness entries to be zero (e.g., $c_{24} = c_{15} = 0$) and others to be identical (e.g., $c_{22} = c_{11}$). One obtains,

$$[c] = \begin{bmatrix} c_{12} & c_{12} & c_{13} & 0 & 0 & c_{16} \\ c_{12} & c_{11} & c_{13} & 0 & 0 & -c_{16} \\ c_{33} & c_{13} & c_{33} & 0 & 0 & 0 \\ 0 & 0 & 0 & c_{44} & 0 & 0 \\ 0 & 0 & 0 & 0 & c_{44} & 0 \\ c_{16} & -c_{16} & 0 & 0 & 0 & c_{66} \end{bmatrix}.$$

Starting next from the latter stiffness tensor (equation (11.6.1)) and imposing the condition in equation (11.4) yields, for the generator operation 2,

$$\begin{bmatrix} c_{11} & c_{13} & c_{12} & 0 & -c_{16} & 0 \\ c_{12} & c_{13} & c_{11} & 0 & c_{16} & 0 \\ c_{13} & c_{33} & c_{13} & 0 & 0 & 0 \\ 0 & 0 & 0 & c_{44} & 0 & 0 \\ 0 & 0 & 0 & 0 & 0 & -c_{44} \\ c_{16} & 0 & -c_{16} & 0 & -c_{66} & 0 \end{bmatrix} = \begin{bmatrix} c_{11} & c_{12} & c_{13} & 0 & 0 & c_{16} \\ c_{13} & c_{13} & c_{33} & 0 & 0 & 0 \\ c_{12} & c_{11} & c_{13} & 0 & 0 & -c_{16} \\ 0 & 0 & 0 & c_{44} & 0 & 0 \\ -c_{16} & c_{16} & 0 & 0 & 0 & -c_{66} \\ 0 & 0 & 0 & 0 & -c_{44} & 0 \end{bmatrix}.$$

Again, comparing the two matrices forces $c_{16} = 0$ and others to be identical (e.g., $c_{33} = c_{11}$). Eventually, one obtains,

$$[c] = \begin{bmatrix} c_{11} & c_{12} & c_{12} & 0 & 0 & 0 \\ c_{12} & c_{11} & c_{12} & 0 & 0 & 0 \\ c_{12} & c_{12} & c_{11} & 0 & 0 & 0 \\ 0 & 0 & 0 & c_{44} & 0 & 0 \\ 0 & 0 & 0 & 0 & c_{44} & 0 \\ 0 & 0 & 0 & 0 & 0 & c_{44} \end{bmatrix}.$$

11.7 Transformation properties of the piezoelectric and permittivity tensors

In order to apply group theory to reduce the piezoelectric strain and stress matrices and the permittivity tensor to the most general form allowed by symmetry for any crystal class, we first need to derive the coordinate transformation properties of the same quantities, i.e., $[d]$, $[e]$, $[\epsilon]$.

11.7.1 Transformation properties of the piezoelectric strain tensor

If the electric field is zero, the electric displacement transforms as,

$$[D'] = [a][D] = [a][d][T] = [a][d][M]^{-1}[T'] = [a][d][N^T][T'] = [d'][T'], \quad (11.7)$$

where the transformation matrices $[a]$, $[M]$, and $[N]$, and the expression $[M]^{-1} = [N^T]$ from chapter 2 were used. Thus,

$$[d'] = [a][d][N]^T. \quad (11.8)$$

This is the transformation law of the piezoelectric strain tensor.

11.7.2 Transformation properties of the piezoelectric stress tensor

If the electric field is zero, the electric displacement transforms as,

$$[D'] = [a][D] = [a][e][S] = [a][e][N]^{-1}[S'] = [a][e][M^T][S'] = [e'][S'], \quad (11.9)$$

where the expression $[N]^{-1} = [M^T]$ was used. Thus,

$$[e'] = [a][e][M]^T. \quad (11.10)$$

This is the transformation law of the piezoelectric stress tensor.

11.7.3 Transformation properties of the permittivity tensor

Since the electric displacement and the electric field transform as vectors, they obey the following transformation law,

$$[D'] = [a][D], \quad (11.11)$$

$$[E'] = [a][E].\tag{11.12}$$

Thus, if the stress tensor is zero,

$$[D'] = [a][D] = [a][\epsilon][E] = [a][\epsilon][a]^{-1}[a][E] = [a][\epsilon][a]^T[a][E] = [\epsilon'][E'],\tag{11.13}$$

whereby,

$$[\epsilon'] = [a][\epsilon][a]^T.\tag{11.14}$$

This is the transformation law of the permittivity tensor. The above transformation law applies to both $[\epsilon^S]$ and $[\epsilon^T]$.

11.7.4 The piezoelectric stress tensor for the tetragonal 4 class

Let us now derive the general form of the piezoelectric stress tensor for the tetragonal class 4 allowed by group symmetry. There is only one generator for the 4 crystal class according to section 11.5 so the coordinate transformation matrix $[a]$ is,

$$\text{Class 4} \quad \text{Gen. matr.} \quad 4 \rightarrow [a] = \begin{bmatrix} 0 & 1 & 0 \\ -1 & 0 & 0 \\ 0 & 0 & 1 \end{bmatrix}.$$

The $[M]$ matrix is obtained from the $[a]$ matrix in the usual way,

$$[M^{(4)}] = \begin{bmatrix} 0 & 1 & 0 & 0 & 0 & 0 \\ 1 & 0 & 0 & 0 & 0 & 0 \\ 0 & 0 & 1 & 0 & 0 & 0 \\ 0 & 0 & 0 & 0 & -1 & 0 \\ 0 & 0 & 0 & 1 & 0 & 0 \\ 0 & 0 & 0 & 0 & 0 & -1 \end{bmatrix}.$$

Then, the piezoelectric stress matrix is, using equation (11.10),

$$[e'] = [a][e][M]^T = \begin{bmatrix} 0 & 1 & 0 \\ -1 & 0 & 0 \\ 0 & 0 & 1 \end{bmatrix} \begin{bmatrix} e_{x1} & e_{x2} & e_{x3} & e_{x4} & e_{x5} & e_{x6} \\ e_{y1} & e_{y2} & e_{y3} & e_{y4} & e_{y5} & e_{y6} \\ e_{z1} & e_{z2} & e_{z3} & e_{z4} & e_{z5} & e_{z6} \end{bmatrix} \begin{bmatrix} 0 & 1 & 0 & 0 & 0 & 0 \\ 1 & 0 & 0 & 0 & 0 & 0 \\ 0 & 0 & 1 & 0 & 0 & 0 \\ 0 & 0 & 0 & 0 & 1 & 0 \\ 0 & 0 & 0 & -1 & 0 & 0 \\ 0 & 0 & 0 & 0 & 0 & -1 \end{bmatrix}$$

$$= \begin{bmatrix} e_{y1} & e_{y2} & e_{y3} & e_{y4} & e_{y5} & e_{y6} \\ -e_{x1} & -e_{x2} & -e_{x3} & -e_{x4} & -e_{x5} & -e_{x6} \\ e_{z1} & e_{z2} & e_{z3} & e_{z4} & e_{z5} & e_{z6} \end{bmatrix} \begin{bmatrix} 0 & 1 & 0 & 0 & 0 & 0 \\ 1 & 0 & 0 & 0 & 0 & 0 \\ 0 & 0 & 1 & 0 & 0 & 0 \\ 0 & 0 & 0 & 0 & 1 & 0 \\ 0 & 0 & 0 & -1 & 0 & 0 \\ 0 & 0 & 0 & 0 & 0 & -1 \end{bmatrix}$$

$$= \begin{bmatrix} e_{y2} & e_{y1} & e_{y3} & -e_{y5} & e_{y4} & -e_{y6} \\ -e_{x2} & -e_{x1} & -e_{x3} & e_{x5} & -e_{x4} & e_{x6} \\ e_{z2} & e_{z1} & e_{z3} & -e_{z5} & e_{z4} & -e_{z6} \end{bmatrix} = \begin{bmatrix} e_{x1} & e_{x2} & e_{x3} & e_{x4} & e_{x5} & e_{x6} \\ e_{y1} & e_{y2} & e_{y3} & e_{y4} & e_{y5} & e_{y6} \\ e_{z1} & e_{z2} & e_{z3} & e_{z4} & e_{z5} & e_{z6} \end{bmatrix},$$

where, for the last equality, it was used that $[e'] = [e]$ for any point-group symmetry. Comparing the two last 3×6 matrices, the following identities apply,

$$e_{x1} = e_{y2} = 0,$$
$$e_{y1} = e_{x2} = 0,$$
$$e_{x3} = e_{y3} = 0,$$
$$e_{x6} = e_{y6} = 0,$$
$$e_{z4} = e_{z5} = 0,$$
$$e_{z6} = 0,$$
$$e_{x4} = -e_{y5},$$
$$e_{y4} = e_{x5},$$
$$e_{z1} = e_{z2},$$

and the most general expression for $[e]$, allowed by tetragonal 4 point group symmetry, becomes,

$$\begin{bmatrix} 0 & 0 & 0 & e_{x4} & e_{x5} & 0 \\ 0 & 0 & 0 & e_{x5} & -e_{x4} & 0 \\ e_{z1} & e_{z1} & e_{z3} & 0 & 0 & 0 \end{bmatrix}.$$

Since, for the class 4 generator matrix, the matrices $[M^{(4)}]$ and $[N^{(4)}]$ are equal, it immediately follows from equation (11.8) that the most general expression for the piezoelectric strain tensor $[d]$, allowed by tetragonal 4 point-group symmetry, also is,

$$[d'] = [a][d][N]^T = \begin{bmatrix} 0 & 0 & 0 & d_{x4} & d_{x5} & 0 \\ 0 & 0 & 0 & d_{x5} & -d_{x4} & 0 \\ d_{z1} & d_{z1} & d_{z3} & 0 & 0 & 0 \end{bmatrix}.$$

11.8 Transformation of the permittivity tensor

From the expressions for the differential energy dU and the three enthalpies,

$$dU = \Theta d\sigma + E_i dD_i + T_{ik} dS_{ik},$$
$$H = U - E_i D_i - T_{ik} S_{ik}, \quad dH = \Theta d\sigma - D_i dE_i - S_{ik} dT_{ik},$$
$$\tilde{H} = U - T_{ik} S_{ik}, \quad d\tilde{H} = \Theta d\sigma + E_i dD_i - S_{ik} dT_{ik},$$
$$\tilde{\tilde{H}} = U - E_i D_i, \quad d\tilde{\tilde{H}} = \Theta d\sigma - D_i dE_i + T_{ik} dS_{ik},$$

it is easy to show using thermodynamics that the permittivity matrix is symmetric. For example,

$$\epsilon_{ij}^S = \left(\frac{\partial D_i}{\partial E_j}\right)|_S = -\left(\frac{\partial^2 \tilde{H}}{\partial E_j \partial E_i}\right)|_S = -\left(\frac{\partial^2 \tilde{H}}{\partial E_i \partial E_j}\right)|_S = \left(\frac{\partial D_j}{\partial E_i}\right)|_S = \epsilon_{ji}^S, \quad (11.15)$$

where the expression for the elastic enthalpy $\tilde{H}$ and the interchangeability of partial derivatives were used. From the expression for the enthalpy H, it is obtained that,

$$\epsilon_{ij}^T = \left(\frac{\partial D_i}{\partial E_j}\right)|_T = -\left(\frac{\partial^2 H}{\partial E_j \partial E_i}\right)|_T = -\left(\frac{\partial^2 H}{\partial E_i \partial E_j}\right)|_T = \left(\frac{\partial D_j}{\partial E_i}\right)|_T = \epsilon_{ji}^T, \quad (11.16)$$

so the symmetry of the permittivity tensor does not depend on whether strain or stress is fixed, and the superscript T (or S) will be left out below.

11.8.1 The permittivity tensor for the tetragonal 4 class

Let us next derive the permittivity tensor for the tetragonal class 4 allowed by group symmetry. Since the coordinate transformation matrix $[a]$ is,

$$\text{Class 4} \quad \text{Gen. matr.} \quad 4 \rightarrow [a] = \begin{bmatrix} 0 & 1 & 0 \\ -1 & 0 & 0 \\ 0 & 0 & 1 \end{bmatrix},$$

we have, using equation (11.14) and the fact that the permittivity must be symmetric,

$$[\epsilon'] = [a][\epsilon][a]^T = \begin{bmatrix} 0 & 1 & 0 \\ -1 & 0 & 0 \\ 0 & 0 & 1 \end{bmatrix} \begin{bmatrix} \epsilon_{xx} & \epsilon_{xy} & \epsilon_{xz} \\ \epsilon_{xy} & \epsilon_{yy} & \epsilon_{yz} \\ \epsilon_{xz} & \epsilon_{yz} & \epsilon_{zz} \end{bmatrix} \begin{bmatrix} 0 & -1 & 0 \\ 1 & 0 & 0 \\ 0 & 0 & 1 \end{bmatrix}$$

$$= \begin{bmatrix} \epsilon_{xy} & \epsilon_{yy} & \epsilon_{yz} \\ -\epsilon_{xx} & -\epsilon_{xy} & -\epsilon_{xz} \\ \epsilon_{xz} & \epsilon_{yz} & \epsilon_{zz} \end{bmatrix} \begin{bmatrix} 0 & -1 & 0 \\ 1 & 0 & 0 \\ 0 & 0 & 1 \end{bmatrix} = \begin{bmatrix} \epsilon_{yy} & -\epsilon_{xy} & \epsilon_{yz} \\ -\epsilon_{xy} & \epsilon_{xx} & -\epsilon_{xz} \\ \epsilon_{yz} & -\epsilon_{xz} & \epsilon_{zz} \end{bmatrix} = \begin{bmatrix} \epsilon_{xx} & \epsilon_{xy} & \epsilon_{xz} \\ \epsilon_{xy} & \epsilon_{yy} & \epsilon_{yz} \\ \epsilon_{xz} & \epsilon_{yz} & \epsilon_{zz} \end{bmatrix},$$

where in the last step the symmetry $[\epsilon'] = [\epsilon]$ was imposed. Comparing the two last matrices forces,

$$\epsilon_{xx} = \epsilon_{yy},$$
$$\epsilon_{xz} = \epsilon_{yz} = 0,$$
$$\epsilon_{xy} = 0$$

and the most general expression for $[\epsilon]$, allowed by tetragonal 4 point-group symmetry, becomes,

$$\begin{bmatrix} \epsilon_{xx} & 0 & 0 \\ 0 & \epsilon_{xx} & 0 \\ 0 & 0 & \epsilon_{zz} \end{bmatrix}.$$

It turns out that the above form for the permittivity tensor holds for any of the tetragonal classes as the reader can easily verify.

11.8.2 The permittivity tensor for the monoclinic $\frac{2}{m}$ class

Let us next derive the permittivity tensor for the monoclinic class $\frac{2}{m}$ allowed by group symmetry. For $\frac{2}{m}$ there are two generator matrices 2 and m. Consider first the coordinate transformation matrix $[a^{(2)}]$,

$$\text{Class } \frac{2}{m} \quad \text{Gen. matr. } 2 \rightarrow [a^{(2)}] = \begin{bmatrix} -1 & 0 & 0 \\ 0 & 1 & 0 \\ 0 & 0 & -1 \end{bmatrix}.$$

Then,

$$[\epsilon'] = [a][\epsilon][a]^T = \begin{bmatrix} -1 & 0 & 0 \\ 0 & 1 & 0 \\ 0 & 0 & -1 \end{bmatrix} \begin{bmatrix} \epsilon_{xx} & \epsilon_{xy} & \epsilon_{xz} \\ \epsilon_{xy} & \epsilon_{yy} & \epsilon_{yz} \\ \epsilon_{xz} & \epsilon_{yz} & \epsilon_{zz} \end{bmatrix} \begin{bmatrix} -1 & 0 & 0 \\ 0 & 1 & 0 \\ 0 & 0 & -1 \end{bmatrix}$$

$$= \begin{bmatrix} -\epsilon_{xx} & -\epsilon_{xy} & -\epsilon_{xz} \\ \epsilon_{xy} & \epsilon_{yy} & \epsilon_{yz} \\ -\epsilon_{xz} & -\epsilon_{yz} & -\epsilon_{zz} \end{bmatrix} \begin{bmatrix} -1 & 0 & 0 \\ 0 & 1 & 0 \\ 0 & 0 & -1 \end{bmatrix} = \begin{bmatrix} \epsilon_{xx} & -\epsilon_{xy} & \epsilon_{xz} \\ -\epsilon_{xy} & \epsilon_{yy} & -\epsilon_{yz} \\ \epsilon_{xz} & -\epsilon_{yz} & \epsilon_{zz} \end{bmatrix} = \begin{bmatrix} \epsilon_{xx} & \epsilon_{xy} & \epsilon_{xz} \\ \epsilon_{xy} & \epsilon_{yy} & \epsilon_{yz} \\ \epsilon_{xz} & \epsilon_{yz} & \epsilon_{zz} \end{bmatrix},$$

where the last equality expresses $[e'] = [e]$. Comparing the two matrices yields.

$$\epsilon_{xy} = 0, \tag{11.17}$$

$$\epsilon_{yz} = 0, \tag{11.18}$$

but no further restrictions result. Thus, the permittivity matrix is simplified to,

$$[\epsilon] = \begin{bmatrix} \epsilon_{xx} & 0 & \epsilon_{xz} \\ 0 & \epsilon_{yy} & 0 \\ \epsilon_{xz} & 0 & \epsilon_{zz} \end{bmatrix}. \tag{11.19}$$

Using next the generator matrix for m,

$$a^{(m)} = \begin{bmatrix} 1 & 0 & 0 \\ 0 & -1 & 0 \\ 0 & 0 & 1 \end{bmatrix}$$

on equation (11.19) does not restrict the general form of the permittivity matrix any further. Hence, the permittivity matrix for the monoclinic $\frac{2}{m}$ class allowed by symmetry becomes

$$[\epsilon] = \begin{bmatrix} \epsilon_{xx} & 0 & \epsilon_{xz} \\ 0 & \epsilon_{yy} & 0 \\ \epsilon_{xz} & 0 & \epsilon_{zz} \end{bmatrix}.$$

11.9 Coupled electromagnetic and mechanical fields in piezoelectric materials

In this section, coupled electromagnetic and mechanical waves in a homogeneous medium will be described. The theory is based on the electromagnetic field equations,

$$-\nabla \times \mathbf{E} = \frac{\partial \mathbf{B}}{\partial t}, \tag{11.20}$$

$$\nabla \times \mathbf{H} = \frac{\partial \mathbf{D}}{\partial t} + \mathbf{J}_c + \mathbf{J}_s, \tag{11.21}$$

and the acoustic field equations,

$$\frac{\partial T_{ij}}{\partial x_j} = \rho \frac{\partial^2 u_i}{\partial t^2} - \mathbf{F}, \tag{11.22}$$

$$S_{ij} = \frac{1}{2}\left(\frac{\partial u_i}{\partial x_j} + \frac{\partial u_j}{\partial x_i}\right). \tag{11.23}$$

Here, $\mathbf{E}, \mathbf{D}, \mathbf{H}, \mathbf{B}$ are the electric field, the electric displacement, the electric displacement, the magnetic intensity, and the magnetic field, respectively. The terms $\mathbf{J}_c, \mathbf{J}_s$, and $\mathbf{F}$ are the conduction current density, the source current density, and the (external) force density, respectively. Finally, $T_{ij}, S_{ij}, u_i,$ *rho* are the stress tensor, the strain tensor, the mechanical displacement vector, and the mass density, respectively. It is customary to assume linearity and non-magnetic media,

$$J_{c,i} = \sigma_{ij} E_j, \tag{11.24}$$

$$B_i = \mu_0 H_i, \tag{11.25}$$

and the usual piezoelectric stress equations,

$$T_{ij} = c_{ijkl} S_{kl} - e_{kij} E_k, \tag{11.26}$$

$$D_i = \epsilon_{ij}^S E_j + P_{\text{spon},i} + e_{ijk} S_{jk}, \tag{11.27}$$

where $c_{ijkl}, e_{ijk}, \epsilon_{ij}, P_{\text{spon},i}$ are the stiffness tensor, the permittivity tensor, the piezoelectric stress tensor, and the spontaneous polarization, respectively. Assume all currents $\mathbf{J}_c, \mathbf{J}_s$ and the external force $\mathbf{F}$ can be neglected. Taking the time derivative of equation (11.22) and using equation (11.26) yield,

$$c_{ijkl}\frac{\partial^2 S_{kl}}{\partial x_j \partial t} - e_{kij}\frac{\partial^2 E_k}{\partial x_j \partial t} = \rho \frac{\partial^3 u_i}{\partial t^3}. \tag{11.28}$$

Next, taking the curl of equation (11.20) and using equations (11.21),

$$-\boldsymbol{\nabla} \times \boldsymbol{\nabla} \times \mathbf{E} = \mu_0 \frac{\partial(\boldsymbol{\nabla} \times \mathbf{H})}{\partial t} = \mu_0 \frac{\partial^2 \mathbf{D}}{\partial t^2} \Rightarrow$$

$$-\boldsymbol{\nabla} \times \boldsymbol{\nabla} \times \mathbf{E} = \mu_0 \epsilon_{ij} \frac{\partial^2 E_j}{\partial t^2} + \mu_0 e_{ijk} \frac{\partial^2 S_{jk}}{\partial t^2},$$

(11.29)

where in the last step, equation (11.27) was used, and the fact that the spontaneous polarization is constant in time. Notice that equations (11.28) and (11.29) constitute a coupled set of vector equations in the displacement **u** and the electric field **E** keeping in mind that the strain tensor S_{ij} is given by equation (11.23). Once they are solved, the electric displacement, magnetic intensity, and magnetic field can be found from equation (11.27) and equations (11.20) and (11.21).

11.10 The quasistatic approximation

A simple decoupling can be made by splitting the electric field into rotational and irrotational parts,

$$\mathbf{E} = \mathbf{E}^{(rot)} + \mathbf{E}^{(irrot)},$$
$$\mathbf{E}^{(rot)} = \boldsymbol{\nabla} \times \mathbf{A},$$
$$\mathbf{E}^{(irrot)} = -\boldsymbol{\nabla}\phi,$$

(11.30)

where **A** and φ are the vector and scalar potentials, respectively. Note the mathematical identities,

$$\boldsymbol{\nabla} \cdot \mathbf{E}^{(rot)} = \boldsymbol{\nabla} \cdot \boldsymbol{\nabla} \times \mathbf{A} = 0,$$
$$\boldsymbol{\nabla} \times \mathbf{E}^{(irrot)} = -\boldsymbol{\nabla} \times \boldsymbol{\nabla}\phi = 0.$$

(11.31)

In nearly all cases the electromagnetic coupling to mechanical fields can be neglected due to the large differences in typical speeds of light versus mechanical wave speeds (about five orders of magnitude difference in most materials). In such cases, for piezoelectric studies, the rotational part of the electric field can be neglected, the so-called quasistatic approximation, i.e.,

$$\mathbf{E} = -\boldsymbol{\nabla}\phi,$$

(11.32)

and the coupled equations, equations (11.28) and (11.29), simplify to a set of coupled equations in the scalar potential and the displacement **u**,

$$c_{ijkl} \frac{\partial^2 S_{kl}}{\partial x_j \partial t} + e_{kij} \frac{\partial^3 \phi}{\partial x_j \partial x_k \partial t} = \rho \frac{\partial^3 u_i}{\partial t^3},$$

(11.33)

$$-\mu_0 \epsilon_{ij} \frac{\partial^3 \phi}{\partial x_j \partial t^2} + \mu_0 e_{ijk} \frac{\partial^2 S_{jk}}{\partial t^2} = 0.$$

(11.34)

Equations (11.33) and (11.34) constitute a coupled set of equations in the displacement **u** and the scalar potential φ, again keeping in mind that the strain tensor S_{ij} is uniquely determined from the displacement **u**. Equations (11.33) and (11.34) apply in the quasistatic approximation.

11.11 Solving the coupled field equations for a strain wave in a cubic material

In this section, the coupled fields due to an x-polarized displacement propagating along the y direction in a cubic material will be discussed,

$$\mathbf{u} = \hat{x}\frac{k}{\rho\omega^2}\cos(\omega t - ky),\qquad(11.35)$$

that is, the only non-zero strain component is,

$$S_6 = 2S_{xy} = \left(\frac{\partial u_x}{\partial y} + \frac{\partial u_y}{\partial x}\right) = \frac{\partial u_x}{\partial y}$$

$$= \frac{k^2}{\rho\omega^2}\sin(\omega t - ky).\qquad(11.36)$$

Writing,

$$-\nabla \times \mathbf{E} = \mu_0\frac{\partial \mathbf{H}}{\partial t},\qquad(11.37)$$

out for each component gives,

$$-\frac{\partial E_z}{\partial y} = \mu_0\frac{\partial H_x}{\partial t},\qquad(11.38)$$

$$0 = \mu_0\frac{\partial H_y}{\partial t},\qquad(11.39)$$

$$\frac{\partial E_x}{\partial y} = \mu_0\frac{\partial H_z}{\partial t}.\qquad(11.40)$$

Similarly, writing,

$$\nabla \times \mathbf{H} = \frac{\partial \mathbf{D}}{\partial t},\qquad(11.41)$$

out gives, for each component,

$$\frac{\partial H_z}{\partial y} = \frac{\partial D_x}{\partial t},\qquad(11.42)$$

$$0 = \frac{\partial D_y}{\partial t},\qquad(11.43)$$

$$-\frac{\partial H_x}{\partial y} = \frac{\partial D_z}{\partial t}.\qquad(11.44)$$

Evidently, the only free-field (no source) solutions are,

$$H_y = D_y = 0, \tag{11.45}$$

except for time-independent contributions that are of no importance in addition to couplings between two sets E_z, H_x, D_z and E_x, H_z, D_x. For the first set, equations (11.38) and (11.44) must be solved,

$$-\frac{\partial E_z}{\partial y} = \mu_0 \frac{\partial H_x}{\partial t}, \tag{11.46}$$

$$-\frac{\partial H_x}{\partial y} = \frac{\partial D_z}{\partial t} = \epsilon_{xx}^S \frac{\partial E_z}{\partial t} + e_{x4} \frac{\partial S_6}{\partial t}, \tag{11.47}$$

where the piezoelectric constitutive equation was used and the piezoelectric stress tensor for cubic materials,

$$[e] = \begin{bmatrix} 0 & 0 & 0 & e_{x4} & 0 & 0 \\ 0 & 0 & 0 & 0 & e_{x4} & 0 \\ 0 & 0 & 0 & 0 & 0 & e_{x4} \end{bmatrix}.$$

Differentiating equation (11.46) with respect to y and using the result in equation (11.47) yield,

$$\frac{\partial^2 E_z}{\partial y^2} - \mu_0 \epsilon_{xx}^S \frac{\partial^2 E_z}{\partial t^2} = \mu_0 e_{x4} \frac{\partial^2 S_6}{\partial t^2} = -\frac{\mu_0 e_{x4} k^2}{\rho} \sin(\omega t - ky), \tag{11.48}$$

with the solution,

$$E_z = -\frac{\mu_0 e_{x4} \omega^2}{\mu_0 \epsilon_{xx}^S \omega^2 - k^2} S_6 = -\frac{\mu_0 e_{x4} k^2}{\rho(\mu_0 \epsilon_{xx}^S \omega^2 - k^2)} \sin(\omega t - ky). \tag{11.49}$$

For the stress component T_6, it follows using equation (11.49),

$$T_6 = c_{44}^E S_6 - e_{x4} E_z = \left(c_{44}^E + \frac{\mu_0 e_{x4}^2 \omega^2}{\mu_0 \epsilon_{xx}^S \omega^2 - k^2} \right) S_6. \tag{11.50}$$

Finally, from the elastic equations and equation (11.50),

$$\frac{\partial T_6}{\partial y} = \frac{\partial T_{xy}}{\partial y} = -\rho \omega^2 u_x,$$

$$\frac{\partial T_6}{\partial y} = \left(c_{44}^E + \frac{\mu_0 e_{x4}^2 \omega^2}{\mu_0 \epsilon_{xx}^S \omega^2 - k^2} \right) \frac{\partial S_6}{\partial y} = \left(c_{44}^E + \frac{\mu_0 e_{x4}^2 \omega^2}{\mu_0 \epsilon_{xx}^S \omega^2 - k^2} \right) \frac{\partial^2 u_x}{\partial y^2} \tag{11.51}$$

$$= -\left(c_{44}^E + \frac{\mu_0 e_{x4}^2 \omega^2}{\mu_0 \epsilon_{xx}^S \omega^2 - k^2} \right) k^2 u_x,$$

i.e., the coupled dispersion equation,

$$\left(c_{44}^E + \frac{\mu_0 e_{x4}^2 \omega^2}{\mu_0 \epsilon_{xx}^S \omega^2 - k^2} \right) k^2 = \rho \omega^2, \tag{11.52}$$

is obtained. The latter expression can be written as,

$$\left(\rho \omega^2 - c_{44}^E k^2 \right)\left(\mu_0 \epsilon_{xx}^S \omega^2 - k^2 \right) = \mu_0 e_{x4}^2 \omega^2 k^2, \tag{11.53}$$

and it is now evident that, in the absence of piezoelectricity, two dispersion equations for uncoupled electromagnetic and elastic modes are found,

$$\mu_0 \epsilon_{xx}^S \omega^2 - k^2 = 0, \text{ (electromagnetic mode)}, \tag{11.54}$$

$$\rho \omega^2 - c_{44}^E k^2 = 0 \text{ (elastic mode)}. \tag{11.55}$$

If piezoelectricity is present, electromagnetic and elastic modes couple weakly, as the reader can easily verify by plotting the dispersion curves calculated from equation (11.53) with and without the piezoelectric term on the right-hand side.

Chapter 12

Piezoelectric reciprocal systems coupled to fluids

Many classical applications of piezoelectricity involve coupling between a piezo-electric system and a fluid embedding it for sensor or actuator purposes. In this chapter, a detailed description of such systems will be presented. In the first part of chapter 12, the governing equations are derived for sound (or ultrasound) waves propagating in fluids. Then, using thermodynamics, general constitutive laws for piezoelectric media are obtained and applied to one-dimensional piezoelectric transducer systems.

12.1 Wave motion in fluids

Consider wave motion in a homogeneous fluid and let $\mathbf{v}$ be the fluid particle velocity in the presence of the wave. The fluid mass density is ρ.

12.2 The equation of continuity

Let us now, in rectangular coordinates, write for the mass flow in and out of a small volume $dV = dxdydz$ due to particle flow along the x direction,

$$\left[\rho v_x - \left(\rho v_x + \frac{\partial(\rho v_x)}{\partial x}dx\right)\right]dydz = -\frac{\partial(\rho v_x)}{\partial x}dV. \tag{12.1}$$

Similar expressions for the flows in and out of the volume along the y and z directions contribute to the net flux, i.e., the total rate of mass increase in dV is,

$$-\left(\frac{\partial(\rho v_x)}{\partial x} + \frac{\partial(\rho v_y)}{\partial y} + \frac{\partial(\rho v_z)}{\partial z}\right)dV = -\nabla \cdot (\rho\mathbf{v})dV. \tag{12.2}$$

Evidently, mass conservation now requires,

$$\frac{\partial m}{\partial t} = \frac{\partial \rho}{\partial t}dV = -\nabla \cdot (\rho\mathbf{v})dV, \tag{12.3}$$

doi:10.1088/978-0-7503-5557-5ch12

where dm is the mass of dV, or,

$$\frac{\partial \rho}{\partial t} + \boldsymbol{\nabla} \cdot (\rho \mathbf{v}) = 0. \tag{12.4}$$

This equation is the equation of continuity.

12.3 The Euler equation

Consider the fluid element dV moves with the fluid. The net force component along the x direction, df_x, on the element is,

$$df_x = \left[\mathcal{P} - \left(\mathcal{P} + \frac{\partial \mathcal{P}}{\partial x} dx \right) \right] dy dz = -\frac{\partial \mathcal{P}}{\partial x} dV, \tag{12.5}$$

where $\mathcal{P}$ is the instantaneous pressure. Neglecting gravitational effects, the total force including contributions from force components along the y and z directions is,

$$d\mathbf{f} = \mathbf{a} dm = \mathbf{a} \rho dV = -\boldsymbol{\nabla} \mathcal{P} dV, \tag{12.6}$$

where $\mathbf{a}$ is the acceleration. Since the fluid element moves in time, the particle velocity is a function of both space and time. Hence,

$$\mathbf{a} = \frac{\partial \mathbf{v}}{\partial t} + v_x \frac{\partial \mathbf{v}}{\partial x} + v_y \frac{\partial \mathbf{v}}{\partial y} + v_z \frac{\partial \mathbf{v}}{\partial z} = \frac{\partial \mathbf{v}}{\partial t} + (\mathbf{v} \cdot \boldsymbol{\nabla}) \mathbf{v}. \tag{12.7}$$

Now, since,

$$\mathcal{P} = \mathcal{P}_0 + p, \tag{12.8}$$

where $\mathcal{P}_0$ is the equilibrium pressure obeying $\boldsymbol{\nabla} \mathcal{P}_0 = 0$ and p is the acoustic pressure, the equation

$$\mathbf{a} dm = \mathbf{a} \rho dV = \left(\frac{\partial \mathbf{v}}{\partial t} + (\mathbf{v} \cdot \boldsymbol{\nabla}) \mathbf{v} \right) \rho dV = -\boldsymbol{\nabla} p dV, \tag{12.9}$$

is obtained, or,

$$\frac{\partial \mathbf{v}}{\partial t} + (\mathbf{v} \cdot \boldsymbol{\nabla}) \mathbf{v} = -\frac{\boldsymbol{\nabla} p}{\rho}. \tag{12.10}$$

In the linear acoustics regime, the particle velocity $\mathbf{v}$, the mass density variation $\rho - \rho_0$, and the acoustic pressure p are small quantities such that products of them can be neglected. Then,

$$\frac{\partial \mathbf{v}}{\partial t} = -\frac{1}{\rho_0} \boldsymbol{\nabla} p, \tag{12.11}$$

which is the Euler equation. The Euler equation and the equation of continuity can be solved together to find the fluid velocity and acoustic pressure in the whole fluid domain subject to specified initial and boundary conditions.

12.4 Wave propagation in fluids

Since our aim is to develop an ultrasonic piezoelectric transducer model, it is important to understand how waves in a solid material couple to a fluid. Ultrasonic transducer applications in, e.g., instrumentation and the health sector, rely on the coupling of an ultrasonic transducer to a surrounding fluid. To address such applications, wave propagation in fluids will now be addressed.

The mass-continuity and non-viscous equations-of-motion read from equations (12.4) and (12.10)

$$\frac{\partial \rho}{\partial t} + \boldsymbol{\nabla} \cdot (\rho \mathbf{v}) = 0, \tag{12.12}$$

$$\frac{\partial \mathbf{v}}{\partial t} + (\mathbf{v} \cdot \boldsymbol{\nabla})\mathbf{v} = -\frac{\boldsymbol{\nabla} p}{\rho}, \tag{12.13}$$

where ρ, $\mathbf{v}$, p, and t denote the fluid mass density, fluid particle velocity, pressure, and time, respectively.

In the case of a small wave disturbance, a first-order perturbative approach gives,

$$\rho = \rho_0 + \tilde{\rho}, \tag{12.14}$$

$$p = p_0 + \tilde{p}, \tag{12.15}$$

$$\mathbf{v} = \tilde{\mathbf{v}}, \tag{12.16}$$

where variables with subscript 0 denote equilibrium values in the absence of a disturbance, and tilde variables ($\tilde{}$) denote changes in these variables in the presence of a disturbance. The perturbative assumption is,

$$\left|\tilde{\rho}\right| < < \rho_0, \tag{12.17}$$

$$\left|\tilde{p}\right| < < p_0. \tag{12.18}$$

It was assumed that the fluid is quiet in the absence of a disturbance, i.e., $\mathbf{v}_0 = 0$ although the generalization to the case with a background flow can be carried out and is of practical importance in many areas such as flow measurement and medical applications.

Let us look for 1D plane wave solutions in the form,

$$\tilde{\rho}(z, t) = \bar{\rho} \exp{(\pm ikz - i\omega t)}, \tag{12.19}$$

$$\tilde{p}(z, t) = \bar{p} \exp{(\pm ikz - i\omega t)}, \tag{12.20}$$

$$\tilde{\mathbf{v}}(z, t) = \bar{\mathbf{v}} \exp{(\pm ikz - i\omega t)}. \tag{12.21}$$

Inserting into equation (12.13) yields, keeping terms linear in the small tilde quantities only,

$$-i\omega \rho_0 \tilde{v}_z = -\frac{\partial \tilde{p}}{\partial z} = \mp ik\tilde{p} \implies \tag{12.22}$$

$$\tilde{v}_z = \pm \frac{k}{\omega} \frac{1}{\rho_0} \tilde{p} = \pm \frac{1}{\rho_0 c} \tilde{p}$$

$$\equiv \pm \frac{\tilde{p}}{Z} = \mp \frac{\tilde{T}}{Z}, \tag{12.23}$$

or,

$$\bar{v}_z \equiv \pm \frac{\tilde{p}}{Z} = \mp \frac{\tilde{T}}{Z}, \tag{12.24}$$

where $\tilde{T} = -\tilde{p}$, since stress T is a positive constant (stiffness) multiplied by strain, with reference to the stress–strain relations, and strain is negative in the presence of a positive pressure. The upper (lower) sign in the latter expressions corresponds to forward (backward) wave propagation. It was also used that the fluid speed-of-sound c is given by $\frac{\omega}{k}$ which must not be confused with the particle velocity. Finally, the fluid impedance $Z \equiv \rho_0 c$ was introduced.

12.5 Piezoelectric constitutive equations

In this section, the general form of the piezoelectric constitutive equations for a solid will be determined using thermodynamic considerations.

In a previous section, the mechanical part of the free energy, equation (1.28), was derived,

$$dF^{\mathrm{mech}} = T_{ik} dS_{ik} - \sigma d\Theta. \tag{12.25}$$

Adding contributions from electrical (dF^{elec}) and magnetic fields (dF^{mag}), the total free energy can be written as,

$$dF = dF^{\mathrm{mech}} + dF^{\mathrm{elec}} + dF^{\mathrm{mag}} = T_{ik} dS_{ik} + E_i dD_i + H_i dB_i - \sigma d\Theta, \tag{12.26}$$

where **B** and **H** denote the magnetic field and the magnetic field intensity, respectively. It will prove useful to introduce the elastic enthalpy defined as,

$$\tilde{H} = F - E_i D_i, \tag{12.27}$$

i.e.,

$$d\tilde{H} = dF - D_i dE_i - E_i dD_i = T_{ik} dS_{ik} - D_i dE_i + H_i dB_i - \sigma d\Theta. \tag{12.28}$$

It now follows from the Chain Rule that,

$$e_{ikl} \equiv \left(\frac{\partial D_i}{\partial S_{kl}} \right)_E = -\left(\frac{\partial^2 \tilde{H}}{\partial S_{kl} \partial E_i} \right) = -\left(\frac{\partial^2 \tilde{H}}{\partial E_i \partial S_{kl}} \right) = -\left(\frac{\partial T_{kl}}{\partial E_i} \right)_S. \tag{12.29}$$

With reference to the discussion in section 11.10 magnetic field effects will be neglected. This is reasonable since the speed of light is many orders of magnitude higher than the mechanical wave speed. Hence,

$$D_i = \epsilon_{ij} E_j + e_{ikl} S_{kl}, \tag{12.30}$$

$$T_{kl} = \lambda_{klij} S_{ij} - e_{ikl} E_i. \tag{12.31}$$

Notice that the *same* piezoelectric coefficient e_{ikl} appears in the two constitutive equations except from a sign change. This result, known as an Onsager equation, relies on reversible thermodynamic arguments and assumes that no loss terms appear in the thermodynamic potentials F and $\tilde{\tilde{H}}$. Here, and in the following, consider the thermodynamic equilibrium case to correspond to all variables E_i, D_i, T_{ij}, S_{ij} being zero. This allows us to write to first order in the variables,

$$D_i = \epsilon_{ij} E_j + e_{ikl} S_{kl},$$

instead of,

$$dD_i = \epsilon_{ij} dE_j + e_{ikl} dS_{kl},$$

etc.

Another piezoelectric coefficient, the piezoelectric h coefficient, will be useful,

$$h_{ikl} \equiv -\left(\frac{\partial E_i}{\partial S_{kl}}\right)_D = -\left(\frac{\partial^2 F}{\partial S_{kl} \partial D_i}\right) = -\left(\frac{\partial^2 F}{\partial D_i \partial S_{kl}}\right) = -\left(\frac{\partial T_{kl}}{\partial D_i}\right)_S, \tag{12.32}$$

thus,

$$T_{kl} = \lambda_{klij} S_{ij} - h_{ikl} D_i. \tag{12.33}$$

Further, the identity (for a fixed zero strain case—note the superscript),

$$E_i \equiv \beta_{ij}^S D_j = \beta_{ij}^S \epsilon_{jk}^S E_k, \tag{12.34}$$

defining the permittivity and the inverse permittivity, yields,

$$\delta_{ik} = \beta_{ij}^S \epsilon_{jk}^S, \tag{12.35}$$

where δ_{ik} is a Kronecker delta. Similarly, for a fixed zero stress case,

$$\delta_{ik} = \beta_{ij}^T \epsilon_{jk}^T. \tag{12.36}$$

12.5.1 1D versions of the piezoelectric constitutive equations

To make a computationally fast 1D model, 1D versions of the constitutive equations are required. They are,

$$D = \epsilon^S E + eS, \tag{12.37}$$

$$T = \lambda^E S - eE, \tag{12.38}$$

$$T = \lambda^D S - hD, \tag{12.39}$$

$$E = \beta^S D - hS. \tag{12.40}$$

Suffixes (i, j, k, l) on variables and coefficients are omitted here since they all correspond to the relevant 1D coordinate, say, z. Note the appearance of superscripts S and E on ϵ and λ in the first two equations. As mentioned earlier, for piezoelectric materials, it is important to specify which field is kept fixed subject to a change. For the first two equations, the superscripts signify that the permittivity and the stiffness are calculated at constant strain and electric field, respectively. Indeed, the permittivity and stiffness have different values evaluated at fixed stress or strain. Similarly, they have different values at fixed electric field or electric displacement. For the stiffness as an example,

$$\lambda^E = \left(\frac{\partial T}{\partial S}\right)_E \neq \left(\frac{\partial T}{\partial S}\right)_D = \lambda^D, \qquad (12.41)$$

etc.

12.6 Equation-of-motion in the 1D solid case

Consider an infinitesimal solid slab as shown in figure 12.1. The cross-sectional area is A, the thickness is dz, and the mass density of the solid is ρ. Newton's second law then yields,

$$\rho A dz \frac{\partial v_z}{\partial t} = F_z - F_{z+dz} = A p_z - A p_{z+dz} = -A T_z + A T_{z+dz} = A \frac{\partial T}{\partial z} dz, \quad (12.42)$$

where p is the pressure, T is the stress, and v_z is the particle velocity along the z direction. In obtaining the third equality, it was used that $T = -p$. Thus,

$$\rho \frac{\partial v_z}{\partial t} = \frac{\partial T}{\partial z}. \qquad (12.43)$$

Consider next a microscopic view of the particles that constitute the solid (figure 12.2). In the presence of a dynamic disturbance, the individual solid particles move a distance ξ_z that depends on time. The particle velocity is given by,

$$v_z = \frac{\partial \xi_z}{\partial t}, \qquad (12.44)$$

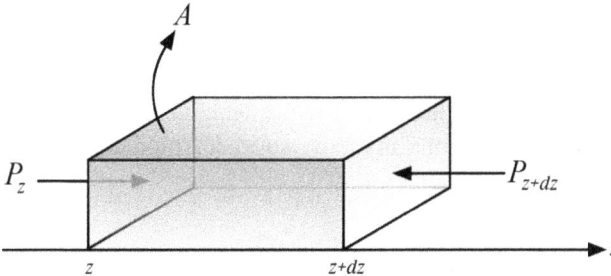

Figure 12.1. Solid slab under pressure (tension).

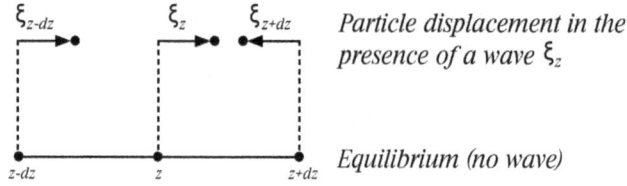

Particle displacement in the presence of a wave ξ_z

Equilibrium (no wave)

Figure 12.2. Strain wave

so,

$$\frac{\partial v_z}{\partial z} = \frac{\partial \left(\frac{\partial \xi_z}{\partial t} \right)}{\partial z} = \frac{\partial}{\partial t} \frac{\partial \xi_z}{\partial z} = \frac{\partial S}{\partial t}. \tag{12.45}$$

Since $T = \lambda^D S - hD$ from equation (12.39), the two dynamic equations can be reshaped to two partial differential equations in two variables v_z, S,

$$\rho \frac{\partial v_z}{\partial t} = \lambda^D \frac{\partial S}{\partial z}, \tag{12.46}$$

$$\frac{\partial v_z}{\partial z} = \frac{\partial S}{\partial t}, \tag{12.47}$$

where the Maxwell–Poisson equation was used to argue that $\frac{\partial D}{\partial z} = 0$, see the discussion below [equation (12.51)]. Differentiating the first equation with respect to time and using the second equation yield,

$$\frac{\partial^2 v_z}{\partial t^2} = \frac{\lambda^D}{\rho} \frac{\partial^2 S}{\partial t \partial z} = \frac{\lambda^D}{\rho} \frac{\partial^2 S}{\partial z \partial t} = \frac{\lambda^D}{\rho} \frac{\partial^2 v_z}{\partial z^2}, \tag{12.48}$$

and, similarly, differentiating the second equation with respect to time,

$$\frac{\partial^2 S}{\partial t^2} = \frac{\lambda^D}{\rho} \frac{\partial^2 S}{\partial z^2}. \tag{12.49}$$

Hence, both particle speed and strain obey the familiar one-dimensional wave equation where the wave speed (speed-of-sound) is given by,

$$c = \sqrt{\frac{\lambda^D}{\rho}}. \tag{12.50}$$

Since the problem we address involves mechanical and electrical fields, the governing mechanical equations must be supplemented by the Maxwell–Poisson equation,

$$\frac{\partial D}{\partial z} = \rho_e^{free} = 0, \tag{12.51}$$

where ρ_e^{free} is the free electrical charge in the solid. Assuming the piezoelectric material is a dielectric (applicable in most transducer applications), the free electrical charge can be set to zero. Another electrical equation is required since the piezoelectric system is connected to an external electric circuit represented by an electric impedance, i.e., Kirchhoff's law,

$$V_{\text{exc}}(t) = Z_e I + \int_{z_L}^{z_R} E_z(z, t)dz, \tag{12.52}$$

where V_{exc}, Z_e, I denote the excitation voltage, the electrical impedance connected to the piezoelectric transducer (see figure 12.4), and the current, respectively, and z_L (z_R) denotes the left (right) side of the piezoelectric material.

12.7 Mono-frequency case

The transducer problem is considerably simplified if mono-frequency conditions apply whereby all variables depend on time as $\exp(-i\omega t)$. Then, equations (12.48) and (12.49) can be written as,

$$\frac{\partial^2 v_z}{\partial z^2} + k^2 v_z = 0, \tag{12.53}$$

$$\frac{\partial^2 S}{\partial z^2} + k^2 S = 0, \tag{12.54}$$

where the wavenumber,

$$k = \frac{\omega}{c}, \tag{12.55}$$

has been introduced. The general solution to equations (12.53) and (12.54) is,

$$v_z(z, t) = v_z^A \exp(-ikz - i\omega t) + v_z^B \exp(ikz - i\omega t), \tag{12.56}$$

$$S(z, t) = S^A \exp(-ikz - i\omega t) + S^B \exp(ikz - i\omega t). \tag{12.57}$$

A simple connection exists between v_z^A and S^A (v_z^B and S^B) by use of equation (12.47),

$$\frac{\partial v_z}{\partial z} = \frac{\partial S}{\partial t},$$

namely,

$$-ikv_z^A = -i\omega S^A, \tag{12.58}$$

$$ikv_z^B = -i\omega S^B, \tag{12.59}$$

or,

$$\frac{v_z^A}{S^A} = \frac{\omega}{k} = c = \frac{\rho c^2}{\rho c} = \frac{\lambda^D}{Z_a}, \tag{12.60}$$

$$\frac{v_z^B}{S^B} = -\frac{\omega}{k} = -\frac{\lambda^D}{Z_a},$$

(12.61)

where the acoustic impedance of the material, $Z_a = \rho_0 c$, was introduced.

12.7.1 Recasting Kirchhoff's law

Taking the time derivative of equation (12.52) yields,

$$\frac{\partial V_{\text{exc}}}{\partial t} = Z_e \frac{\partial I}{\partial t} + \int_{z_L(t)}^{z_R(t)} \frac{\partial E}{\partial t} dz + E(z_R(t), t)\frac{dz_R}{dt} - E(z_L(t), t)\frac{dz_L}{dt}.$$

(12.62)

Since the only current contribution through a dielectric material is the displacement current, the current can be written as,

$$I = A\frac{\partial D}{\partial t}.$$

(12.63)

Employing the constitutive relation equation (12.40) and the fact that D is constant in space, yields,

$$
\begin{aligned}
\frac{\partial V_{\text{exc}}}{\partial t} &= AZ_e \frac{\partial^2 D}{\partial t^2} + \int_{z_L(t)}^{z_R(t)} \left(\beta^S \frac{\partial D}{\partial t} - h\frac{\partial S}{\partial t} \right) dz + E(z_R(t), t)\frac{dz_R}{dt} - E(z_L(t), t)\frac{dz_L}{dt} \\
&= AZ_e \frac{\partial^2 D}{\partial t^2} + \beta^S \frac{\partial D}{\partial t}l - h\int_{z_L(t)}^{z_R(t)} \frac{\partial v_z}{\partial z} dz + E(z_R(t), t)\frac{dz_R}{dt} - E(z_L(t), t)\frac{dz_L}{dt} \\
&= AZ_e \frac{\partial^2 D}{\partial t^2} + \beta^S \frac{\partial D}{\partial t}l - h(v_z(z_R(t)) - v_z(z_L(t))) + E(z_R(t), t)\frac{dz_R}{dt} - E(z_L(t), t)\frac{dz_L}{dt}.
\end{aligned}
$$

(12.64)

The latter two terms can be shown to be negligibly small. Compare, e.g., the fourth and second terms,

$$\frac{\left| E(z_R(t), t)\dfrac{dz_R}{dt} \right|}{\left| \beta^S \dfrac{\partial D}{\partial t}l \right|} \approx \frac{|\tilde{z}_R|}{l} << 1,$$

(12.65)

since the displacement amplitude $\tilde{z}_R$ is much smaller than the piezoelectric material thickness l for typical frequencies and applications. In deriving this result, it was used that $\beta^S \epsilon \approx 1$. Hence,

$$\frac{\partial V_{\text{exc}}}{\partial t} = AZ_e \frac{\partial^2 D}{\partial t^2} + \beta^S \frac{\partial D}{\partial t}l - h(v_z(z_R(t)) - v_z(z_L(t))),$$

(12.66)

or, in the mono-frequency case,

$$-i\omega V_{\text{exc}} = -AZ_e \omega^2 D - i\omega\beta^S Dl - h(v_z(z_R) - v_z(z_L)).$$

(12.67)

12.8 A one-dimensional model of a classical piezoelectric transmitter

Consider a piezoelectric material of length l and cross-sectional area A immersed in a fluid (figure 12.3). The fluid that faces the piezoelectric material on its left side might be different than the fluid on its right side. This situation is common in applications since the wave energy is intentionally either absorbed (or emitted) from one side of the transducer only.

Continuity of pressure (or normal stress) and particle velocity applies everywhere in the piezoelectric structure and at the interfaces with the fluids. Applying these boundary conditions to the left side of the piezoelectric material, acting as a transmitter, yields,

$$T_L = \lambda^D(S^A + S^B) - hD, \tag{12.68}$$

$$\frac{T_L}{Z_a^L} = \lambda^D\left(\frac{S^A}{Z_a} - \frac{S^B}{Z_a}\right), \tag{12.69}$$

where equations (12.24), (12.58) and (12.59) were used to obtain equation (12.69). The plus sign $(+\frac{T_L}{Z_a^L})$ on the left-hand side of equation (12.69) corresponds to backward traveling waves in the fluid coupled to the left since pressure waves are generated in the piezoelectric material by virtue of electrical excitation through the application of a voltage.

Applying the boundary conditions to the right side of the piezoelectric material yields,

$$T_R = \lambda^D(S^A \exp(-ikl) + S^B \exp(ikl)) - hD, \tag{12.70}$$

$$-\frac{T_R}{Z_a^R} = \lambda^D\left(\frac{S^A}{Z_a} \exp(-ikl) - \frac{S^B}{Z_a} \exp(ikl)\right). \tag{12.71}$$

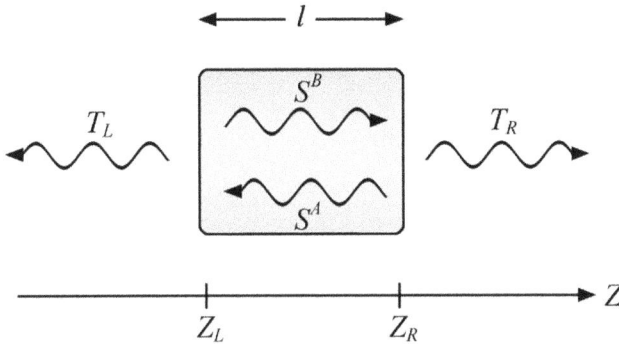

Figure 12.3. Piezoelectric transmitter.

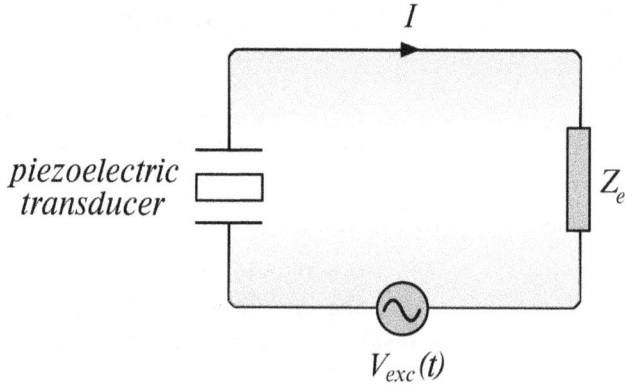

Figure 12.4. Transmitter circuit.

Finally, Kirchhoff's law leads to,

$$-i\omega\tilde{V}_{exc} = -AZ_e\omega^2 D - i\omega\frac{l}{\epsilon^S}D - h[v_z(z_R) - v_z(z_L)] = -AZ_e\omega^2 D - i\omega\frac{l}{\epsilon^S}D - h\left[-\frac{T_R}{Z_a^R} - \frac{T_L}{Z_a^L}\right], \quad (12.72)$$

where,

$$V_{exc}(t) = \tilde{V}_{exc} \exp(-i\omega t), \quad (12.73)$$

is the excitation voltage across the piezoelectric transmitter and the electric impedance (see figure 12.4).

The combined set of boundary conditions consists of 5 equations in the 5 unknowns S^A, S^B, T_L, T_R, D. They can be written in matrix form as,

$$\begin{bmatrix} \lambda^D & \lambda^D & -1 & 0 & -h \\ \dfrac{\lambda^D}{Z_a} & -\dfrac{\lambda^D}{Z_a} & -\dfrac{1}{Z_a^L} & 0 & 0 \\ \lambda^D\exp(-ikl) & \lambda^D\exp(ikl) & 0 & -1 & -h \\ \dfrac{\lambda^D}{Z_a}\exp(-ikl) & -\dfrac{\lambda^D}{Z_a}\exp(ikl) & 0 & \dfrac{1}{Z_a^R} & 0 \\ 0 & 0 & \dfrac{h}{Z_a^L} & \dfrac{h}{Z_a^R} & -AZ_e\omega^2 - i\omega\dfrac{l}{\epsilon^S} \end{bmatrix}\begin{bmatrix} S^A \\ S^B \\ T_L \\ T_R \\ D \end{bmatrix} = \begin{bmatrix} 0 \\ 0 \\ 0 \\ 0 \\ -i\omega\tilde{V}_{exc} \end{bmatrix}. \quad (12.74)$$

This matrix equation is of the form,

$$\mathbf{A}\mathbf{x} = \mathbf{b}, \quad (12.75)$$

where $\mathbf{A}$ is a 5×5 matrix and $\mathbf{x}$, $\mathbf{b}$ are 5×1 vectors. It can be solved easily numerically to find $\mathbf{x}$,

$$\mathbf{x} = \mathbf{A}^{-1}\mathbf{b}. \quad (12.76)$$

12.9 A one-dimensional model of a classical piezoelectric receiver

A piezoelectric receiver is usually either physically the same transducer as the piezoelectric transmitter or a separate physical component constructed in the same way (intentionally at least, except for possible variations during fabrication). Such a double transducer configuration is called a reciprocal transducer system.

The input to the receiver is an acoustic (or ultrasonic) wave. Due to the piezoelectric properties of the receiver, the generated acoustic (or ultrasonic) wave in the receiver piezoelectric material induces an oscillating electric field that can be registered by coupling an electric impedance to the piezoelectric material (or, more correctly, coupling an electric impedance to the two electrodes on the piezoelectric material).

The receiver is schematically shown in figure 12.5. Applying continuity in pressure (stress) and velocity at the ends of the piezoelectric material (and discarding the negligible influence of the electrodes) leads to the four equations,

$$T_I + T_B = \lambda^D(S^C + S^D) - hD, \tag{12.77}$$

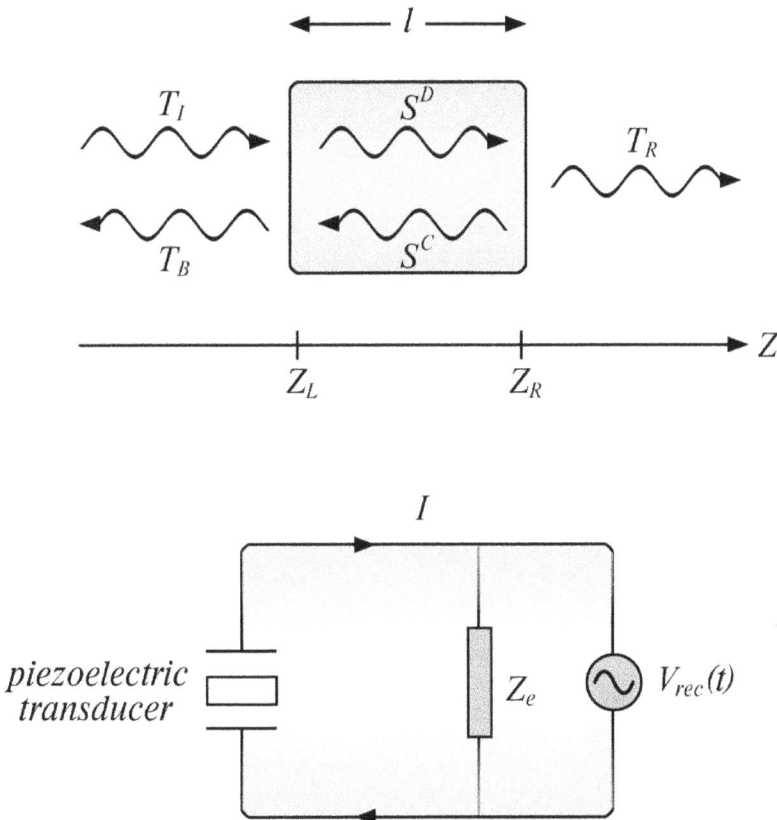

Figure 12.5. Piezoelectric receiver.

$$-\frac{T_I}{Z_a^L} + \frac{T_B}{Z_a^L} = \lambda^D\left(\frac{S^C}{Z_a} - \frac{S^D}{Z_a}\right), \tag{12.78}$$

$$T_R = \lambda^D(S^C \exp(-ikl) + S^D \exp(ikl)) - hD, \tag{12.79}$$

$$-\frac{T_R}{Z_a^R} = \lambda^D\left(\frac{S^C}{Z_a} \exp(-ikl) - \frac{S^D}{Z_a} \exp(ikl)\right), \tag{12.80}$$

where T_I, T_B, T_R, S^C, S^D, D, Z_a^L, Z_a, Z_a^R, l denote the incoming stress wave, the reflected stress wave, the stress wave generated on the right aperture of the receiver, the strain amplitude associated with the backward propagating wave in the piezoelectric material, the strain amplitude associated with the forward propagating wave in the piezoelectric material, the acoustic impedance of the left medium, the acoustic impedance of the piezoelectric material, the acoustic impedance of the right medium, and the piezoelectric layer thickness, respectively.

Differentiating Kirchhoff's law in time yields,

$$0 = -AZ_e\omega^2 D - i\omega\frac{l}{\epsilon^S}D - h\left[-\frac{T_R}{Z_a^R} - \left(-\frac{T_I}{Z_a^L} + \frac{T_B}{Z_a^L}\right)\right], \tag{12.81}$$

where A and Z_e denote the piezoelectric material cross-sectional area and the electric impedance, respectively.

Writing this set of equations in matrix form yields,

$$\begin{bmatrix} \lambda^D & \lambda^D & -1 & 0 & -h \\ \dfrac{\lambda^D}{Z_a} & -\dfrac{\lambda^D}{Z_a} & -\dfrac{1}{Z_a^L} & 0 & 0 \\ \lambda^D\exp(-ikl) & \lambda^D\exp(ikl) & 0 & -1 & -h \\ \dfrac{\lambda^D}{Z_a}\exp(-ikl) & -\dfrac{\lambda^D}{Z_a}\exp(ikl) & 0 & \dfrac{1}{Z_a^R} & 0 \\ 0 & 0 & \dfrac{h}{Z_a^L} & \dfrac{h}{Z_a^R} & -AZ_e\omega^2 - i\omega\dfrac{l}{\epsilon^S} \end{bmatrix} \begin{bmatrix} S^C \\ S^D \\ T_B \\ T_R \\ D \end{bmatrix} = \begin{bmatrix} T_I \\ -\dfrac{T_I}{Z_a^L} \\ 0 \\ 0 \\ h\dfrac{T_I}{Z_a^L} \end{bmatrix}. \tag{12.82}$$

Solving the above system of 5 equations in 5 unknowns: S^C, S^D, T_B, T_R, D for the input T_I gives us the current I_{rec} and the voltage V_{rec} across the electric impedance,

$$I_{\text{rec}}(\omega, t) = A\frac{\partial D}{\partial t} = -i\omega AD \exp(-i\omega t), \tag{12.83}$$

$$V_{\text{rec}}(\omega, t) = Z_e I(\omega, t) = -i\omega Z_e AD \exp(-i\omega t). \tag{12.84}$$

Note that for a reciprocal transducer system, the combination of the piezoelectric transmitter and receiver equation systems allows us to determine the receiver voltage in equation (12.84) from the transmitter excitation voltage given by equation (12.73). The coupling,

$$V_{\text{rec}}(\omega) = H(\omega)V_{\text{exc}}(\omega), \tag{12.85}$$

Table 12.1. Model parameters for a reciprocal transducer system operating in water at 1 MHz.

Parameter	Value	Unit
Driving frequency $\omega_0/(2\pi)$:	1	MHz
Applied voltage amplitude $\tilde{V}_e$:	10	V
Transducer thickness l	1.96	mm
Transducer area A	380	mm^2
Piezoelectric h coefficient	1.42×10^9	F^{2}V m^{-3}
Piezoelectric ϵ^S	$1440\epsilon_0$	F m^{-1}
Piezoelectric λ^D	1.19×10^{11}	Pa
Piezoelectric mass density ρ	7750	kg m^{-3}
Water acoustic impedance Z_a	1.5×10^6	kg (m^{-2} s^{-1})
Air acoustic impedance Z_a	440	kg (m^{-2} s^{-1})
Electric impedance Z_e	100	Ω

introduces $H(\omega)$ as the reciprocal transducer system transfer function. From knowledge of $H(\omega)$, information about the fluid medium between the transmitter and receiver can be extracted such as its properties, scattering particles or objects, flow, temperature variations, etc. A reciprocal transducer system finds application in numerous areas.

In table 12.1, typical parameters of a piezoelectric transducer system operating at 1 MHz. are listed. One side of the transmitter (receiver) faces air—the other side faces water.

12.10 General time-dependent excitation of a reciprocal piezoelectric transducer system

Since the calculation of the signals above is a linear problem, it is easy to determine the response of a piezoelectric reciprocal transducer system to a general time-dependent excitation voltage using Fourier analysis.

To this purpose, the Fourier transforms,

$$V_{\text{exc}}(t) = \frac{1}{\sqrt{2\pi}} \int_{-\infty}^{\infty} V_{\text{exc}}(\omega)\exp(-i\omega t)d\omega, \qquad (12.86)$$

$$V_{\text{exc}}(\omega) = \frac{1}{\sqrt{2\pi}} \int_{-\infty}^{\infty} V_{\text{exc}}(t)\exp(i\omega t)dt, \qquad (12.87)$$

will be used, where $V_{\text{exc}}(\omega)$ is the Fourier transform of $V_{\text{exc}}(t)$.

12.10.1 Example: Fourier transform of a sinus burst

Consider the following time-dependent excitation,

$$V_{\text{exc}}(t) = V_0 \sin(\omega_0 t)\Theta\left(|t| \leqslant \frac{8\pi}{\omega_0}\right), \qquad (12.88)$$

where $\Theta(\cdot)$ denotes Heaviside's function. This signal is an 8-period sinus burst of driving frequency $f_0 = \frac{\omega_0}{2\pi}$.

12.10.2 Exercise

Prove that the Fourier transform becomes,

$$V_{\text{exc}}(\omega) = i\sqrt{\frac{2}{\pi}}\, V_0 \left[\frac{\sin\left((\omega_0 - \omega)(8\pi/\omega_0)\right)}{2(\omega_0 - \omega)} - \frac{\sin\left((\omega_0 + \omega)(8\pi/\omega_0)\right)}{2(\omega_0 + \omega)} \right], \quad (12.89)$$

Hint: Use,

$$\sin(\omega_0 t) = \frac{1}{2i}[\exp(i\omega_0 t) - \exp(-i\omega_0 t)]. \quad (12.90)$$

12.10.3 Exercise

Prove that the time-dependent receiver voltage due to the excitation voltage in equation (12.88) can be found by computing the inverse Fourier transform,

$$V_{\text{rec}}(t) = \frac{1}{\sqrt{2\pi}} \int_{-\infty}^{\infty} H(\omega) V_{\text{exc}}(\omega) \exp(-i\omega t) d\omega, \quad (12.91)$$

where $H(\omega)$ is the transfer function in equation (12.85) and $V_{\text{exc}}(\omega)$ is the Fourier transform obtained in equation (12.89). Clearly this prescription can be carried out for *any* time-dependent excitation voltage as long as the excitation is small, i.e., a linear analysis suffices.

Chapter 13

Three-dimensional axisymmetric piezoelectric vibrations

Sometimes, when a more detailed understanding of all the resonances of piezoelectric systems is important, a full three-dimensional analysis of piezoelectric systems is necessary. In chapter 13, a model for axisymmetric three-dimensional piezoelectric systems is given and a finite-difference scheme is provided to determine the displacements and eigenfrequencies subject to a complete set of boundary conditions and initial conditions. Results for a typical transducer setup (PZT-5 material with or without a matching layer) and parameters corresponding to air flow measurement are presented.

13.1 Piezoelectric cylindrical rod—$C_{6v}(6\ mm)$ crystal symmetry

Consider a piezoelectric cylindrical rod. Assume that the crystal symmetry is $C_{6v}(6\ mm)$ which is common for many important piezoelectric materials such as the commercial ferroelectric ceramic PZT-5H. The general piezoelectric constitutive equations are,

$$T_{ij} = c_{ijkl}^{E} S_{kl} - e_{mij} E_{m}, \tag{13.1}$$

$$D_{i} = e_{ikl} S_{kl} + \epsilon_{im} E_{m}, \tag{13.2}$$

From the strain result,

$$S_{kl} = \frac{1}{2}\left(\frac{\partial u_{k}}{\partial x_{l}} + \frac{\partial u_{l}}{\partial x_{k}}\right), \tag{13.3}$$

the electric field,

$$E_{m} = -\frac{\partial \phi}{\partial x_{m}}, \tag{13.4}$$

where φ is the electric potential, the elastic equations,

doi:10.1088/978-0-7503-5557-5ch13

$$\rho \frac{\partial^2 u_i}{\partial t^2} = \frac{\partial T_{ij}}{\partial x_j}, \tag{13.5}$$

and the Maxwell–Poisson equation,

$$\frac{\partial D_i}{\partial x_i} = 0, \tag{13.6}$$

where **D** is the electric displacement, the complete set of governing equations is obtained.

Neglecting torsional modes, the following simplification can be made,

$$u_\theta = 0, \tag{13.7}$$

$$D_\theta = 0, \tag{13.8}$$

$$\frac{\partial}{\partial \theta} = 0, \tag{13.9}$$

corresponding to full rotational symmetry.

The constitutive equations for the stress components are in cylindrical coordinates,

$$T_{rr} = c_{11}^E \frac{\partial u_r}{\partial r} + c_{12}^E \frac{u_r}{r} + c_{13}^E \frac{\partial u_z}{\partial z} + e_{31} \frac{\partial \phi}{\partial z}, \tag{13.10}$$

$$T_{\theta\theta} = c_{12}^E \frac{\partial u_r}{\partial r} + c_{11}^E \frac{u_r}{r} + c_{13}^E \frac{\partial u_z}{\partial z} + e_{31} \frac{\partial \phi}{\partial z}, \tag{13.11}$$

$$T_{zz} = c_{13}^E \frac{\partial u_r}{\partial r} + c_{13}^E \frac{u_r}{r} + c_{33}^E \frac{\partial u_z}{\partial z} + e_{33} \frac{\partial \phi}{\partial z}, \tag{13.12}$$

$$T_{rz} = c_{44}^E \left(\frac{\partial u_r}{\partial z} + \frac{\partial u_z}{\partial r} \right) + e_{15} \frac{\partial \phi}{\partial r}, \tag{13.13}$$

$$T_{\theta z} = T_{r\theta} = 0. \tag{13.14}$$

Similarly, the constitutive equations for the electric displacement components are,

$$D_r = e_{15} \left(\frac{\partial u_r}{\partial z} + \frac{\partial u_z}{\partial r} \right) - \epsilon_{11}^S \frac{\partial \phi}{\partial r}, \tag{13.15}$$

$$D_\theta = 0, \tag{13.16}$$

$$D_z = e_{31} \left(\frac{\partial u_r}{\partial r} + \frac{u_r}{r} \right) + e_{33} \frac{\partial u_z}{\partial z} - \epsilon_{33}^S \frac{\partial \phi}{\partial z}. \tag{13.17}$$

The governing elastic and electric equations are,

$$\rho\frac{\partial^2 u_z}{\partial t^2} = \frac{\partial T_{rz}}{\partial r} + \frac{T_{rz}}{r} + \frac{\partial T_{zz}}{\partial z}, \tag{13.18}$$

$$\rho\frac{\partial^2 u_r}{\partial t^2} = \frac{\partial T_{rr}}{\partial r} + \frac{\partial T_{rz}}{\partial z} + \frac{T_{rr} - T_{\theta\theta}}{r}, \tag{13.19}$$

$$\frac{\partial D_r}{\partial r} + \frac{D_r}{r} + \frac{\partial D_z}{\partial z} = 0. \tag{13.20}$$

13.1.1 Boundary conditions

If the piezoelectric rod is finite with end facets located at $z = 0$ and $z = L$, stress-free boundaries require,

$$T_{zz} = T_{rz} = 0, \text{ at } z = 0, \tag{13.21}$$

$$T_{zz} = T_{rz} = 0, \text{ at } z = L. \tag{13.22}$$

Assuming stress-free boundaries at $r = a$, where a is the cylinder radius, implies,

$$T_{rr} = T_{rz} = 0, \text{ at } r = a. \tag{13.23}$$

If the end facets are covered with metallic electrodes, the electric boundary conditions are,

$$\phi = \phi_L, \text{ at } z = 0, \tag{13.24}$$

$$\phi = \phi_R, \text{ at } z = L, \tag{13.25}$$

where ϕ_L and ϕ_R are the electric potentials at $z = 0$ and $z = L$, respectively.

Finally, an electric boundary condition must be specified at $r = a$. Assuming no currents can be transported through the circumferential area of the piezoelectric rod, the normal displacement current density component $J_r = \frac{\partial D_r}{\partial t}$ vanishes at $r = a$,

$$D_r(r = a) = 0. \tag{13.26}$$

Alternative boundary conditions corresponding to open-circuit conditions, that is (no metallic coverage on the) end facets, are,

$$D_z(z = 0) = 0, \tag{13.27}$$

$$D_z(z = L) = 0, \tag{13.28}$$

so that the current density component $J_z = \frac{\partial D_z}{\partial t}$ vanishes.

Other types of boundary conditions can be relevant depending on the application.

In addition to the above, it follows from symmetry and boundedness of physical fields, by inspection of the elastic equations and constitutive equations, that

$$u_r = 0, \quad \text{if } r = 0, \tag{13.29}$$

$$\frac{\partial u_r}{\partial r} = 0, \quad \text{if } r = 0, \tag{13.30}$$

$$T_{rz} = 0, \quad \text{if } r = 0, \tag{13.31}$$

$$D_r = 0, \quad \text{if } r = 0. \tag{13.32}$$

Note that the latter two boundary equations are equivalent to

$$E_r = 0, \quad \text{if } r = 0, \tag{13.33}$$

$$\frac{\partial u_r}{\partial z} + \frac{\partial u_z}{\partial r} = 0, \quad \text{if } r = 0. \tag{13.34}$$

13.1.2 Initial conditions

Initial conditions can be chosen to correspond to a quiet situation, i.e.,

$$u_r = u_z = 0, \tag{13.35}$$

$$E_r = E_z = 0. \tag{13.36}$$

An external voltage $V_{\text{appl}} = \phi(z = L) - \phi(z = 0) = \phi_R - \phi_L$ may be invoked at a time $t > 0$ to excite vibrations. We will typically consider V_{appl} as a given function of time.

13.2 Effective one-dimensional spatial model

It is possible to make an effective one-dimensional implementation in the spatial coordinates. First let us attempt solutions in the form [1],

$$u_z(r, z, t) = F(r)f(z, t), \tag{13.37}$$

$$\phi(r, z, t) = F(r)g(z, t), \tag{13.38}$$

$$u_r(r, z, t) = \frac{\partial F}{\partial r}h(z, t), \tag{13.39}$$

$$E_z(r, z, t) = -F(r)\frac{\partial g(z, t)}{\partial z}. \tag{13.40}$$

One finds,

$$\frac{d^2F}{dr^2} + \frac{1}{r}\frac{dF}{dr} + k_r^2 F_r = 0, \tag{13.41}$$

where k_r is a separation constant. An approximative expression for k_r can be found by invoking the free boundary condition, $T_{rz} = 0$, at $r = a$. We note that really both

$T_{rr} = T_{rz} = 0$ when $r = a$ in the case of a free transducer surface. The full set of boundary conditions can, however, not be fulfilled if separation-of-variables in r and z is attempted.

Demanding $T_{rz}(r = a) = 0$ yields,

$$F(r) = J_0(k_r r) = J_0\left(\frac{3.83}{a} r\right), \tag{13.42}$$

by choosing the first zero (≈ 3.83) of $\frac{dJ_0(x)}{dx} = -J_1$.

Further, let us rewrite equation (13.18),

$$\rho\frac{\partial^2 u_z}{\partial t^2} = c_{44}\left(\frac{\partial^2 u_r}{\partial z\partial r} + \frac{\partial^2 u_z}{\partial r^2}\right) + e_{15}\frac{\partial^2 \phi}{\partial r^2} + c_{44}\left(\frac{\partial^2 u_r}{\partial z}/r + \frac{\partial u_z}{\partial r}/r\right) + e_{15}\frac{\partial\phi}{\partial r}/r + c_{13}\frac{\partial^2 u_r}{\partial z\partial r}$$

$$+ c_{13}\frac{\partial u_r}{\partial z}/r + c_{33}\frac{\partial^2 u_z}{\partial z^2} + e_{33}\frac{\partial^2 \phi}{\partial z^2}\frac{\partial c_{13}}{\partial z}\frac{\partial u_r}{\partial r} + \frac{\partial c_{13}}{\partial z}u_r/r + \frac{\partial c_{33}}{\partial z}\frac{\partial u_z}{\partial z} + \frac{\partial e_{33}}{\partial z}\frac{\partial\phi}{\partial r}, \tag{13.43}$$

as

$$\rho\frac{\partial^2 f}{\partial t^2} = -k_r^2 c_{44}(h'(z) + f(z)) - k_r^2 e_{15}g(z) - k_r^2 c_{13}h'(z) + c_{33}f''(z) + e_{33}g''(z) - k_r^2 c'_{13}h(z) \tag{13.44}$$

$$+ c'_{33}f'(z) + e'_{33}g'(z),$$

where the prime denotes $\frac{\partial}{\partial z}$ and a division by $F(r)$ has been made.

Similarly, from equation (13.20),

$$0 = e_{15}\left(\frac{\partial^2 u_r}{\partial z\partial r} + \frac{\partial^2 u_z}{\partial r^2}\right) - \varepsilon_{11}\frac{\partial^2 \phi}{\partial r^2} + e_{15}\left(\frac{\partial^2 u_r}{\partial z}/r + \frac{\partial u_z}{\partial r}/r\right) - \varepsilon_{11}\frac{\partial\phi}{\partial r}/r$$

$$+ e_{31}\left(\frac{\partial^2 u_r}{\partial z\partial r} + \frac{\partial u_r}{\partial z}/r\right) + e_{33}\frac{\partial^2 u_z}{\partial z^2} - \varepsilon_{33}\frac{\partial^2 \phi}{\partial z^2} + e'_{31}\left(\frac{\partial u_r}{\partial r} + u_r/r\right) + e'_{33}\frac{\partial u_z}{\partial z} - \varepsilon'_{33}\frac{\partial\phi}{\partial z}, \tag{13.45}$$

and using equations (13.37)–(13.39) followed by a division of $F(r)$,

$$0 = -k_r^2 e_{15}h'(z) - k_r^2 e_{15}f(z) + \varepsilon_{11}k_r^2 g(z) - e_{31}k_r^2 h'(z) + e_{33}f''(z) - \varepsilon_{33}g''(z) \tag{13.46}$$

$$- e'_{31}k_r^2 h(z) + e'_{33}f'(z) - \varepsilon'_{33}g'(z).$$

From equation (13.19) we obtain,

$$\rho\frac{\partial^2 u_r}{\partial t^2} = c_{11}\frac{\partial^2 u_r}{\partial r^2} + c_{12}\left(\frac{\partial u_r}{\partial r}/r - u_r/r^2\right) + c_{13}\frac{\partial^2 u_z}{\partial r\partial z} + e_{31}\frac{\partial^2 \phi}{\partial r\partial z} + c'_{44}\left(\frac{\partial u_r}{\partial z} + \frac{\partial u_z}{\partial r}\right)$$

$$+ e'_{15}\frac{\partial\phi}{\partial r} + c_{44}\left(\frac{\partial^2 u_r}{\partial z^2} + \frac{\partial^2 u_z}{\partial r\partial z}\right) + e_{15}\frac{\partial^2 \phi}{\partial r\partial z} + (c_{11} - c_{12})\left(\frac{\partial u_r}{\partial r}/r - u_r/r^2\right). \tag{13.47}$$

Hence,

$$\rho\frac{\partial^2 h}{\partial t^2} = -k_r^2 c_{11}h(z) + c_{13}f'(z) + e_{31}g'(z) + c_{44}(h''(z) + f'(z)) \tag{13.48}$$

$$+ c'_{44}(h'(z) + f(z)) + e'_{15}g(z) + e_{15}g'(z).$$

13.2.1 Effective one-dimensional boundary conditions

The effective one-dimensional boundary conditions can be found from,

$$\phi(z = 0, t) = -V_{\text{appl}}/2, \tag{13.49}$$

$$\phi(z = L, t) = V_{\text{appl}}/2, \tag{13.50}$$

$$T_{rz}(z = 0, t) = T_{rz}(z = L, t) = 0, \tag{13.51}$$

$$T_{zz}(z = 0, t) = T_{zz}(z = L, t) = 0, \tag{13.52}$$

assuming vacuum environment, and are,

$$g(z = 0, t) = -V_{\text{appl}}/2, \tag{13.53}$$

$$g(z = L, t) = V_{\text{appl}}/2, \tag{13.54}$$

$$c_{44}(h'(z = 0, t) + f(z = 0, t)) + e_{15}g'(z = 0, t) = 0, \tag{13.55}$$

$$c_{44}(h'(z = L, t) + f(z = L, t)) + e_{15}g'(z = L, t) = 0, \tag{13.56}$$

$$-k_r^2 c_{13}(z = 0)h(z = 0, t) + c_{33}(z = 0)f'(z = 0, t) + e_{33}(z = 0)g'(z = 0, t) = 0 \tag{13.57}$$

$$-k_r^2 c_{13}(z = L)h(z = L, t) + c_{33}(z = L)f'(z = L, t) + e_{33}(z = L)g'(z = L, t) = 0. \tag{13.58}$$

These boundary conditions are implemented using the one-dimensional finite difference time domain (FDTD) method, i.e., derivative approximations near the boundaries are approximated as,

$$h'(z = 0, t) = \frac{h(2, t) - h(1, t)}{\Delta z}, \tag{13.59}$$

$$h'(z = L, t) = \frac{h(N, t) - h(N - 1, t)}{\Delta z}, \tag{13.60}$$

etc.

We note that the effective one-dimensional spatial model captures three-dimensional transducer effects including buckling, axial and radial displacements, and the influence of varying material parameters.

13.2.2 Transducer immersed in a fluid medium

If the transducer is immersed in a fluid medium at $z = 0$ and $z = L$, equations (13.51) and (13.52) must be replaced by,

$$T_{rz}(z = 0, t) = T_{rz}(z = L, t) = 0, \tag{13.61}$$

$$T_{zz}(z = 0, t) = Z_{\text{fluid}}\frac{\partial u_z}{\partial t}\bigg|_{z=0}, \tag{13.62}$$

$$T_{zz}(z = L, t) = -Z_{\text{fluid}}\frac{\partial u_z}{\partial t}\bigg|_{z=L}, \tag{13.63}$$

where $Z_{\text{fluid}} = \rho_{\text{fluid}}c_{\text{fluid}}$ is the acoustic impedance of the fluid.

Recasting equations (13.57) and (13.58) to appropriately correct for a fluid medium gives,

$$- k_r^2 c_{13}(z = 0)h(z = 0, t) + c_{33}(z = 0)f'(z = 0, t) + e_{33}(z = 0)g'(z = 0, t)$$
$$= Z_{\text{fluid}}\frac{\partial f(z, t)}{\partial t}\bigg|_{z=0}, \tag{13.64}$$

$$- k_r^2 c_{13}(z = L)h(z = L, t) + c_{33}(z = L)f'(z = L, t) + e_{33}(z = L)g'(z = L, t)$$
$$= -Z_{\text{fluid}}\frac{\partial f(z, t)}{\partial t}\bigg|_{z=L}. \tag{13.65}$$

13.3 Full two-dimensional numerical implementation

Numerical implementation in two dimensions is done using a FDTD scheme used earlier which is convenient for solving practical transducer problems.

Hence, we introduce N_i, N_j discretization points along the r and z directions, respectively. We also introduce N_k points for the time discretization.

Then we use, neglecting piezoelectric terms just to illustrate the calculational method,

$$\frac{\partial u_r}{\partial r} = \frac{u_r(i + 1, j, k) - u_r(i, j, k)}{\Delta_r}, \quad r(i) = 0 + \frac{(i - 1)}{(N_i - 1)}R, \quad i = 1, \dots, N_i, \tag{13.66}$$

$$\frac{\partial u_r}{\partial z} = \frac{u_r(i, j + 1, k) - u_r(i, j, k)}{\Delta_z}, \quad z(j) = 0 + \frac{(j - 1)}{(N_j - 1)}L, \quad j = 1, \dots, N_j, \tag{13.67}$$

and similarly discretizations for u_z and φ. It should be pointed out that for all plots shown, all piezoelectric terms are included in the calculations.

13.3.1 Discretizing boundary conditions

We now give the discretization equations for the boundary conditions; for simplicity in the absence of piezoelectric effects. Imposing $T_{rr} = 0$ when $i = N_i$ we get from equations (13.10) and (13.23),

$$c_{11}(j)\frac{u_r(N_i + 1, j, k) - u_r(N_i, j, k)}{\Delta_r} + c_{12}(j)\frac{u_r(N_i, j, k)}{R} + c_{13}(j)\frac{u_z(N_i, j + 1, k) - u_z(N_i, j, k)}{\Delta_z} = 0,$$
$$\to u_r(N_i + 1, j, k) = \alpha(u_r, u_z, \phi), \tag{13.68}$$

where $\alpha(u_r, u_z, \phi)$ is a function of internal grid points only, and Δ_r and Δ_z are the grid element size along the r and z directions, respectively. Note that the coefficients

c_{ij}, etc are functions of j allowing us to consider stacked materials such as a piezoelectric material, a bonding layer, a matching layer.

Imposing instead $T_{rz} = 0$ when $i = N_i$ we get from equations (13.13) and (13.23),

$$\frac{u_r(N_i, j+1, k) - u_r(N_i, j, k)}{\Delta_z} + \frac{u_z(N_i + 1, j, k) - u_z(N_i, j, k)}{\Delta_r} = 0,$$

$$\rightarrow u_z(N_i + 1, j, k) = \beta(u_r, u_z, \phi),$$
(13.69)

where $\beta(u_r, u_z, \phi)$ is a function of internal grid points only.

Imposing $T_{rz} = 0$ when $j = 1$ we get from equations (13.13) and (13.21),

$$\frac{u_r(i, 1, k) - u_r(i, 0, k)}{\Delta_z} + \frac{u_z(i+1, 1, k) - u_z(i, 1, k)}{\Delta_r} = 0,$$

$$\rightarrow u_r(i, 0, k) = \gamma(u_r, u_z, \phi),$$
(13.70)

where $\gamma(u_r, u_z, \phi)$ is a function of internal grid points only.

Imposing instead $T_{zz} = 0$ when $j = 1$ we get from equations (13.12) and (13.21),

$$c_{13}\frac{u_r(i+1, 1, k) - u_r(i, 1, k)}{\Delta_r} + c_{13}\frac{u_r(i, 1, k)}{r(i)} + c_{33}\frac{u_z(1, j, k) - u_z(0, j, k)}{\Delta_z} = 0,$$

$$\rightarrow u_z(0, j, k) = \delta(u_r, u_z, \phi),$$
(13.71)

where $\delta(u_r, u_z, \phi)$ is a function of internal grid points only.

Imposing $u_r = 0$, $\frac{\partial u_r}{\partial r} = 0$, $E_r = 0$, $\frac{\partial u_r}{\partial z} + \frac{\partial u_z}{\partial r} = 0$ when $r = 0$ yield,

$$\begin{aligned}
u_r(1, j, k) &= 0, \\
u_r(2, j, k) &= u_r(1, j, k) = 0, \\
\phi(2, j, k) &= \phi(1, j, k), \\
u_z(2, j, k) &= u_z(1, j, k).
\end{aligned}$$
(13.72)

For the external force acting only on the upper surface $z = L$ we have,

$$F_{ext}(i, j, k) = F_{ext}(i, N_j, k) \cdot (j = N_j).$$
(13.73)

where $(j = N_j)$ is equal to 1 if $j = N_j$ else 0.

13.3.2 Discretizing initial conditions

Initial conditions are simply

$$u_z(i, j, 0) = u_r(i, j, 0) = \phi(i, j, 0) = 0.$$
(13.74)

13.3.3 Discretizing time variations

The quantity $\frac{\partial^2 \mathbf{u}}{\partial t^2}$ is discretized as,

$$\frac{\partial^2 u_z}{\partial t^2} = \frac{u_z(i, j, k+1) - 2u_z(i, j, k) + u_z(i, j, k-1)}{\Delta_t^2},$$
(13.75)

$$\frac{\partial^2 u_r}{\partial t^2} = \frac{u_r(i, j, k+1) - 2u_r(i, j, k) + u_r(i, j, k-1)}{\Delta_t^2}, \tag{13.76}$$

where Δ_t is the discretization time step.

13.3.4 Relevant 2D output

A numerical expression of the pressure p at the transducer aperture ($z = N_j$) can be found as, assuming azimuthal symmetry,

$$p(z = N_j, k) = \frac{2\pi}{\pi a^2} \int_0^a T_{zz}(r, z = N_j, t) r \, dr$$

$$= \frac{2\pi}{\pi a^2} \sum_{i=1}^{N_j-1} \frac{T_{zz}(i+1, z = N_j, k) + T_{zz}(i, z = N_j, k)}{2} \frac{r(i+1) + r(i)}{2} (r(i+1) - r(i)). \tag{13.77}$$

13.4 Numerical results using the effective one-dimensional model

In this section, u_z, u_r and the electric field components E_z, E_r are computed as a function of z and time for different electric excitation frequencies for a single PZT-5 material transducer (parameters are given in table 13.1). An applied voltage of 1 V (-0.5 V and 0.5 V on the two electrodes) is assumed. The excitation occurs from 2 periods until 5 periods at the driving frequency. The radius R is 0.012 m and the thickness L is 0.01 m. In figure 13.1, a schematic drawing of the axisymmetric transducer is shown. Figure 13.2 shows the variation in time (first axis) and along the z coordinate (second axis) of the functions $g(z, t)$, $e(z, t) = -\frac{\partial g}{\partial z}$, $f(z, t)$, and $h(z, t)$ that uniquely describe the dependencies of the electric potential φ, electric field component E_z, axial displacement u_z, and radial displacement u_r with z and t (refer to equations (13.37)–(13.40)).

13.4.1 Example 1: frequency 240 kHz for a transducer composed of a PZT-5 layer

In this example, we consider a transducer composed of a piezoceramic layer (0.00182 m) and an excitation frequency of 240 kHz. The radius R is assumed equal to 0.0045 m as in the preceding example. Results for the axial and radial displacements are shown in figures 13.3 and 13.4 at a fixed but arbitrary time.

In figure 13.5, the time-dependent average pressure at the transducer aperture is plotted. The acoustic pressure is computed as $Re(i\omega Z_{\text{air}} u_z(z = L, r = 0))$.

13.4.2 Example 2: frequency 240 kHz for a transducer composed of a PZT-5 layer and epoxy matching layer

In example 2, a transducer composed of a piezoceramic layer (0.00182 m) and an epoxy matching layer of thickness 0.0027 m is considered, and the excitation frequency is 240 kHz. The total transducer thickness is then 0.004 52 m. The radius R is assumed equal to 0.0045 m as in the preceding example. Results for the axial and radial displacements are shown in figures 13.6 and 13.7. All calculations in

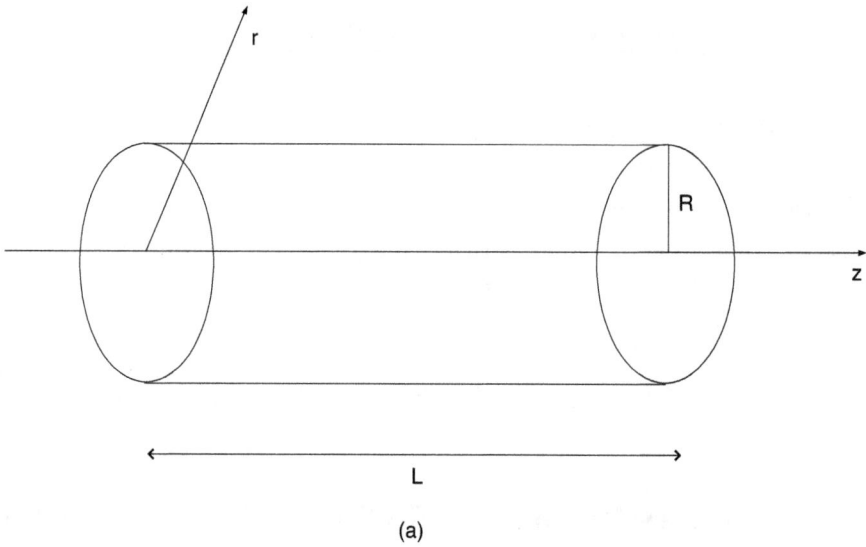

(a)

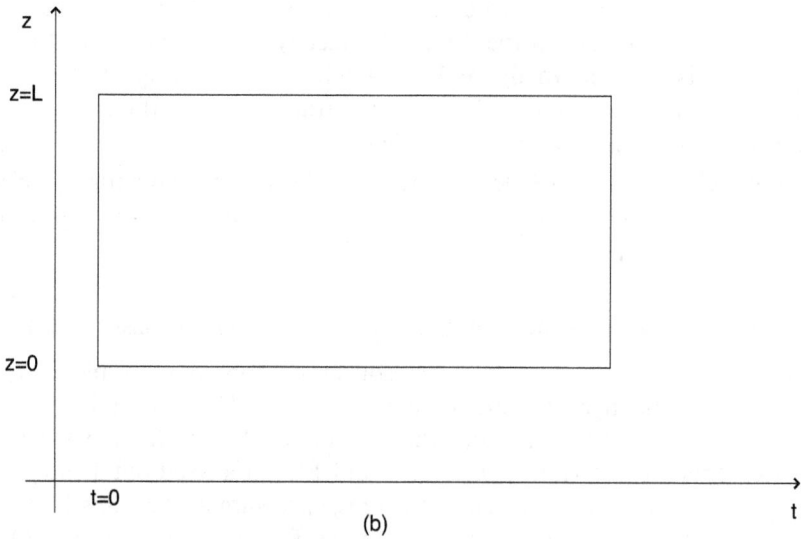

(b)

Figure 13.1. Schematic drawing of the axisymmetric two-dimensional transducer. The subplot (a) shows the real three-dimensional transducer domain using cylindrical coordinates z and r. Subplot (b) shows the $t - z$ domain plotted in figure 13.2. The first axis is time (t) and the second axis is the z coordinate. The r coordinate dependence is not shown in the latter figures since they are known Bessel functions according to the theory of Schnabel [1].

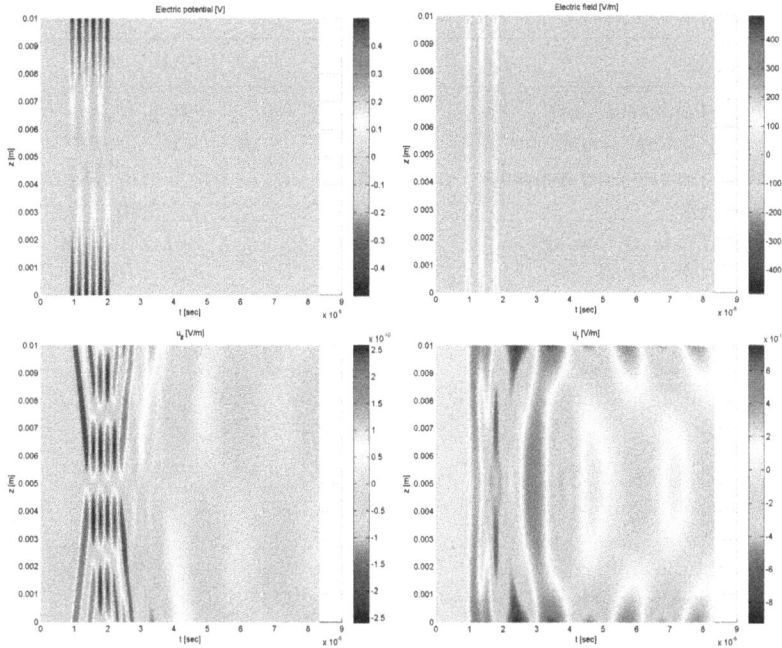

Figure 13.2. Computation of the potential factor g (upper-left), the electric field factor $e = -\frac{\partial g}{\partial z}$ (upper-right), the axial displacement factor f (lower-left), and the radial displacement factor h (lower-right). The driving frequency is 240 kHz and excitation occurs from 2 to 5 periods. The transducer has dimensions $R = 0.012$ m and $L = 0.01$ m and consists of PZT-5 solely.

Table 13.1. Two-dimensional transducer physical properties.

Parameter	PZT-5
c_{11} (PZT-5)	12.1×10^{10} Pa
c_{12} (PZT-5)	7.55×10^{10} Pa
c_{13} (PZT-5)	7.52×10^{10} Pa
c_{33} (PZT-5)	11.1×10^{10} Pa
c_{44} (PZT-5)	2.11×10^{10} Pa
e_{15} (PZT-5)	12.3 C m^{-2}
e_{31} (PZT-5)	-5.4 C m^{-2}
e_{33} (PZT-5)	15.8 C m^{-2}
ϵ_{11} (PZT-5)	8.11×10^{-9} F m^{-1}
ϵ_{33} (PZT-5)	7.35×10^{-9} F m^{-1}
ρ (PZT-5)	7800 kg m^{-3}
E (E module—window material)	3.23×10^{9} Pa
σ (Poisson ratio—window material)	0.3
ρ (air)	1.38 kg m^{-3}

(*Continued*)

Table 13.1. (*Continued*)

Parameter	PZT-5
c (speed of sound—air)	340 m s^{-1}
Driving voltage amplitude	1 V
$f = \omega/(2\pi)$ (driving frequency)	2.4×10^5 Hz
a (radius)	4.5×10^{-3} m
L_p (PZT-5 layer thickness)	1.82×10^{-3} m
L (total thickness)	4.52×10^{-3} m
k_r	$3.83/a$

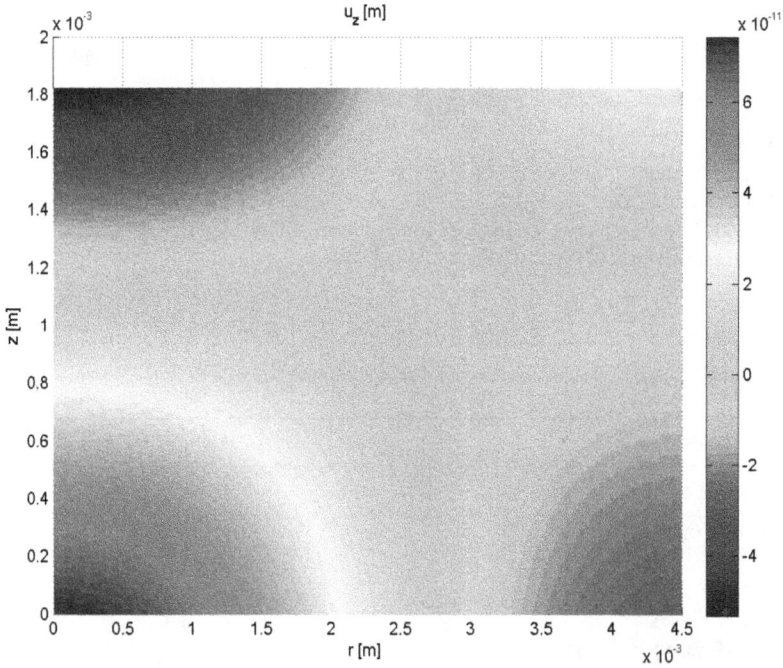

Figure 13.3. Plot of the axial displacement u_z at a specified but arbitrary time versus the spatial coordinates z and r. The frequency is 240 kHz and the transducer radius R is 0.0045 m. The transducer consists of a layer of PZT-5 (thickness 0.001 82 m).

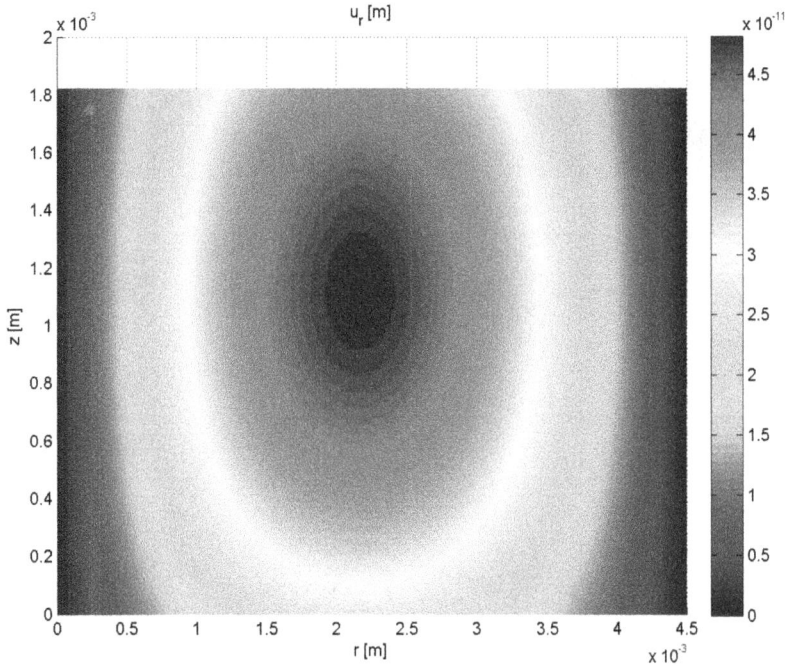

Figure 13.4. Plot of the radial displacement u_r at a specified but arbitrary time versus the spatial coordinates z and r. The frequency is 240 kHz and the transducer radius R is 0.0045 m. The transducer consists of a layer of PZT-5 (thickness 0.001 82 m).

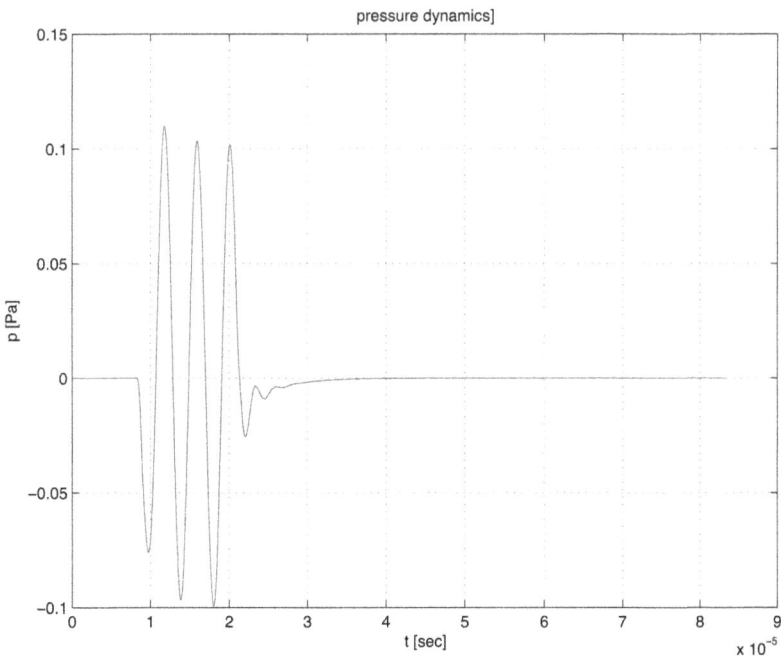

Figure 13.5. Plot of the average pressure p, defined by equation (13.77), as a function of time. The frequency is 240 kHz and the transducer radius R is 0.0045 m. The transducer consists a layer of PZT-5 (thickness 0.001 82 m).

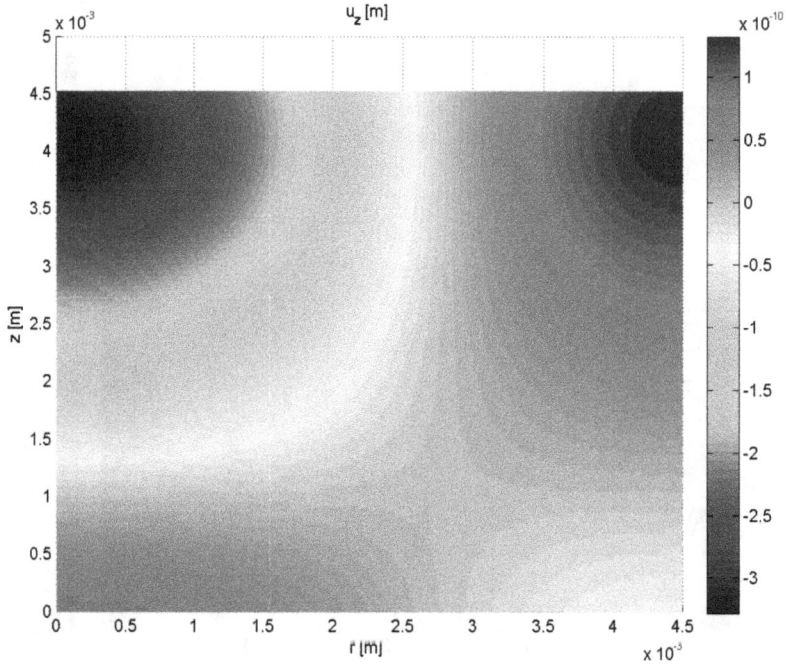

Figure 13.6. Plot of the axial displacement u_z at a specified but arbitrary time versus the spatial coordinates z and r. The frequency is 240 kHz, the transducer radius R is 0.0045 m and the total length $L = 0.00452$ m. The transducer is composed of a layer of PZT-5 (thickness 0.001 82 m) and an epoxy matching layer of thickness 0.0027 m.

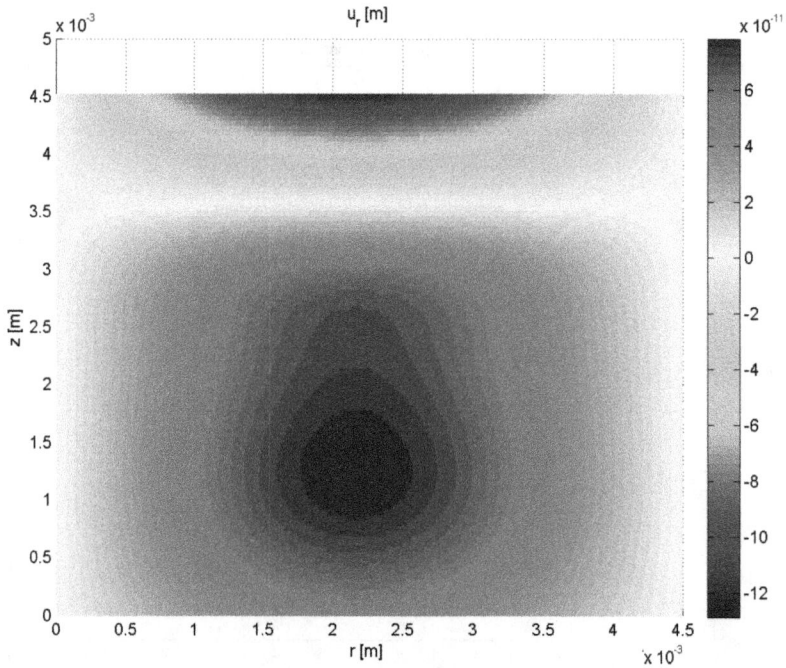

Figure 13.7. Plot of the radial displacement u_r at a specified but arbitrary time versus the spatial coordinates z and r. The frequency is 240 kHz, the transducer radius R is 0.0045 m and the total length $L = 0.00452$ m. The transducer is composed of a layer of PZT-5 (thickness 0.001 82 m) and an epoxy matching layer of thickness 0.0027 m.

Example 2 correspond to electrodes placed at the PZT-5 layer ends, i.e., one electrode is attached between the PZT-5 layer and the window material.

In figure 13.8, the time-dependent average pressure at the transducer aperture is plotted. Observe a substantially higher pressure amplitude as a consequence of the mechanical vibration amplification due to a near quarter-wavelength resonance of the window and the PZT-5 layers. The acoustic pressure is computed as $Re(i\omega Z_{air} u_z(z = L, r = 0))$.

In figure 13.9 the electric potential is plotted. Note that the electric potential is independent of z in the matching layer.

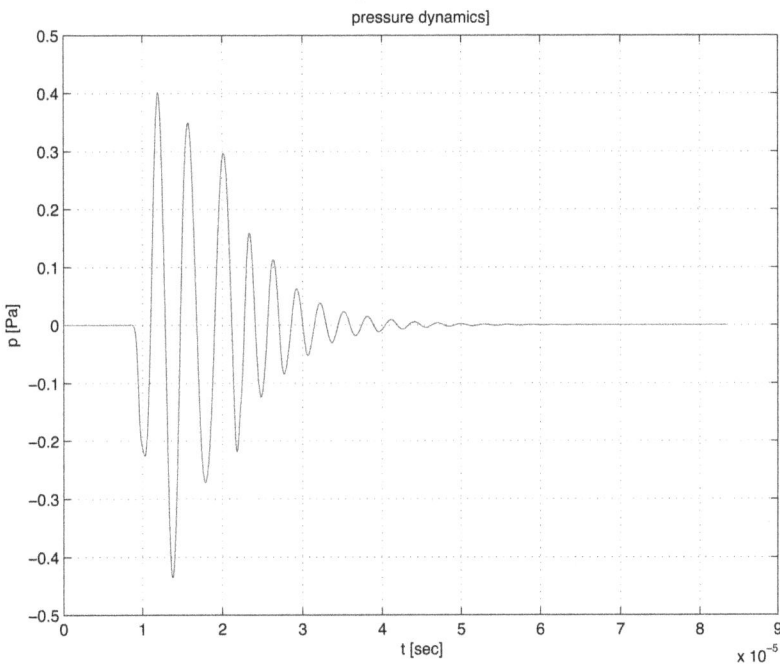

Figure 13.8. Plot of the average pressure p, defined by equation (13.77), as a function of time. The frequency is 240 kHz, the transducer radius R is 0.0045 m and the total length $L = 0.00452$ m. The transducer is composed of a layer of PZT-5 (thickness 0.001 82 m) and an epoxy matching layer of thickness 0.0027 m.

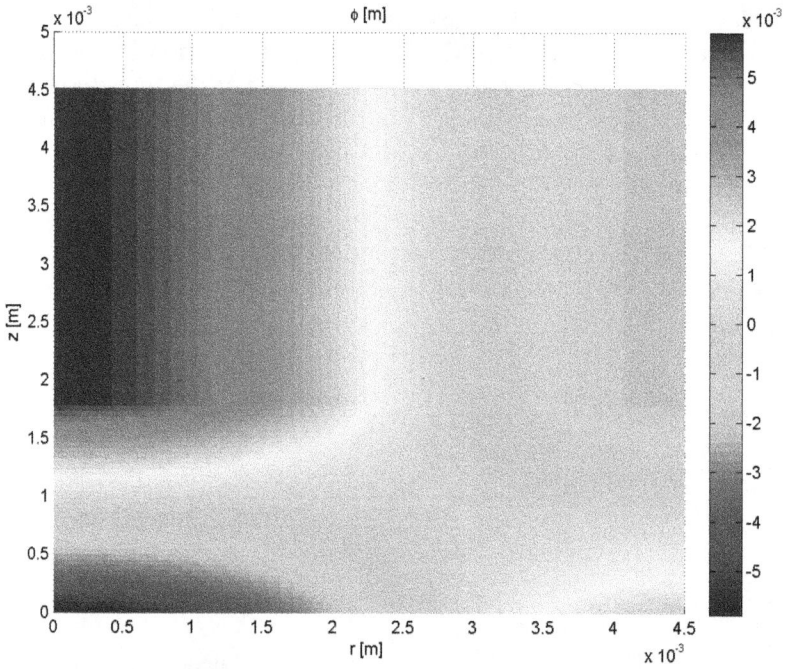

Figure 13.9. Plot of the electric potential φ at a specified but arbitrary time versus the spatial coordinates z and r. The frequency is 240 kHz, the transducer radius R is 0.0045 m and the total length $L = 0.00452$ m. The transducer is composed of a layer of PZT-5 (thickness 0.001 82 m) and an epoxy matching layer of thickness 0.0027 m.

Reference

[1] Schnabel P 1978 Dispersion of thickness vibrations of piezoceramic disk resonators *IEEE Trans. Sonics Ultrason.* **SU-25** 16–24

Chapter 14

Flexoelectricity and electrostriction

In addition to piezoelectricity, two other important electromechanical coupling effects will be discussed. Flexoelectricity and electrostriction are mechanisms that occasionally play an important role. This chapter will briefly introduce the two effects and the corresponding constitutive equations will be derived. Flexoelectricity describes the local coupling between strain gradients and the electric displacement or the electric field. There is also a converse effect that is a local coupling between electric field gradients and mechanical stress or strain. Strain gradients can be expected to be important in some applications of highly flexible materials such as two-dimensional materials. Electrostriction is a local coupling between squares of electric field components and mechanical stress or strain and is important in applications where moderate to high electric fields are present. The two effects, flexoelectricity and electrostriction, can exist in all materials, both centrosymmetric and non-centrosymmetric structures, as they are described by four-rank tensor coefficients.

14.1 Flexoelectricity and symmetry properties of hexagonal crystals

In order to derive the constitutive equations for flexoelectricity, let us start out by writing down the expression for the total energy differential,

$$dU = \Theta d\sigma + T_{ij} dS_{ij} + E_i dD_i, \tag{14.1}$$

where U is the total energy and Θ, σ, T_{ik}, S_{ik}, E_i, D_i are the temperature, entropy, stress tensor, strain tensor, electric field, and the electric displacement, respectively. Since the electric enthalpy is defined by $H_E = U - E_l D_l$,

$$dH_E = \Theta d\sigma + T_{ij} dS_{ij} - D_i dE_i. \tag{14.2}$$

From the latter equation it follows that,

$$T_{ij} = \frac{\partial H_E}{\partial S_{ij}}, \tag{14.3}$$

$$D_i = -\frac{\partial H_E}{\partial E_i}. \tag{14.4}$$

Equation (14.1) will be needed when deriving the general symmetry-allowed form of the converse flexoelectric coefficient ν_{ijlk} [1] given by,

$$T_{ij} = -\nu_{ijkl}\frac{\partial E_l}{\partial x_k}, \tag{14.5}$$

and equations (14.3) and (14.4) to determine the general symmetry-allowed form of the direct flexoelectric coefficient μ_{ijkl},

$$D_l = \mu_{ijkl}\frac{\partial S_{ij}}{\partial x_k}. \tag{14.6}$$

14.1.1 Relation between μ_{ijkl} and ν_{ijlk}

It follows from equation (14.2), assuming isentropic conditions $d\sigma = 0$,

$$
\begin{aligned}
d\left(\frac{\partial H_E}{\partial x_k}\right) &= \frac{\partial}{\partial x_k}dH_E = \frac{\partial}{\partial x_k}(T_{ij}dS_{ij} - D_l dE_l)\\
&= T_{ij}d\frac{\partial S_{ij}}{\partial x_k} + \frac{\partial T_{ij}}{\partial x_k}dS_{ij} - \frac{\partial D_l}{\partial x_k}dE_l - D_l d\frac{\partial E_l}{\partial x_k},
\end{aligned} \tag{14.7}
$$

so,

$$\mu_{ijkl} = \frac{\partial D_l}{\partial\left(\frac{\partial S_{ij}}{\partial x_k}\right)} = -\frac{\partial^2\left(\frac{\partial H_E}{\partial x_k}\right)}{\partial\left(\frac{\partial S_{ij}}{\partial x_k}\right)\partial\left(\frac{\partial E_l}{\partial x_k}\right)} = -\frac{\partial^2\left(\frac{\partial H_E}{\partial x_k}\right)}{\partial\left(\frac{\partial E_l}{\partial x_k}\right)\partial\left(\frac{\partial S_{ij}}{\partial x_k}\right)} = -\frac{\partial T_{ij}}{\partial\left(\frac{\partial E_l}{\partial x_k}\right)} = \nu_{ijkl}, \tag{14.8}$$

and it is proven that the direct and converse flexoelectric coefficients are the same.

14.1.2 Hexagonal coordinates

In the following, the c direction of the hexagonal crystal is, unless otherwise is mentioned, along the z direction. Let us use the standard trick for hexagonal crystals [2] and introduce the complex rotated variables ξ and η replacing the in-plane coordinates x and y,

$$\xi = x + iy, \tag{14.9}$$

$$\eta = x - iy. \tag{14.10}$$

The hexagonal lattice transforms into itself upon a rotation of 60 degrees around the z axis. Under this symmetry operation the new coordinates ξ and η transform according to,

$$\xi \rightarrow \xi e^{i\pi/3}, \tag{14.11}$$

$$\eta \rightarrow \eta e^{-i\pi/3}. \tag{14.12}$$

Hence, for the total energy (and enthalpy) to be unchanged upon a symmetry operation, only the same number of ξ's and η's are allowed in each term contributing to the total energy (and enthalpy).

14.2 Converse flexoelectricity

With the latter rule, it is now easy to write down the possible terms to the differential free energy allowed by symmetry contributing to converse flexoelectricity,

$$
\begin{aligned}
dU = & -\tilde{\nu}_1 \frac{\partial E_z}{\partial z} dS_{zz} - \tilde{\nu}_2 \frac{\partial E_\eta}{\partial \xi} dS_{zz} - \tilde{\nu}_3 \frac{\partial E_\xi}{\partial \eta} dS_{zz} \\
& - 2\tilde{\nu}_4 \frac{\partial E_z}{\partial \eta} dS_{z\xi} - 2\tilde{\nu}_5 \frac{\partial E_\eta}{\partial z} dS_{z\xi} - 2\tilde{\nu}_6 \frac{\partial E_z}{\partial \xi} dS_{z\eta} \\
& - 2\tilde{\nu}_7 \frac{\partial E_\xi}{\partial z} dS_{z\eta} - 2\tilde{\nu}_8 \frac{\partial E_z}{\partial z} dS_{\xi\eta} - \tilde{\nu}_9 \frac{\partial E_\eta}{\partial \eta} dS_{\xi\xi} \\
& - \tilde{\nu}_{10} \frac{\partial E_\xi}{\partial \xi} dS_{\eta\eta} - 2\tilde{\nu}_{11} \frac{\partial E_\eta}{\partial \xi} dS_{\xi\eta} - 2\tilde{\nu}_{12} \frac{\partial E_\xi}{\partial \eta} dS_{\xi\eta},
\end{aligned} \tag{14.13}
$$

where 12 unspecified coefficients $\{\tilde{\nu}_1, \ldots, \tilde{\nu}_{12}\}$ are introduced. To simplify the above expression, it is used that the mirror operation in the x–z plane is a symmetry, i.e., $y \rightarrow -y$ equivalent to $\xi \rightarrow \eta$ and $\eta \rightarrow \xi$. Then, $\tilde{\nu}_2 = \tilde{\nu}_3$ and $\tilde{\nu}_4 = \tilde{\nu}_6$, etc,

$$
\begin{aligned}
dU = & -\tilde{\nu}_1 \frac{\partial E_z}{\partial z} dS_{zz} - \tilde{\nu}_2 \left(\frac{\partial E_\eta}{\partial \xi} + \frac{\partial E_\xi}{\partial \eta} \right) dS_{zz} \\
& - 2\tilde{\nu}_4 \left(\frac{\partial E_z}{\partial \eta} dS_{z\xi} + \frac{\partial E_z}{\partial \xi} dS_{z\eta} \right) - 2\tilde{\nu}_5 \left(\frac{\partial E_\eta}{\partial z} dS_{z\xi} + \frac{\partial E_\xi}{\partial z} dS_{z\eta} \right) \\
& - 2\tilde{\nu}_8 \frac{\partial E_z}{\partial z} dS_{\xi\eta} - \tilde{\nu}_9 \left(\frac{\partial E_\eta}{\partial \eta} dS_{\xi\xi} + \frac{\partial E_\xi}{\partial \xi} dS_{\eta\eta} \right) - 2\tilde{\nu}_{11} \left(\frac{\partial E_\eta}{\partial \xi} dS_{\xi\eta} + \frac{\partial E_\xi}{\partial \eta} dS_{\xi\eta} \right).
\end{aligned} \tag{14.14}
$$

At this point, notice that $\tilde{\nu}_1, \tilde{\nu}_2, \tilde{\nu}_4, \tilde{\nu}_5, \tilde{\nu}_8, \tilde{\nu}_9, \tilde{\nu}_{11}$ must be real numbers to ensure dU is a real number for any symmetry-allowed strain configuration. The standard coordinates x and y can be reintroduced in equation (14.14) replacing the complex variables ξ and η by noting that,

$$\partial_\xi = \partial_x + i\partial_y, \tag{14.15}$$

$$\partial_\eta = \partial_x - i\partial_y, \tag{14.16}$$

$$S_{\xi\xi} = S_{xx} - S_{yy} + 2iS_{xy}, \tag{14.17}$$

$$S_{\eta\eta} = S_{xx} - S_{yy} - 2iS_{xy}, \tag{14.18}$$

$$S_{\xi\eta} = S_{xx} + S_{yy}, \tag{14.19}$$

$$S_{z\xi} = S_{zx} + iS_{zy}, \tag{14.20}$$

$$S_{z\eta} = S_{zx} - iS_{zy}, \tag{14.21}$$

From equations (14.14), (14.15)–(14.21) it is found that,

$$
\begin{aligned}
dU = &-\tilde{\nu}_1\frac{\partial E_z}{\partial z}dS_{zz} - \tilde{\nu}_2\left(\frac{\partial E_x}{\partial x} - i\frac{\partial E_y}{\partial x} + i\frac{\partial E_x}{\partial y} + \frac{\partial E_y}{\partial y} + c.\,c.\right)dS_{zz} \\
&- 2\tilde{\nu}_4\left(\left(\frac{\partial E_z}{\partial x} - i\frac{\partial E_z}{\partial y}\right)(dS_{zx} + idS_{zy}) + c.\,c.\right) - 2\tilde{\nu}_5\left(\left(\frac{\partial E_x}{\partial z} - i\frac{\partial E_y}{\partial z}\right)(dS_{zx} + idS_{zy}) + c.\,c.\right) \\
&- 2\tilde{\nu}_8\frac{\partial E_z}{\partial z}d(S_{xx} + S_{yy}) - \tilde{\nu}_9\left(\left(\frac{\partial E_x}{\partial x} - i\frac{\partial E_y}{\partial x} - i\frac{\partial E_x}{\partial y} - \frac{\partial E_y}{\partial y}\right)(dS_{xx} - dS_{yy} + 2idS_{xy}) + c.\,c.\right) \\
&- 2\tilde{\nu}_{11}\left(\left(\frac{\partial E_x}{\partial x} - i\frac{\partial E_y}{\partial x} + i\frac{\partial E_x}{\partial y} + \frac{\partial E_y}{\partial y}\right)(dS_{xx} + dS_{yy}) + c.\,c.\right),
\end{aligned}
\tag{14.22}
$$

or,

$$
\begin{aligned}
dU = &-\left(\tilde{\nu}_1\frac{\partial E_z}{\partial z} + 2\tilde{\nu}_2\left(\frac{\partial E_x}{\partial x} + \frac{\partial E_y}{\partial y}\right)\right)dS_{zz} \\
&- \left(4\tilde{\nu}_4\frac{\partial E_z}{\partial x} + 4\tilde{\nu}_5\frac{\partial E_x}{\partial z}\right)dS_{zx} - \left(4\tilde{\nu}_4\frac{\partial E_z}{\partial y} + 4\tilde{\nu}_5\frac{\partial E_y}{\partial z}\right)dS_{zy} \\
&- \left(2\tilde{\nu}_8\frac{\partial E_z}{\partial z} + 2\tilde{\nu}_9\left(\frac{\partial E_x}{\partial x} - \frac{\partial E_y}{\partial y}\right) + 4\tilde{\nu}_{11}\left(\frac{\partial E_x}{\partial x} + \frac{\partial E_y}{\partial y}\right)\right)dS_{xx} \\
&- \left(2\tilde{\nu}_8\frac{\partial E_z}{\partial z} - 2\tilde{\nu}_9\left(\frac{\partial E_x}{\partial x} - \frac{\partial E_y}{\partial y}\right) + 4\tilde{\nu}_{11}\left(\frac{\partial E_x}{\partial x} + \frac{\partial E_y}{\partial y}\right)\right)dS_{yy} \\
&- 4\tilde{\nu}_9\left(\frac{\partial E_y}{\partial x} + \frac{\partial E_x}{\partial y}\right)dS_{xy} \\
= &-\left(\tilde{\nu}_1\frac{\partial E_z}{\partial z} + 2\tilde{\nu}_2\left(\frac{\partial E_x}{\partial x} + \frac{\partial E_y}{\partial y}\right)\right)dS_3 \\
&- \left(2\tilde{\nu}_4\frac{\partial E_z}{\partial x} + 2\tilde{\nu}_5\frac{\partial E_x}{\partial z}\right)dS_5 - \left(2\tilde{\nu}_4\frac{\partial E_z}{\partial y} + 2\tilde{\nu}_5\frac{\partial E_y}{\partial z}\right)dS_4 \\
&- \left(2\tilde{\nu}_8\frac{\partial E_z}{\partial z} + 2\tilde{\nu}_9\left(\frac{\partial E_x}{\partial x} - \frac{\partial E_y}{\partial y}\right) + 4\tilde{\nu}_{11}\left(\frac{\partial E_x}{\partial x} + \frac{\partial E_y}{\partial y}\right)\right)dS_1 \\
&- \left(2\tilde{\nu}_8\frac{\partial E_z}{\partial z} - 2\tilde{\nu}_9\left(\frac{\partial E_x}{\partial x} - \frac{\partial E_y}{\partial y}\right) + 4\tilde{\nu}_{11}\left(\frac{\partial E_x}{\partial x} + \frac{\partial E_y}{\partial y}\right)\right)dS_2 \\
&- 2\tilde{\nu}_9\left(\frac{\partial E_y}{\partial x} + \frac{\partial E_x}{\partial y}\right)dS_6.
\end{aligned}
\tag{14.23}
$$

14.2.1 Flexoelectricity and $T_I - \frac{\partial E_i}{\partial x_j}$ relations

With the above expression for the differential of the internal energy,

$$T_1 = \frac{\partial U}{\partial S_1} = -2\tilde{v}_8 \frac{\partial E_z}{\partial z} - 2(\tilde{v}_9 + 2\tilde{v}_{11}) \frac{\partial E_x}{\partial x} - 2(2\tilde{v}_{11} - \tilde{v}_9) \frac{\partial E_y}{\partial y},$$

$$T_2 = \frac{\partial U}{\partial S_2} = -2\tilde{v}_8 \frac{\partial E_z}{\partial z} - 2(2\tilde{v}_{11} - \tilde{v}_9) \frac{\partial E_x}{\partial x} - 2(\tilde{v}_9 + 2\tilde{v}_{11}) \frac{\partial E_y}{\partial y},$$

$$T_3 = \frac{\partial U}{\partial S_3} = -\tilde{v}_1 \frac{\partial E_z}{\partial z} - 2\tilde{v}_2 \frac{\partial E_x}{\partial x} - 2\tilde{v}_2 \frac{\partial E_y}{\partial y},$$

$$T_4 = \frac{\partial U}{\partial S_4} = -2\tilde{v}_4 \frac{\partial E_z}{\partial y} - 2\tilde{v}_5 \frac{\partial E_y}{\partial z},$$ (14.24)

$$T_5 = \frac{\partial U}{\partial S_5} = -2\tilde{v}_4 \frac{\partial E_z}{\partial x} - 2\tilde{v}_5 \frac{\partial E_x}{\partial z},$$

$$T_6 = \frac{\partial U}{\partial S_6} = -2\tilde{v}_9 \frac{\partial E_y}{\partial x} - 2\tilde{v}_9 \frac{\partial E_x}{\partial y}.$$

In matrix form the above equations can be written as,

$$
v_{6\times9} = \begin{bmatrix}
2(\tilde{v}_9 + 2\tilde{v}_{11}) & 0 & 0 & 0 & 2(2\tilde{v}_{11} - \tilde{v}_9) & 0 & 0 & 0 & 2\tilde{v}_8 \\
2(2\tilde{v}_{11} - \tilde{v}_9) & 0 & 0 & 0 & 2(\tilde{v}_9 + 2\tilde{v}_{11}) & 0 & 0 & 0 & 2\tilde{v}_8 \\
2\tilde{v}_2 & 0 & 0 & 0 & 2\tilde{v}_2 & 0 & 0 & 0 & \tilde{v}_1 \\
0 & 0 & 0 & 0 & 0 & 2\tilde{v}_5 & 0 & 2\tilde{v}_4 & 0 \\
0 & 0 & 2\tilde{v}_5 & 0 & 0 & 0 & 2\tilde{v}_4 & 0 & 0 \\
0 & 2\tilde{v}_9 & 0 & 2\tilde{v}_9 & 0 & 0 & 0 & 0 & 0
\end{bmatrix}, \quad (14.25)
$$

where the matrix dimension 6×9 describes the flexoelectric coupling between the six stress components using Voigt notation $(T_1, T_2, T_3, T_4, T_5, T_6)$ and the nine electric field gradient components $(\frac{\partial E_x}{\partial x}, \frac{\partial E_x}{\partial y}, \frac{\partial E_x}{\partial z}, \frac{\partial E_y}{\partial x}, \frac{\partial E_y}{\partial y}, \frac{\partial E_y}{\partial z}, \frac{\partial E_z}{\partial x}, \frac{\partial E_z}{\partial y}, \frac{\partial E_z}{\partial z})$. Evidently, there are only seven independent constants, a result that follows immediately from equation (14.14). A further simplification is made by renaming the above entries to,

$$
v_{6\times9} = \begin{bmatrix}
v_1 & 0 & 0 & 0 & v_2 & 0 & 0 & 0 & v_3 \\
v_2 & 0 & 0 & 0 & v_1 & 0 & 0 & 0 & v_3 \\
v_4 & 0 & 0 & 0 & v_4 & 0 & 0 & 0 & v_5 \\
0 & 0 & 0 & 0 & 0 & v_6 & 0 & v_7 & 0 \\
0 & 0 & v_6 & 0 & 0 & 0 & v_7 & 0 & 0 \\
0 & \frac{1}{2}(v_1 - v_2) & 0 & \frac{1}{2}(v_1 - v_2) & 0 & 0 & 0 & 0 & 0
\end{bmatrix}. \quad (14.26)
$$

14.2.2 Flexoelectricity and $S_l - D_l$ relations

In order to establish a relation between flexoelectric coefficients that couple strain S_{ij} and electric displacement D_l it is convenient to use the enthalpy,

$$H = U - T_{ij}S_{ij} - D_l E_l, \tag{14.27}$$

which yields the differential enthalpy,

$$dH = \Theta d\sigma - S_{ij}dT_{ij} - D_l dE_l. \tag{14.28}$$

It follows from equation (14.28), assuming isentropic conditions $d\sigma = 0$,

$$
\begin{aligned}
d\left(\frac{\partial H}{\partial x_k}\right) &= \frac{\partial}{\partial x_k}dH = -\frac{\partial}{\partial x_k}(S_{ij}dT_{ij} + D_l dE_l) \\
&= -S_{ij}d\frac{\partial T_{ij}}{\partial x_k} - \frac{\partial S_{ij}}{\partial x_k}dT_{ij} - \frac{\partial D_l}{\partial x_k}dE_l - D_l d\frac{\partial E_l}{\partial x_k},
\end{aligned} \tag{14.29}
$$

so,

$$
f_{ijkl} = \frac{\partial D_l}{\partial\left(\dfrac{\partial T_{ij}}{\partial x_k}\right)} = -\frac{\partial^2\left(\dfrac{\partial H}{\partial x_k}\right)}{\partial\left(\dfrac{\partial T_{ij}}{\partial x_k}\right)\partial\left(\dfrac{\partial E_l}{\partial x_k}\right)} = -\frac{\partial^2\left(\dfrac{\partial H}{\partial x_k}\right)}{\partial\left(\dfrac{\partial E_l}{\partial x_k}\right)\partial\left(\dfrac{\partial T_{ij}}{\partial x_k}\right)} = \frac{\partial S_{ij}}{\partial\left(\dfrac{\partial E_l}{\partial x_k}\right)}, \tag{14.30}
$$

and the following relations are obtained,

$$
\begin{aligned}
S_{ij} &= f_{ijkl}\frac{\partial E_l}{\partial x_k}, \\
D_l &= f_{ijkl}\frac{\partial T_{ij}}{\partial x_k}.
\end{aligned} \tag{14.31}
$$

Note that the direct and converse flexoelectric coefficients for the $S_{ij} - D_l$ relations are similar in structure to the direct and converse flexoelectric coefficients for the T_{ij} and D_l relations except for the change in sign relating S_{ij} to $\frac{\partial E_l}{\partial x_k}$ versus T_{ij} to $\frac{\partial E_l}{\partial x_k}$.

Similar to the expression relating the free energy to the strain and electric field gradient, an expression for the enthalpy as a function of the stress and the electric field gradient can be written down,

$$
\begin{aligned}
dH = {}&-\tilde{f}_1\frac{\partial E_z}{\partial z}dT_{zz} - \tilde{f}_2\frac{\partial E_\eta}{\partial \xi}dT_{zz} - \tilde{f}_3\frac{\partial E_\xi}{\partial \eta}dT_{zz} \\
&-2\tilde{f}_4\frac{\partial E_z}{\partial \eta}dT_{z\xi} - 2\tilde{f}_5\frac{\partial E_\eta}{\partial z}dT_{z\xi} - 2\tilde{f}_6\frac{\partial E_z}{\partial \xi}dT_{z\eta} \\
&-2\tilde{f}_7\frac{\partial E_\xi}{\partial z}dT_{z\eta} - 2\tilde{f}_8\frac{\partial E_z}{\partial z}dT_{\xi\eta} - \tilde{f}_9\frac{\partial E_\eta}{\partial \eta}dT_{\xi\xi} \\
&-\tilde{f}_{10}\frac{\partial E_\xi}{\partial \xi}dT_{\eta\eta} - 2\tilde{f}_{11}\frac{\partial E_\eta}{\partial \xi}dT_{\xi\eta} - 2\tilde{f}_{12}\frac{\partial E_\xi}{\partial \eta}dT_{\xi\eta}.
\end{aligned} \tag{14.32}
$$

Repeating steps that led to equation (14.23),

$$dH = -\left(\tilde{f}_1\frac{\partial E_z}{\partial z} + 2\tilde{f}_2\left(\frac{\partial E_x}{\partial x} + \frac{\partial E_y}{\partial y}\right)\right)dT_3$$

$$-\left(4\tilde{f}_4\frac{\partial E_z}{\partial x} + 4\tilde{f}_5\frac{\partial E_x}{\partial z}\right)dT_5 - \left(4\tilde{f}_4\frac{\partial E_z}{\partial y} + 4\tilde{f}_5\frac{\partial E_y}{\partial z}\right)dT_4$$

$$-\left(2\tilde{f}_8\frac{\partial E_z}{\partial z} + 2\tilde{f}_9\left(\frac{\partial E_x}{\partial x} - \frac{\partial E_y}{\partial y}\right) + 4\tilde{f}_{11}\left(\frac{\partial E_x}{\partial x} + \frac{\partial E_y}{\partial y}\right)\right)dT_1 \qquad (14.33)$$

$$-\left(2\tilde{f}_8\frac{\partial E_z}{\partial z} - 2\tilde{f}_9\left(\frac{\partial E_x}{\partial x} - \frac{\partial E_y}{\partial y}\right) + 4\tilde{f}_{11}\left(\frac{\partial E_x}{\partial x} + \frac{\partial E_y}{\partial y}\right)\right)dT_2$$

$$-4\tilde{f}_9\left(\frac{\partial E_y}{\partial x} + \frac{\partial E_x}{\partial y}\right)dT_6.$$

Hence,

$$S_1 = -\frac{\partial H}{\partial T_1} = 2\tilde{f}_8\frac{\partial E_z}{\partial z} + 2(\tilde{f}_9 + 2\tilde{f}_{11})\frac{\partial E_x}{\partial x} + 2(2\tilde{f}_{11} - \tilde{f}_9)\frac{\partial E_y}{\partial y},$$

$$S_2 = -\frac{\partial H}{\partial T_2} = 2\tilde{f}_8\frac{\partial E_z}{\partial z} + 2(2\tilde{f}_{11} - \tilde{f}_9)\frac{\partial E_x}{\partial x} + 2(\tilde{f}_9 + 2\tilde{f}_{11})\frac{\partial E_y}{\partial y},$$

$$S_3 = -\frac{\partial H}{\partial T_3} = \tilde{f}_1\frac{\partial E_z}{\partial z} + 2\tilde{f}_2\frac{\partial E_x}{\partial x} + 2\tilde{f}_2\frac{\partial E_y}{\partial y},$$

$$S_4 = -\frac{\partial H}{\partial T_4} = 4\tilde{f}_4\frac{\partial E_z}{\partial y} + 4\tilde{f}_5\frac{\partial E_y}{\partial z}, \qquad (14.34)$$

$$S_5 = -\frac{\partial H}{\partial T_5} = 4\tilde{f}_4\frac{\partial E_z}{\partial x} + 4\tilde{f}_5\frac{\partial E_x}{\partial z},$$

$$S_6 = -\frac{\partial H}{\partial T_6} = 4\tilde{f}_9\frac{\partial E_y}{\partial x} + 4\tilde{f}_9\frac{\partial E_x}{\partial y}.$$

14.2.3 Flexoelectricity and $S_I - \frac{\partial E_i}{\partial x_j}$ relations

In matrix form, equations (14.34) are,

$$f_{6\times 9} = \begin{bmatrix} 2(\tilde{f}_9 + 2\tilde{f}_{11}) & 0 & 0 & 0 & 2(2\tilde{f}_{11} - \tilde{f}_9) & 0 & 0 & 0 & 2\tilde{f}_8 \\ 2(2\tilde{f}_{11} - \tilde{f}_9) & 0 & 0 & 0 & 2(\tilde{f}_9 + 2\tilde{f}_{11}) & 0 & 0 & 0 & 2\tilde{f}_8 \\ 2\tilde{f}_2 & 0 & 0 & 0 & 2\tilde{f}_2 & 0 & 0 & 0 & \tilde{f}_1 \\ 0 & 0 & 0 & 0 & 0 & 4\tilde{f}_5 & 0 & 4\tilde{f}_4 & 0 \\ 0 & 0 & 4\tilde{f}_5 & 0 & 0 & 0 & 4\tilde{f}_4 & 0 & 0 \\ 0 & 4\tilde{f}_9 & 0 & 4\tilde{f}_9 & 0 & 0 & 0 & 0 & 0 \end{bmatrix}, \qquad (14.35)$$

where the matrix dimension 6×9 describes coupling between the six strain components in Voigt notation ($S_1, S_2, S_3, S_4, S_5, S_6$) and the nine electric field gradient components ($\frac{\partial E_x}{\partial x}, \frac{\partial E_x}{\partial y}, \frac{\partial E_x}{\partial z}, \frac{\partial E_y}{\partial x}, \frac{\partial E_y}{\partial y}, \frac{\partial E_y}{\partial z}, \frac{\partial E_z}{\partial x}, \frac{\partial E_z}{\partial y}, \frac{\partial E_z}{\partial z}$). A further simplification is made by renaming the above entries to,

$$f_{6\times9} = \begin{bmatrix} f_1 & 0 & 0 & 0 & f_2 & 0 & 0 & 0 & f_3 \\ f_2 & 0 & 0 & 0 & f_1 & 0 & 0 & 0 & f_3 \\ f_4 & 0 & 0 & 0 & f_4 & 0 & 0 & 0 & f_5 \\ 0 & 0 & 0 & 0 & 0 & f_6 & 0 & f_7 & 0 \\ 0 & 0 & f_6 & 0 & 0 & 0 & f_7 & 0 & 0 \\ 0 & f_1 & -f_2 & 0 & f_1 & -f_2 & 0 & 0 & 0 & 0 \end{bmatrix}. \tag{14.36}$$

Note that the latter result agrees with the result on p. 7 of reference [3] (but there is a typo for the entry μ_{21} in their matrix which should be μ_{15}).

14.2.3.1 Compiling flexoelectric equations

Let us write down the full set of flexoelectric equations,

$$T_{ij} = -\nu_{ijkl}\frac{\partial E_l}{\partial x_k}, \qquad D_l = \nu_{ijkl}\frac{\partial S_{ij}}{\partial x_k},$$

$$S_{ij} = f_{ijkl}\frac{\partial E_l}{\partial x_k}, \qquad D_l = f_{ijkl}\frac{\partial T_{ij}}{\partial x_k}. \tag{14.37}$$

14.3 Electrostriction and symmetry properties of hexagonal crystals

Another important electromechanical phenomenon is electrostriction. Electrostriction refers to coupling between stress or strain components to products of electric field components. This effect, like flexoelectricity, exists in all materials as it does not require the unit cell to be centrosymmetric.

14.3.1 Electrostriction and $T_I - E_i E_j$ relations

The contribution from electrostrictive effects to the differential energy of a hexagonal crystal can be written as,

$$\begin{aligned} dU = & -\tilde{g}_1 S_{zz} E_z^2 - 2\tilde{g}_2 S_{\xi\eta} E_z^2 - 4\tilde{g}_3 S_{\xi z} E_\eta E_z \\ & - 4\tilde{g}_4 S_{\eta z} E_\xi E_z - 2\tilde{g}_5 S_{zz} E_\xi E_\eta - \tilde{g}_6 S_{\xi\xi} E_\eta^2 \\ & - \tilde{g}_7 S_{\eta\eta} E_\xi^2 - 4\tilde{g}_8 S_{\xi\eta} E_\xi E_\eta \\ = & -\tilde{g}_1 S_{zz} E_z^2 - 2\tilde{g}_2 (S_{xx}+S_{yy})E_z^2 - 4\tilde{g}_3((S_{xz}+iS_{yz})(E_x E_z - iE_y E_z) + c.c.) \\ & - 2\tilde{g}_5 S_{zz}(E_x+iE_y)(E_x-iE_y) - \tilde{g}_6((S_{xx}-S_{yy}+2iS_{xy})(E_x - iE_y)^2 + c.c.) \\ & - 4\tilde{g}_8(S_{xx}+S_{yy})(E_x+iE_y)(E_x-iE_y) \\ & - \left(2\tilde{g}_6\left(E_x^2 - E_y^2\right) + 4\tilde{g}_8\left(E_x^2 + E_y^2\right) + 2\tilde{g}_2 E_z^2\right)S_{xx} \\ & - \left(-2\tilde{g}_6\left(E_x^2 - E_y^2\right) + 4\tilde{g}_8\left(E_x^2 + E_y^2\right) + 2\tilde{g}_2 E_z^2\right)S_{yy} \\ & - \left(\tilde{g}_1 E_z^2 + 4\tilde{g}_5\left(E_x^2 + E_y^2\right)\right)S_{zz} \\ = & -8\tilde{g}_3 E_y E_z S_{yz} - 8\tilde{g}_3 E_x E_z S_{xz} - 8\tilde{g}_6 E_x E_y S_{xy}. \end{aligned} \tag{14.38}$$

where, in obtaining the second quality, it was used that dU must be a real number.

From the latter general expression, the following stress contributions due to electrostriction are obtained,

$$T_1 = \frac{\partial U}{\partial S_1} = -(2\tilde{g}_6 + 4\tilde{g}_8)E_x^2 - (-2\tilde{g}_6 + 4\tilde{g}_8)E_y^2 - 2\tilde{g}_2 E_z^2,$$

$$T_2 = \frac{\partial U}{\partial S_2} = -(-2\tilde{g}_6 + 4\tilde{g}_8)E_x^2 - (2\tilde{g}_6 + 4\tilde{g}_8)E_y^2 - 2\tilde{g}_2 E_z^2,$$

$$T_3 = \frac{\partial U}{\partial S_3} = -\tilde{g}_1 E_z^2 - 4\tilde{g}_5 E_x^2 - 4\tilde{g}_5 E_y^2,$$

$$T_4 = \frac{\partial U}{\partial S_4} = -4\tilde{g}_3 E_y E_z, \tag{14.39}$$

$$T_5 = \frac{\partial U}{\partial S_5} = -4\tilde{g}_3 E_x E_z,$$

$$T_6 = \frac{\partial U}{\partial S_6} = -4\tilde{g}_6 E_x E_y.$$

In matrix form the above equations are,

$$g_{6\times6} = -\begin{bmatrix} 2\tilde{g}_6 + 4\tilde{g}_8 & -2\tilde{g}_6 + 4\tilde{g}_8 & 2\tilde{g}_2 & 0 & 0 & 0 \\ -2\tilde{g}_6 + 4\tilde{g}_8 & 2\tilde{g}_6 + 4\tilde{g}_8 & 2\tilde{g}_2 & 0 & 0 & 0 \\ 4\tilde{g}_5 & 4\tilde{g}_5 & \tilde{g}_1 & 0 & 0 & 0 \\ 0 & 0 & 0 & 4\tilde{g}_3 & 0 & 0 \\ 0 & 0 & 0 & 0 & 4\tilde{g}_3 & 0 \\ 0 & 0 & 0 & 0 & 0 & 4\tilde{g}_6 \end{bmatrix}, \tag{14.40}$$

where the matrix dimension 6×6 describes coupling between the six stress tensor components in Voigt notation (T_1, T_2, T_3, T_4, T_5, T_6) and the six products of electric field components (E_x^2, E_y^2, E_z^2, $E_y E_z$, $E_x E_z$, $E_x E_y$). A further simplification is made by renaming the above entries to,

$$g_{6\times6} = -\begin{bmatrix} \tilde{g}_{11} & \tilde{g}_{12} & \tilde{g}_{13} & 0 & 0 & 0 \\ \tilde{g}_{12} & \tilde{g}_{11} & \tilde{g}_{13} & 0 & 0 & 0 \\ \tilde{g}_{31} & \tilde{g}_{31} & \tilde{g}_{33} & 0 & 0 & 0 \\ 0 & 0 & 0 & \tilde{g}_{44} & 0 & 0 \\ 0 & 0 & 0 & 0 & \tilde{g}_{44} & 0 \\ 0 & 0 & 0 & 0 & 0 & \tilde{g}_{11} - \tilde{g}_{12} \end{bmatrix}. \tag{14.41}$$

14.3.2 Electrostriction and $S_I - E_i E_j$ relations

The contribution from electrostrictive effects to the differential enthalpy of a hexagonal crystal can be written as,

$$dH = -\tilde{h}_1 T_{zz} E_z^2 - 2\tilde{h}_2 T_{\xi\eta} E_z^2 - 4\tilde{h}_3 T_{\xi z} E_\eta E_z$$
$$\quad - 4\tilde{h}_4 T_{\eta z} E_\xi E_z - 2\tilde{h}_5 T_{zz} E_\xi E_\eta - \tilde{h}_6 T_{\xi\xi} E_\eta^2$$
$$\quad - \tilde{h}_7 T_{\eta\eta} E_\xi^2 - 4\tilde{h}_8 T_{\xi\eta} E_\xi E_\eta$$
$$\quad = -\tilde{h}_1 T_{zz} E_z^2 - 2\tilde{h}_2 (T_{xx} + T_{yy}) E_z^2 - 4\tilde{h}_3 ((T_{xz} + iT_{yz})(E_x E_z - iE_y E_z) + c.\ c.)$$
$$\quad - 2\tilde{h}_5 T_{zz} (E_x + iE_y)(E_x - iE_y) - \tilde{h}_6 ((T_{xx} - T_{yy} + 2iT_{xy})(E_x - iE_y)^2 + c.\ c.)$$
$$\quad - 4\tilde{h}_8 (T_{xx} + T_{yy})(E_x + iE_y)(E_x - iE_y) \tag{14.42}$$
$$\quad = -\left(2\tilde{h}_6 (E_x^2 - E_y^2) + 4\tilde{h}_8 (E_x^2 + E_y^2) + 2\tilde{h}_2 E_z^2\right) T_{xx}$$
$$\quad - \left(-2\tilde{h}_6 (E_x^2 - E_y^2) + 4\tilde{h}_8 (E_x^2 + E_y^2) + 2\tilde{h}_2 E_z^2\right) T_{yy}$$
$$\quad - \left(\tilde{h}_1 E_z^2 + 2\tilde{h}_5 (E_x^2 + E_y^2)\right) T_{zz}$$
$$\quad - 8\tilde{h}_3 E_y E_z T_{yz} - 8\tilde{h}_3 E_x E_z T_{xz} - 8\tilde{h}_6 E_x E_y T_{xy},$$

where it was used that dH must be a real number.

From the latter general expression we obtain the following stress contributions due to electrostriction,

$$S_1 = -\frac{\partial H}{\partial T_1} = (2\tilde{h}_6 + 4\tilde{h}_8) E_x^2 + (-2\tilde{h}_6 + 4\tilde{h}_8) E_y^2 + 2\tilde{h}_2 E_z^2,$$

$$S_2 = -\frac{\partial H}{\partial T_2} = (-2\tilde{h}_6 + 4\tilde{h}_8) E_x^2 + (2\tilde{h}_6 + 4\tilde{h}_8) E_y^2 + 2\tilde{h}_2 E_z^2,$$

$$S_3 = -\frac{\partial H}{\partial T_3} = \tilde{h}_1 E_z^2 + 2\tilde{h}_5 E_x^2 + 2\tilde{h}_5 E_y^2,$$

$$S_4 = -\frac{\partial H}{\partial T_4} = 8\tilde{h}_3 E_y E_z, \tag{14.43}$$

$$S_5 = -\frac{\partial H}{\partial T_5} = 8\tilde{h}_3 E_x E_z,$$

$$S_6 = -\frac{\partial H}{\partial T_6} = 8\tilde{h}_6 E_x E_y.$$

In matrix form the above equations are,

$$h_{6\times6} = \begin{bmatrix} 2\tilde{h}_6 + 4\tilde{h}_8 & -2\tilde{h}_6 + 4\tilde{h}_8 & 2\tilde{h}_2 & 0 & 0 & 0 \\ -2\tilde{h}_6 + 4\tilde{h}_8 & 2\tilde{h}_6 + 4\tilde{h}_8 & 2\tilde{h}_2 & 0 & 0 & 0 \\ 2\tilde{h}_5 & 2\tilde{h}_5 & \tilde{h}_1 & 0 & 0 & 0 \\ 0 & 0 & 0 & 8\tilde{h}_3 & 0 & 0 \\ 0 & 0 & 0 & 0 & 8\tilde{h}_3 & 0 \\ 0 & 0 & 0 & 0 & 0 & 8\tilde{h}_6 \end{bmatrix}, \tag{14.44}$$

where the matrix dimension 6×6 describes coupling between six strain components in Voigt notation (S_1, S_2, S_3, S_4, S_5, S_6) and six products of electric field components

$(E_x^2, E_y^2, E_z^2, E_y E_z, E_x E_z, E_x E_y)$. A further simplification is made by renaming the above entries to,

$$
h_{6 \times 6} =
\begin{bmatrix}
h_{11} & h_{12} & h_{13} & 0 & 0 & 0 \\
h_{12} & h_{11} & h_{13} & 0 & 0 & 0 \\
h_{31} & h_{31} & h_{33} & 0 & 0 & 0 \\
0 & 0 & 0 & h_{44} & 0 & 0 \\
0 & 0 & 0 & 0 & h_{44} & 0 \\
0 & 0 & 0 & 0 & 0 & 2(h_{11} - h_{12})
\end{bmatrix}.
\tag{14.45}
$$

Note that the latter result agrees with the result (for Q_{ij}) on p 7 of reference [3].

14.3.3 Electrostriction and D_l to $S_{ij} E_k$ relations

In order to establish a relation between electrostrictive coefficients describing coupling between products of strain and electric field components to electric displacement components observe that,

$$
T_{ij} = -g_{ijkl} E_k E_l,
\tag{14.46}
$$

implies,

$$
\frac{\partial^2 T_{ij}}{\partial E_k \partial E_l} = -g_{ijkl} - g_{ijlk} = -2 g_{ijkl},
\tag{14.47}
$$

and,

$$
g_{ijkl} = -\frac{1}{2} \frac{\partial^2 T_{ij}}{\partial E_k \partial E_l} = -\frac{1}{2} \frac{\partial^3 H_E}{\partial E_k \partial E_l \partial S_{ij}} = \frac{1}{2} \frac{\partial^2 D_l}{\partial E_k \partial S_{ij}},
\tag{14.48}
$$

thus,

$$
D_l = 2 g_{ijkl} S_{ij} E_k.
\tag{14.49}
$$

14.3.4 Electrostriction and D_l to $T_{ij} E_k$ relations

In order to establish a relation between electrostrictive coefficients describing coupling between products of stress and electric field components to electric displacement components observe that,

$$
S_{ij} = h_{ijkl} E_k E_l,
\tag{14.50}
$$

implies,

$$
\frac{\partial^2 S_{ij}}{\partial E_k \partial E_l} = h_{ijkl} + h_{ijlk} = 2 h_{ijkl},
\tag{14.51}
$$

and,

$$h_{ijkl} = \frac{1}{2}\frac{\partial^2 S_{ij}}{\partial E_k \partial E_l} = -\frac{1}{2}\frac{\partial^3 H}{\partial E_k \partial E_l \partial T_{ij}} = \frac{1}{2}\frac{\partial^2 D_l}{\partial E_k \partial T_{ij}}, \qquad (14.52)$$

thus,

$$D_l = 2h_{ijkl}T_{ij}E_k. \qquad (14.53)$$

Observe the sign change between equations (14.46) and (14.50).

14.3.4.1 Compiling electrostrictive equations

The above manipulations allow us to obtain the following set of constitutive equations for electrostriction,

$$\begin{aligned} T_{ij} &= -g_{ijkl}E_k E_l, & D_l &= 2g_{ijkl}S_{ij}E_k, \\ S_{ij} &= h_{ijkl}E_k E_l, & D_l &= 2h_{ijkl}T_{ij}E_k. \end{aligned} \qquad (14.54)$$

14.4 Flexoelectricity and piezoelectricity in graphene

In this section, flexoelectricity and piezoelectricity in a graphene membrane will be considered. Consider the membrane fixed at its rim along two edges, $y = -L$ and $y = L$ and free to move along the two other edges $x = -W$ and $x = W$. The z direction is the hexagonal c direction. The graphene membrane is assumed to be inversion asymmetric for example due to the presence of holes of triangular shape or the membrane is bent. In this case, there is an effect of bending on the graphene membrane symmetry properties. As speculated in reference [4], and confirmed by DFT calculations, bending of graphene or presence of triangular holes in graphene changes the symmetry from the non-piezoelectric $6/mmm$ structure to the piezo-electric structure $\bar{6}m2$. In both cases, flexoelectric effects can be expected since the appearance of holes in a membrane or a bent membrane will generate strain gradients. An example is now discussed where a pre-bent graphene membrane subject to an external mechanical force may lead to either flexoelectricity or piezoelectricity.

14.5 Set of dynamic equations for a two-dimensional membrane

Let us take a crude approach and assume a classical form of the current-continuity equations applies to two-dimensional materials, and write down the full set of governing equations (including the Maxwell–Poisson and the elastic equation). Then,

$$\frac{\partial n}{\partial t} = \frac{1}{e}\boldsymbol{\nabla} \cdot \mathbf{J}_n, \qquad (14.55)$$

$$\frac{\partial p}{\partial t} = -\frac{1}{e}\boldsymbol{\nabla} \cdot \mathbf{J}_p, \qquad (14.56)$$

$$\nabla \cdot \mathbf{D} = e(-n(\mathbf{r}) + p(\mathbf{r})), \tag{14.57}$$

$$\rho\frac{\partial^2 u_i}{\partial t^2} = \frac{\partial T_{ij}}{\partial j}, \tag{14.58}$$

where, accounting for both flexoelectricity and converse flexoelectricity,

$$\mathbf{J}_n = eD_n\nabla n + e\mu_n n\mathbf{E}, \tag{14.59}$$

$$\mathbf{J}_p = -eD_p\nabla p + e\mu_p p\mathbf{E}, \tag{14.60}$$

$$D_i = \epsilon_{ij}E_j + P_i + e_{ijk}S_{jk} + \nu_{jkli}\frac{\partial S_{jk}}{\partial x_l}, \tag{14.61}$$

$$T_{ij} = c_{ijkl}S_{kl} - e_{kij}E_k - \nu_{ijkl}\frac{\partial E_l}{\partial x_k}, \tag{14.62}$$

and it will be assumed that there are no charged impurities (or dopants). If spontaneous polarization is absent in the two-dimensional material, P_i is zero. If free carriers are neglected as a simplifying assumption, then the problem reduces to only solving the Maxwell–Poisson equation and the elastic equation. This simplification is made in the two case studies below. Note that $\mathbf{u}$ is the particle displacement connected to the strain tensor by,

$$S_{ij} = \frac{1}{2}\left(\frac{\partial u_i}{\partial x_j} + \frac{\partial u_j}{\partial x_i}\right). \tag{14.63}$$

14.6 Case study 1

Firstly, since graphene is a two-dimensional material of hexagonal symmetry it is a good approximation to use an isotropic form of the membrane equations [5]. Here, a small contribution from electric fields on strains is neglected (stemming from converse piezoelectricity and converse flexoelectricity) but we shall keep the effect of strain and strain gradients on electric fields, i.e., the direct piezoelectric and flexoelectric terms in the constitutive expression for the electric displacement,

$$-\rho\omega^2 u_x = (B + \mu)\frac{\partial^2 u_x}{\partial x^2} + \mu\frac{\partial^2 u_x}{\partial y^2} + B\frac{\partial^2 u_y}{\partial x\partial y} + \rho f_x, \tag{14.64}$$

$$-\rho\omega^2 u_y = (B + \mu)\frac{\partial^2 u_y}{\partial y^2} + \mu\frac{\partial^2 u_y}{\partial x^2} + B\frac{\partial^2 u_x}{\partial x\partial y} + \rho f_y, \tag{14.65}$$

$$-\rho\omega^2 u_z = -\kappa\left(\frac{\partial^2}{\partial x^2} + \frac{\partial^2}{\partial y^2}\right)^2 u_z + \rho f_z, \tag{14.66}$$

where ρ, ω, B and μ, κ are the graphene surface mass density, the angular frequency of the external force, the two-dimensional bulk moduli, and the bending rigidity, respectively. The two-dimensional bulk moduli are given in terms of their three-dimensional analogs and the graphene layer thickness [5].

Consider the force f is along the z direction, and of the form,

$$f_z(y) = F\delta(y)e^{i\omega t}, \tag{14.67}$$

where F is a constant. Due to the symmetry of the problem and fixed ends at $y = \pm L$, it follows from equations (14.64)–(14.66) that u_x, u_y are both zero. Note that $u_z = u_z(y)$ then satisfies,

$$-\rho\omega^2 u_z = -\kappa \frac{\partial^4 u_z}{\partial y^4} + \rho F\delta(y)e^{i\omega t}, \tag{14.68}$$

subject to the boundary conditions,

$$u_z = 0, \ \text{at } y = \pm L,$$
$$\frac{\partial^2 u_z}{\partial y^2} = 0, \ \text{at } y = \pm L, \tag{14.69}$$

corresponding to fixed ends and assuming no bending moments at $y = \pm L$.

The general solution to equation (14.68) is,

$$u_z(y) = A_1 \cosh(ky) + B_1 \cos(ky) + C_1 \sinh(ky) + D_1 \sin(ky), \ y > 0,$$
$$u_z(y) = A_2 \cosh(ky) + B_2 \cos(ky) + C_2 \sinh(ky) + D_2 \sin(ky), \ y < 0,$$

where,

$$k = \left(\frac{\rho\omega^2}{\kappa}\right)^{1/4}. \tag{14.70}$$

A particular solution is sought by requiring,

$$u_z(0-) = u_z(0+),$$
$$\frac{\partial u_z}{\partial y}(0-) = \frac{\partial u_z}{\partial y}(0+),$$
$$\frac{\partial^2 u_z}{\partial y^2}(0-) = \frac{\partial^2 u_z}{\partial y^2}(0+).$$

Integrating equation (14.68) from $y = -\epsilon$ to $y = +\epsilon$ and letting $\epsilon \to 0$ gives the last boundary condition,

$$\kappa\left[\left.\frac{\partial^3 u_z}{\partial y^3}\right|_{0+} - \left.\frac{\partial^3 u_z}{\partial y^3}\right|_{0-}\right] = \rho F e^{i\omega t}. \tag{14.71}$$

The boundary conditions lead to the following expressions between A_1, A_2,..., D_1, D_2,

$$A_1 + B_1 = A_2 + B_2,$$
$$C_1 + D_1 = C_2 + D_2,$$
$$A_1 - B_1 = A_2 - B_2, \tag{14.72}$$
$$C_1 - C_2 + D_2 - D_1 = \frac{\rho F e^{i\omega t}}{\kappa k^3},$$

and a particular solution becomes,

$$u_z(y) = \frac{\rho F e^{i\omega t}}{4\kappa k^3} \sinh(ky) - \frac{\rho F e^{i\omega t}}{4\kappa k^3} \sin(ky), \quad y > 0,$$

$$u_z(y) = -\frac{\rho F e^{i\omega t}}{4\kappa k^3} \sinh(ky) + \frac{\rho F e^{i\omega t}}{4\kappa k^3} \sin(ky), \quad y < 0.$$

By adding a general homogeneous solution,

$$u_z(y) = A \cosh(ky) + B \cos(ky) + C \sinh(ky) + D \sin(ky),$$

to the above particular solution and imposing the four boundary conditions in equation (14.69) determine the coefficients A, B, C, D. The solution is,

$$u_z(y) = -\frac{\rho F e^{i\omega t}}{4\kappa k^3} \frac{\sinh(kL)}{\cosh(kL)} \cosh(ky) + \frac{\rho F e^{i\omega t}}{4\kappa k^3} \frac{\sin(kL)}{\cos(kL)} \cos(ky)$$

$$+ \frac{\rho F e^{i\omega t}}{4\kappa k^3} \sinh(ky) - \frac{\rho F e^{i\omega t}}{4\kappa k^3} \sin(ky), \quad y > 0,$$

$$u_z(y) = -\frac{\rho F e^{i\omega t}}{4\kappa k^3} \frac{\sinh(kL)}{\cosh(kL)} \cosh(ky) + \frac{\rho F e^{i\omega t}}{4\kappa k^3} \frac{\sin(kL)}{\cos(kL)} \cos(ky)$$

$$- \frac{\rho F e^{i\omega t}}{4\kappa k^3} \sinh(ky) + \frac{\rho F e^{i\omega t}}{4\kappa k^3} \sin(ky), \quad y < 0. \tag{14.73}$$

14.6.1 Contribution from flexoelectricity

It is evident from the above result that only the S_{yz} strain component is nonzero and the strain gradient component $S_{yz,y}$ is nonzero too. Since the piezoelectric e tensor for $\bar{6}m2$ has the form,

$$e_{3\times6} = \begin{bmatrix} e_{x1} & -e_{x1} & 0 & 0 & 0 & 0 \\ 0 & 0 & 0 & 0 & 0 & -e_{x1} \\ 0 & 0 & 0 & 0 & 0 & 0 \end{bmatrix}, \tag{14.74}$$

there is no contribution from $e_{ijk} S_{jk}$ to the electric displacement. Indeed, as soon as the graphene layer is bent, the symmetry of the graphene layer changes and it becomes piezoelectric. It was shown in reference [6] that for corrugated graphene the polarization due to flexoelectricity is about an order of magnitude larger than the piezoelectric contribution. Hence, the latter is neglected in the following. Let us check to see if there is a contribution from flexoelectricity. From equation (14.37) and since $S_{yz,y} \neq 0$, there is a contribution $\nu_7 S_{yz,y}$ to D_z.

Thus, the displacement current I^{disp} through half of the $x - y$ membrane surface (assuming electrodes cover half of the membrane area from $y = 0$ to $y = L$), is,

$$
\begin{aligned}
I^{\text{disp}} &= \int_{-W}^{W} \int_{0}^{L} J_z^{\text{disp}} \, dx dy = \int_{-W}^{W} \int_{0}^{L} \frac{\partial D_z}{\partial t} \, dx dy = 2Wi\omega\nu_7 \int_{0}^{L} S_{yz,y}(y) dy \\
&= i\omega\nu_7 W \left[u_z(L) - u_z(0) \right] = -i\omega\nu_7 W u_z(0) = -i\omega\nu_7 W \frac{\rho F e^{i\omega t}}{4\kappa k^3} \left(\frac{\sin(kL)}{\cos(kL)} - \frac{\sinh(kL)}{\cosh(kL)} \right),
\end{aligned} \tag{14.75}
$$

since $S_{yz} = \frac{1}{2} \frac{\partial u_z}{\partial y}$ and equation (14.73) has been used in obtaining the last equality. Hence, we conclude there is a current contribution from flexoelectricity! Assuming the other half of the membrane is covered by an electrode as well (in such a way that there is a small gap between the two electrodes to avoid short-circuiting) and resistors R are connected to each of the membrane halves, the power P harvested by the system is

$$
P = 2RRe(I^{\text{disp}})^2 = 2R \left[\omega\nu_7 W \frac{\rho F \sin(\omega t)}{4\kappa k^3} \left(\frac{\sin(kL)}{\cos(kL)} - \frac{\sinh(kL)}{\cosh(kL)} \right) \right]^2, \tag{14.76}
$$

where $Re(I^{\text{disp}})$ is the real part of the displacement current given by equation (14.75). It appears that for some frequencies (resonances) the harvested power diverges (when either $\cos(kL) = 0$ or $\cosh(kL) = 0$). In practice, due to the presence of losses not included in this analysis, power divergences will not take place, however, the harvested power will take its maximum values near these resonances. It must be emphasized that since no free carriers are assumed there are also no drift or diffusion current contributions through the membrane.

14.7 Case study 2

Consider a graphene membrane of symmetry $\bar{6}m2$ subject to a constant (in time and space) stress T along the $y = \pm L$ edges but free to move along the x direction. In this case it follows from,

$$
\begin{aligned}
T_{xx} &= c_{11} S_{xx} + c_{12} S_{yy} = 0, \\
T_{yy} &= c_{12} S_{xx} + c_{11} S_{yy} = T,
\end{aligned}
$$

that,

$$
\begin{aligned}
S_{yy} &= \frac{T}{c_{11} - \dfrac{c_{12}^2}{c_{11}}}, \\
S_{xx} &= -\frac{c_{12}}{c_{11}} S_{yy} = \frac{T}{c_{12} - \dfrac{c_{11}^2}{c_{12}}}.
\end{aligned} \tag{14.77}
$$

All shear strains are zero and normal strains are constant in time and space so all strain gradients vanish and flexoelectricity is not generated. But we have piezoelectric contributions to the electric displacement,

$$D_x = \epsilon_{xx} E_x + P_x + e_{x1} S_{xx} - e_{x1} S_{yy} = \epsilon_{xx} E_x + T\frac{e_{x1}}{c_{12} - c_{11}}, \tag{14.78}$$

$$D_y = \epsilon_{yy} E_y + P_y - e_{x1} S_{xy} = \epsilon_{yy} E_y, \tag{14.79}$$

since $P_x = P_y = 0$ and equation (14.77) is used to get the last expression for D_x. Neglecting free carriers in the structure and electrically open-circuiting the membrane edges,

$$D_x = D_y = 0, \tag{14.80}$$

everywhere, i.e.,

$$E_x = -e_{x1}\frac{T}{\epsilon_{xx}(c_{12} - c_{11})}, \tag{14.81}$$
$$E_y = 0.$$

Thus a dc voltage V is generated across the x direction, due to piezoelectricity alone, equal to,

$$V = -2We_{x1}\frac{T}{\epsilon_{xx}(c_{12} - c_{11})}. \tag{14.82}$$

References

[1] Shu L, Li F, Huang W, Wei X, Yao X and Jiang X 2014 Relationship between direct and converse flexoelectric coefficients *J. Appl. Phys.* **116** 144105
[2] Landau L D and Lifshitz E M 1986 *Theory of elasticity Course of Theoretical Physics* **vol 7** 3rd edn (Oxford: Heinemann)
[3] Shu L, Wei X and Wang C 2011 Symmetry of flexoelectric coefficients in crystalline medium *J. Appl. Phys.* **110** 104106
[4] Kundalwal S I, Meguid S A and Weng G J 2017 Strain gradient induced polarization in graphene (arXiv:1703.00597)
[5] Droth M and Burkard G 2011 Acoustic phonons and spin relaxation in graphene nanoribbons *Phys. Rev.* **B84** 155404
[6] Tan D, Willatzen M and Wang Z L 2021 Electron transfer in the contact-electrification between corrugated 2d materials: a first-principles study *Nano Energy* **79** 105386

IOP Publishing

Piezoelectricity in Classical and Modern Systems

Morten Willatzen

Chapter 15

Atomistic approach to piezoelectric properties

Approximately 30 years ago, Vanderbilt and King-Smith, and Resta [1–3] suggested a quantum-mechanical method for computing the bulk properties of electric polarization of solids. This method is known as the modern theory of polarization. In chapter 15 the essential details of the modern theory of polarization and consequences related to piezoelectric properties. The second part of the chapter is devoted to earlier important atomistic theories of strain and piezoelectricity in solids.

15.1 Modern theory of polarization

Consider a *finite* crystal confined by a sample volume V. The electric polarization is defined as,

$$\mathbf{P} = \frac{1}{V} \int_V d\mathbf{r} \rho(\mathbf{r}) \mathbf{r}, \tag{15.1}$$

where ρ is the total charge density due to electrons and ions. It is clear that this definition is meaningful only if the volume V is finite to guarantee the integral is bounded. In search of a bulk description of the electric polarization, the electric polarization of the crystal must be insensitive to the surface termination such as surface charges. However, assuming a macroscopic cubic crystal of side length L, prepared such that one face is charged by $\Delta\sigma$ while the opposite face is charged by $-\Delta\sigma$, then the crystal dipole moment has a surface contribution that scales as $\Delta\sigma L^3$, and the integral in equation (15.1) changes by $\Delta\sigma$. In other words, equation (15.1) is surface-termination dependent and therefore not a good representative of the bulk electric polarization. Instead, we suggest to compute changes in the electric polarization,

$$\Delta\mathbf{P} = \mathbf{P}(\lambda = 1) - \mathbf{P}(\lambda = 0) = \frac{1}{V} \int_V d\mathbf{r} \rho_{\lambda=1}(\mathbf{r})\mathbf{r} - \frac{1}{V} \int_V d\mathbf{r} \rho_{\lambda=0}(\mathbf{r})\mathbf{r}, \tag{15.2}$$

doi:10.1088/978-0-7503-5557-5ch15

due to a change in the state of the crystal defined by a continuous parameter λ. Note that the volume V is considered unchanged by the change in λ such that the electric polarization change is effected by a change in the total charge density ρ alone. If, as λ is changed, the surface of V is electrically isolated and the shape and volume of V is unchanged, then any initial surface charge will cancel out in the calculation of $\Delta\mathbf{P}$. Note that many physical properties of interest can be determined as derivatives of $\mathbf{P}$ with respect to a suitably defined λ. This is the case of dielectric susceptibility, piezoelectricity, Born effective charges (for lattice dynamics), and pyroelectricity. Hence, a heuristic approach to determine the latter physical properties is to compute the change in the electric polarization due to a change in the state of the crystal,

$$\Delta\mathbf{P} = \mathbf{P}(\lambda = 1) - \mathbf{P}(\lambda = 0) = \int_0^1 d\lambda \frac{\partial\mathbf{P}(\lambda)}{\partial\lambda}, \tag{15.3}$$

where $\lambda = 0$ represents the initial state and $\lambda = 1$ represents the final state of the crystal.

Using density-functional theory [4], the electronic charge density can be computed as,

$$\rho_{el}(\mathbf{r}) = q_{el}\sum_i f_i |\phi_i(\mathbf{r})|^2, \tag{15.4}$$

where q_{el} is the electron charge, ϕ_i are the (λ-dependent) eigenfunctions of the one-electron Kohn–Sham (KS) Hamiltonian,

$$H_{KS} = T + V_{KS}(\lambda), \tag{15.5}$$

and f_i denote their occupation factors. The latter are assumed to be independent of λ which is a good approximation for dielectrics at $T = 0$ K where the Fermi level is well separated by the occupied bands. Thus, from,

$$\mathbf{P} = \frac{q_{el}}{V}\sum_i \int_V f_i |\phi_i(\mathbf{r})|^2 \mathbf{r} d\mathbf{r} = \frac{q_{el}}{V}\sum_i f_i \langle\phi_i|\mathbf{r}|\phi_i\rangle, \tag{15.6}$$

it is found that,

$$
\begin{aligned}
\frac{\partial\mathbf{P}(\lambda)}{\partial\lambda} &= \frac{q_{el}}{V}\sum_i f_i\left(\langle\frac{\partial\phi_i}{\partial\lambda}|\mathbf{r}|\phi_i\rangle + \langle\phi_i|\mathbf{r}|\frac{\partial\phi_i}{\partial\lambda}\rangle\right) \\
&= \frac{q_{el}}{V}\sum_i f_i\left(\langle\phi_i|\mathbf{r}|\frac{\partial\phi_i}{\partial\lambda}\rangle + c.\,c.\right) \\
&= \frac{q_{el}}{V}\sum_i f_i \sum_{j\neq i}\left(\langle\phi_i|\mathbf{r}|\phi_j\rangle\frac{1}{E_i - E_j}\langle\phi_j|\frac{\partial V_{KS}(\lambda)}{\partial\lambda}|\phi_i\rangle + c.\,c.\right) \\
&= -\frac{i\hbar}{m_{el}}\frac{q_{el}}{V}\sum_i\sum_{j\neq i} f_i \frac{\langle\phi_i|\mathbf{p}|\phi_j\rangle\langle\phi_j|\frac{\partial V_{KS}(\lambda)}{\partial\lambda}|\phi_i\rangle}{(E_i - E_j)^2} + c.\,c.
\end{aligned}
\tag{15.7}
$$

In deriving the steps above, we used the first-order perturbation result,

$$|\delta\phi_i\rangle = \sum_{j\neq i} \frac{\langle\phi_j|\delta V_{KS}(\lambda)|\phi_i\rangle}{E_i - E_j}|\phi_j\rangle.\tag{15.8}$$

It was also used that,

$$[H_{KS}, \mathbf{r}] = -\frac{i\hbar}{m_{el}}\mathbf{p},\tag{15.9}$$

whereby,

$$\langle\phi_i|\mathbf{r}|\phi_j\rangle = -\frac{i\hbar}{m_{el}}\frac{1}{E_i - E_j}\langle\phi_i|\mathbf{p}|\phi_j\rangle.\tag{15.10}$$

A note is appropriate on the first-order perturbation result which requires $|E_i - E_j| > >|\langle\phi_j|\delta V_{KS}(\lambda)|\phi_i\rangle|$. In practice the sum over j is restricted to unoccupied states, and assuming this, the inequality holds well since the occupied states are assumed to be well separated in energy from the unoccupied states. Equation (15.7) can be rewritten as,

$$\frac{\partial \mathbf{P}(\lambda)}{\partial\lambda} = -\frac{i\hbar}{m_{el}}\frac{q_{el}}{V}f\sum_{\mathbf{k}}\sum_{n=1}^{M}\sum_{m=M+1}^{\infty} \frac{\langle\phi_{n\mathbf{k}}|\mathbf{p}|\phi_{m\mathbf{k}}\rangle\langle\phi_{m\mathbf{k}}|\dfrac{\partial V_{KS}(\lambda)}{\partial\lambda}|\phi_{n\mathbf{k}}\rangle}{(E_{n\mathbf{k}} - E_{m\mathbf{k}})^2} + c.\,c.\tag{15.11}$$

where for dielectrics it was used that f is constant for the occupied bands and equal to 2 (summing over the two spin possibilities). Note that the bands $1,\ldots, M$ are assumed occupied while all bands $M + 1$ and above are unoccupied.

Let us now use two commutator relations to rewrite the above equation. Firstly, the cell-periodic Hamiltonian is introduced,

$$\hat{H}_{\mathbf{k}}(\lambda) = \frac{1}{2m_{el}}(-i\hbar\boldsymbol{\nabla} + \hbar\mathbf{k})^2 + V_{KS}(\lambda).\tag{15.12}$$

It then follows that,

$$\left[\frac{\partial}{\partial k_\alpha}, \hat{H}_{\mathbf{k}}(\lambda)\right] = \frac{\hbar}{m_{el}}\left(-i\hbar\frac{\partial}{\partial x_\alpha} + \hbar k_\alpha\right),\tag{15.13}$$

and,

$$\begin{aligned}
\langle\phi_{n\mathbf{k}}|p_\alpha|\phi_{m\mathbf{k}}\rangle &= \langle\phi_{n\mathbf{k}}|-i\hbar\frac{\partial}{\partial x_\alpha}|\phi_{m\mathbf{k}}\rangle = \langle\phi_{n\mathbf{k}}|-i\hbar\frac{\partial}{\partial x_\alpha}|u_{m\mathbf{k}}e^{i\mathbf{k}\cdot\mathbf{r}}\rangle\\
&= \langle\phi_{n\mathbf{k}}|-i\hbar e^{i\mathbf{k}\cdot\mathbf{r}}\left(\frac{\partial}{\partial x_\alpha} + ik_\alpha\right)|u_{m\mathbf{k}}\rangle = \langle\phi_{n\mathbf{k}}|e^{i\mathbf{k}\cdot\mathbf{r}}\left(-i\hbar\frac{\partial}{\partial x_\alpha} + \hbar k_\alpha\right)|u_{m\mathbf{k}}\rangle\\
&= \langle u_{n\mathbf{k}}|\left(-i\hbar\frac{\partial}{\partial x_\alpha} + \hbar k_\alpha\right)|u_{m\mathbf{k}}\rangle.
\end{aligned}\tag{15.14}$$

thus,

$$\langle \phi_{nk} | p_\alpha | \phi_{mk} \rangle = \langle u_{nk} | \frac{m_{el}}{\hbar} \left[\frac{\partial}{\partial k_\alpha}, \hat{H}_k(\lambda) \right] | u_{mk} \rangle. \qquad (15.15)$$

The second commutator relation to be used is,

$$\left[\frac{\partial}{\partial \lambda}, \hat{H}_k(\lambda) \right] = \frac{\partial V_{KS}(\lambda)}{\partial \lambda}, \qquad (15.16)$$

i.e.,

$$\langle \phi_{nk} | \frac{\partial V_{KS}(\lambda)}{\partial \lambda} | \phi_{mk} \rangle = \langle u_{nk} | \frac{\partial V_{KS}(\lambda)}{\partial \lambda} | u_{mk} \rangle = \langle u_{nk} | \left[\frac{\partial}{\partial \lambda}, \hat{H}_k(\lambda) \right] | u_{mk} \rangle. \qquad (15.17)$$

Secondly, equation (15.11) is rewritten as,

$$\frac{\partial \mathbf{P}(\lambda)}{\partial \lambda} = -\frac{i\hbar}{m_{el}} \frac{q_{el}}{V} f \sum_{\mathbf{k}} \sum_{n=1}^{M} \sum_{m=M+1}^{\infty} \frac{\langle \phi_{nk} | \mathbf{p} | \phi_{mk} \rangle \langle \phi_{mk} | \frac{\partial V_{KS}(\lambda)}{\partial \lambda} | \phi_{nk} \rangle - \langle \phi_{mk} | \mathbf{p} | \phi_{nk} \rangle \langle \phi_{nk} | \frac{\partial V_{KS}(\lambda)}{\partial \lambda} | \phi_{mk} \rangle}{(E_{nk} - E_{mk})^2}. \qquad (15.18)$$

Use of the two commutator relations above allows us to rewrite the vector components of the numerator in the last factor after the summation signs,

$$\langle \phi_{nk} | p_\alpha | \phi_{mk} \rangle \langle \phi_{mk} | \frac{\partial V_{KS}(\lambda)}{\partial \lambda} | \phi_{nk} \rangle - \langle \phi_{mk} | p_\alpha | \phi_{nk} \rangle \langle \phi_{nk} | \frac{\partial V_{KS}(\lambda)}{\partial \lambda} | \phi_{mk} \rangle$$

$$= \frac{m_{el}}{\hbar} \langle u_{nk} | \frac{\partial}{\partial k_\alpha} | u_{mk} \rangle (E_{mk} - E_{nk}) \langle u_{mk} | \frac{\partial}{\partial \lambda} | u_{nk} \rangle (E_{nk} - E_{mk})$$

$$- \langle u_{mk} | \frac{\partial}{\partial k_\alpha} | u_{nk} \rangle (E_{nk} - E_{mk}) \langle u_{nk} | \frac{\partial}{\partial \lambda} | u_{mk} \rangle (E_{mk} - E_{nk}) \qquad (15.19)$$

$$= \frac{m_{el}}{\hbar} \langle \frac{\partial u_{nk}}{\partial k_\alpha} | u_{mk} \rangle \langle u_{mk} | \frac{\partial u_{nk}}{\partial \lambda} \rangle (E_{nk} - E_{mk})^2$$

$$- \frac{m_{el}}{\hbar} \langle u_{mk} | \frac{\partial u_{nk}}{\partial k_\alpha} \rangle \langle \frac{\partial u_{nk}}{\partial \lambda} | u_{mk} \rangle (E_{nk} - E_{mk})^2,$$

where partial integration and $\langle u_{mk} | u_{nk} \rangle = \delta_{nm}$ were used in obtaining the second equality, so,

$$\frac{\partial P_\alpha(\lambda)}{\partial \lambda} = -\frac{iq_{el}}{V} f \sum_{\mathbf{k}} \sum_{n=1}^{M} \sum_{m=M+1}^{\infty} \left(\langle \frac{\partial u_{nk}}{\partial k_\alpha} | u_{mk} \rangle \langle u_{mk} | \frac{\partial u_{nk}}{\partial \lambda} \rangle - \langle u_{mk} | \frac{\partial u_{nk}}{\partial k_\alpha} \rangle \langle \frac{\partial u_{nk}}{\partial \lambda} | u_{mk} \rangle \right)$$

$$= -\frac{iq_{el}}{V} f \sum_{\mathbf{k}} \sum_{n=1}^{M} \sum_{m=M+1}^{\infty} \left(\langle \frac{\partial u_{nk}}{\partial k_\alpha} | u_{mk} \rangle \langle u_{mk} | \frac{\partial u_{nk}}{\partial \lambda} \rangle - \langle \frac{\partial u_{nk}}{\partial \lambda} | u_{mk} \rangle \langle u_{mk} | \frac{\partial u_{nk}}{\partial k_\alpha} \rangle \right) \quad (15.20)$$

$$\approx -\frac{iq_{el}}{V} f \sum_{\mathbf{k}} \sum_{n=1}^{M} \left(\langle \frac{\partial u_{nk}}{\partial k_\alpha} | \frac{\partial u_{nk}}{\partial \lambda} \rangle - \langle \frac{\partial u_{nk}}{\partial \lambda} | \frac{\partial u_{nk}}{\partial k_\alpha} \rangle \right).$$

In the last step the approximation,

$$1 \approx \sum_{m=M+1}^{\infty} |u_{mk}\rangle\langle u_{mk}|, \tag{15.21}$$

was used. The change in the electronic polarization is now,

$$\Delta P_\alpha = \int_0^1 d\lambda \frac{\partial P_\alpha(\lambda)}{\partial \lambda} = -\frac{iq_{el}}{V} f \sum_{n=1}^{M} \int_{BZ} \frac{d\mathbf{k}}{\frac{(2\pi)^3}{V}} \int_0^1 d\lambda \left(\langle \frac{\partial u_{n\mathbf{k}}}{\partial k_\alpha} | \frac{\partial u_{n\mathbf{k}}}{\partial \lambda} \rangle - \langle \frac{\partial u_{n\mathbf{k}}}{\partial \lambda} | \frac{\partial u_{n\mathbf{k}}}{\partial k_\alpha} \rangle \right)$$
$$\tag{15.22}$$
$$= -\frac{iq_{el}}{(2\pi)^3} f \sum_{n=1}^{M} \int_{BZ} d\mathbf{k} \int_0^1 d\lambda \left(\langle \frac{\partial u_{n\mathbf{k}}}{\partial k_\alpha} | \frac{\partial u_{n\mathbf{k}}}{\partial \lambda} \rangle - \langle \frac{\partial u_{n\mathbf{k}}}{\partial \lambda} | \frac{\partial u_{n\mathbf{k}}}{\partial k_\alpha} \rangle \right),$$

since there is one $\mathbf{k}$ state for each volume element $\frac{(2\pi)^3}{V}$ in reciprocal space. The integral over $\mathbf{k}$ extends over any primitive cell in reciprocal space.

15.1.1 The one-dimensional system

For a one-dimensional system the integral expression of equation (15.22) reads,

$$\int_{BZ} dk \int_0^1 d\lambda \left(\langle \frac{\partial u_{nk}}{\partial k} | \frac{\partial u_{nk}}{\partial \lambda} \rangle - \langle \frac{\partial u_{nk}}{\partial \lambda} | \frac{\partial u_{nk}}{\partial k} \rangle \right)$$
$$= \int_{BZ} dk \int_0^1 d\lambda \left(\frac{\partial}{\partial k} \langle u_{nk} | \frac{\partial}{\partial \lambda} | u_{nk} \rangle - \langle u_{nk} | \frac{\partial}{\partial \lambda} \frac{\partial}{\partial k} | u_{nk} \rangle \right)$$
$$- \int_{BZ} dk \int_0^1 d\lambda \left(\frac{\partial}{\partial \lambda} \langle u_{nk} | \frac{\partial}{\partial k} | u_{nk} \rangle - \langle u_{nk} | \frac{\partial}{\partial \lambda} \frac{\partial}{\partial k} | u_{nk} \rangle \right) \tag{15.23}$$
$$= \int_{BZ} dk \int_0^1 d\lambda \left(\frac{\partial F_\lambda}{\partial k} - \frac{\partial F_k}{\partial \lambda} \right)$$

where,

$$F_\gamma = \langle u_{nk} | \frac{\partial}{\partial \gamma} | u_{nk} \rangle, \tag{15.24}$$

was introduced. Then,

$$\int_{BZ} dk \int_0^1 d\lambda \left(\frac{\partial F_\lambda}{\partial k} - \frac{\partial F_k}{\partial \lambda} \right) = \int_{BZ} dk \int_0^1 d\lambda \, (\nabla \times \mathbf{F})_{k\lambda} = \oint_C F_\lambda d\lambda + \oint_C F_k dk$$
$$= \oint_C \langle u_{nk} | \frac{\partial}{\partial \lambda} | u_{nk} \rangle d\lambda + \oint_C \langle u_{nk} | \frac{\partial}{\partial k} | u_{nk} \rangle dk, \tag{15.25}$$

where $(\nabla \times \mathbf{F})_{k\lambda}$ is the component of $\nabla \times \mathbf{F}$ perpendicular to the plane formed by k and λ. In deriving the second equality, Stokes' theorem was used.

Combining the above results, the polarization change can be written as,

$$\Delta P = -\frac{iq_{el}}{2\pi} f \sum_{n=1}^{M} \left[\oint_C \langle u_{nk} | \frac{\partial}{\partial \lambda} | u_{nk} \rangle d\lambda + \oint_C \langle u_{nk} | \frac{\partial}{\partial k} | u_{nk} \rangle dk \right] = -\frac{fq_{el}}{2\pi} \sum_{n=1}^{M} \left\{ i \oint_C \sum_{j=1}^{2} \langle u_{nk} | \frac{\partial}{\partial \tau_j} | u_{nk} \rangle d\tau_j \right\} \tag{15.26}$$

where τ is a two-component vector with elements $(\tau_1, \tau_2) = (\lambda, k)$ and the contour of integration C is around the loop in τ space from $(0, \pi/a) \rightarrow (1, \pi/a) \rightarrow (1, -\pi/a) \rightarrow (0, -\pi/a) \rightarrow (0, \pi/a)$. The quantity in curly brackets can be recognized as the change in Berry phase for fictitious adiabatic evolution of the cell-periodic wave function around the loop C. In these circumstances the quantity in curly brackets measures the change in the phase of the wave function at any given real-space point as (λ, k) is taken around C. Given that the cell-periodic parts of the wave function can be chosen to be analytic in k and λ, this change in phase must be an integer multiple of 2π (refer to section 15.2 for a discussion of the Berry phase). We therefore conclude that the polarization per unit length of a one-dimensional system can only change by an integer multiple of fq_{el} for adiabatic changes in the Hamiltonian for which $V_{KS}^{(0)} = V_{KS}^{(1)}$. An analogous result for three-dimensional systems will be derived below.

15.2 Berry phase

With reference to figure 15.1 illustrating the evolution of a complex unit vector around a path, consider the quantity,

$$\phi = \oint i\langle u(\lambda)| \frac{\partial}{\partial\lambda} |u(\lambda)\rangle d\lambda, \tag{15.27}$$

which is the Berry phase. Note that $\langle u(\lambda)| \frac{\partial}{\partial\lambda} |u(\lambda)\rangle$ is purely imaginary since,

$$2Re\langle u(\lambda)| \frac{\partial}{\partial\lambda} |u(\lambda)\rangle = \langle u(\lambda)| \frac{\partial}{\partial\lambda} u(\lambda)\rangle + \langle \frac{\partial}{\partial\lambda} u(\lambda)|u(\lambda)\rangle = \frac{\partial}{\partial\lambda} \langle u(\lambda)|u(\lambda)\rangle = 0. \tag{15.28}$$

The integrand on the right-hand side of equation (15.27) is known as the Berry connection or Berry potential,

$$A(\lambda) = i\langle u(\lambda)| \frac{\partial}{\partial\lambda} |u(\lambda)\rangle = -Im\langle u(\lambda)| \frac{\partial}{\partial\lambda} |u(\lambda)\rangle, \tag{15.29}$$

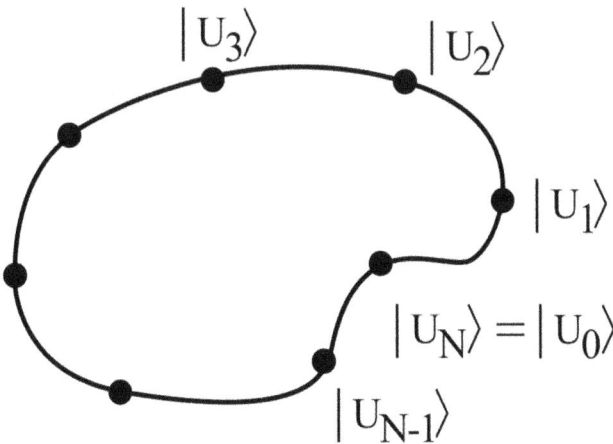

Figure 15.1. Illustration of the evolution of some complex unit vector $|u\rangle$ around a path in parameter space. The first and last points $|u_0\rangle$ and $|u_N\rangle$ are identical and correspond to $\lambda = 0$ and $\lambda = 1$, respectively. Adapted from reference [5] with permission from Cambridge University Press. All rights reserved.

in terms of which the Berry phase is,

$$\phi = \oint A(\lambda)d\lambda. \tag{15.30}$$

Let us understand how these quantities vary under a gauge transformation,

$$|\tilde{u}(\lambda)\rangle = e^{-i\beta(\lambda)}|u(\lambda)\rangle, \tag{15.31}$$

where $\beta(\lambda)$ is a continuous real function of λ. It is found that,

$$\tilde{A}(\lambda) = i\langle\tilde{u}(\lambda)|\frac{\partial}{\partial\lambda}|\tilde{u}(\lambda)\rangle = i\langle u(\lambda)|e^{i\beta(\lambda)}\frac{\partial}{\partial\lambda}e^{-i\beta(\lambda)}|u(\lambda)\rangle = i\langle u(\lambda)|\frac{\partial}{\partial\lambda}|u(\lambda)\rangle + \frac{d\beta}{d\lambda}. \tag{15.32}$$

Hence, the Berry potential is not gauge-invariant. It transforms under a gauge change according to,

$$\tilde{A}(\lambda) = A(\lambda) + \frac{d\beta}{d\lambda}. \tag{15.33}$$

How does the Berry phase change under a gauge transformation? Recall that since $\lambda = 0$ and $\lambda = 1$ label the same state (figure 15.1), we must have $|\tilde{u}(\lambda = 1)\rangle = |\tilde{u}(\lambda = 0)\rangle$ just as $|u(\lambda = 1)\rangle = |u(\lambda = 0)\rangle$. But this implies that,

$$|\tilde{u}(\lambda = 1)\rangle = e^{-i\beta(\lambda=1)}|u(\lambda = 1)\rangle = |\tilde{u}(\lambda = 0)\rangle$$
$$= e^{-i\beta(\lambda=0)}|u(\lambda = 0)\rangle, \tag{15.34}$$

so,

$$e^{-i\beta(\lambda=1)} = e^{-i\beta(\lambda=0)}, \tag{15.35}$$

thus,

$$\beta(\lambda = 1) = \beta(\lambda = 0) + 2\pi l, \tag{15.36}$$

where l is an integer. Then,

$$\int_0^1 \frac{d\beta}{d\lambda}d\lambda = \beta(\lambda = 1) - \beta(\lambda = 0) = 2\pi l, \tag{15.37}$$

so that replacing $A(\lambda)$ by $\tilde{A}(\lambda)$ in equation (15.33) and using equation (15.30) yields,

$$\tilde{\phi} = \phi + 2\pi l. \tag{15.38}$$

That is, the Berry phase φ is gauge-invariant modulo 2π, or in other words, gauge-invariant when regarded as a phase angle! We can think of the Berry phase as the phase that is 'left over' after parallel transport around the loop.

15.3 The three-dimensional system

Let us return to the three-dimensional system. We shall repeat the discussions in reference [1].

The physical implications of equation (15.22) become more transparent by working in a gauge where the wave functions are periodic in reciprocal space, i.e.,

$$\phi_{n\mathbf{k}}(\mathbf{r}) = \phi_{n\mathbf{k}+\mathbf{G}}(\mathbf{r}), \tag{15.39}$$

for all reciprocal lattice vectors $\mathbf{G}$. The cell functions fulfill,

$$u_{n\mathbf{k}}(\mathbf{r}) = e^{i\mathbf{G}\cdot\mathbf{r}} u_{n\mathbf{k}+\mathbf{G}}(\mathbf{r}). \tag{15.40}$$

Notice that the gauge condition of equation (15.40) does not define the phase of the wave functions uniquely. Observe that for the integral factor of equation (15.22), integration by parts yields,

$$\int_{BZ} d\mathbf{k} \int_0^1 d\lambda \left(\langle \frac{\partial u_{n\mathbf{k}}}{\partial k_\alpha} | \frac{\partial u_{n\mathbf{k}}}{\partial \lambda} \rangle - \langle \frac{\partial u_{n\mathbf{k}}}{\partial \lambda} | \frac{\partial u_{n\mathbf{k}}}{\partial k_\alpha} \rangle \right)$$

$$= \int_{BZ} d\mathbf{k} \int_0^1 d\lambda \left(\langle \frac{\partial u_{n\mathbf{k}}}{\partial k_\alpha} | \frac{\partial u_{n\mathbf{k}}}{\partial \lambda} \rangle + \langle u_{n\mathbf{k}} | \frac{\partial}{\partial k_\alpha} \frac{\partial u_{n\mathbf{k}}}{\partial \lambda} \rangle \right) - \int_{BZ} d\mathbf{k} \left[\langle u_{n\mathbf{k}} | \frac{\partial u_{n\mathbf{k}}}{\partial k_\alpha} \rangle \right]_{\lambda=0}^{\lambda=1} \tag{15.41}$$

$$= \int_{BZ} d\mathbf{k} \int_0^1 d\lambda \frac{\partial}{\partial k_\alpha} \langle u_{n\mathbf{k}} | \frac{\partial u_{n\mathbf{k}}}{\partial \lambda} \rangle - \int_{BZ} d\mathbf{k} \left[\langle u_{n\mathbf{k}} | \frac{\partial u_{n\mathbf{k}}}{\partial k_\alpha} \rangle \right]_{\lambda=0}^{\lambda=1},$$

and,

$$\Delta P_\alpha = \frac{ifq_{el}}{(2\pi)^3} \sum_{n=1}^M \int_{BZ} d\mathbf{k} \left\{ \left[\langle u_{n\mathbf{k}} | \frac{\partial u_{n\mathbf{k}}}{\partial k_\alpha} \rangle \right]_{\lambda=0}^{\lambda=1} - \int_0^1 d\lambda \frac{\partial}{\partial k_\alpha} \langle u_{n\mathbf{k}} | \frac{\partial u_{n\mathbf{k}}}{\partial \lambda} \rangle \right\}. \tag{15.42}$$

With the chosen gauge, $\langle u_{n\mathbf{k}} | \frac{\partial u_{n\mathbf{k}}}{\partial \lambda} \rangle$ is periodic in $\mathbf{k}$. The gradient of this quantity integrated over the Brillouin zone (BZ) is zero and the second term in equation (15.42) makes no contribution. Therefore, in the periodic gauge,

$$\Delta \mathbf{P} = \mathbf{P}^{(1)} - \mathbf{P}^{(0)}, \tag{15.43}$$

where,

$$\mathbf{P}^{(\lambda)} = \frac{ifq_{el}}{(2\pi)^3} \sum_{n=1}^M \int_{BZ} d\mathbf{k} \langle u_{n\mathbf{k}}^{(\lambda)} | \frac{\partial u_{n\mathbf{k}}^{(\lambda)}}{\partial k_\alpha} \rangle, \tag{15.44}$$

and the notation clarifies that the cell-periodic part of the wave function depends on λ. The integral on the right-hand side of equation (15.44) is closely connected to the Berry phase of band n [6, 7]. The form of equation (15.44) is conveniently written in terms of the Wannier functions $W_n^{(\lambda)}(\mathbf{r})$ of the occupied bands. The Wannier functions depend on the choice of phases used in the periodic gauge. The Wannier function is defined using,

$$W_n^{(\lambda)}(\mathbf{r} - \mathbf{R}) = \frac{\sqrt{N}\Omega}{(2\pi)^3} \int_{BZ} d\mathbf{k}\, e^{i\mathbf{k}\cdot(\mathbf{r}-\mathbf{R})} u_{n\mathbf{k}}^{(\lambda)}(\mathbf{r}), \tag{15.45}$$

which implies that,

$$u_{n\mathbf{k}}^{(\lambda)}(\mathbf{r}) = \frac{1}{\sqrt{N}} \sum_{\mathbf{R}} e^{-i\mathbf{k}\cdot(\mathbf{r}-\mathbf{R})} W_n^{(\lambda)}(\mathbf{r} - \mathbf{R}), \tag{15.46}$$

where the sum over $\mathbf{R}$ runs over all real-space lattice vectors, $\mathbf{N}$ is the number of unit cells in the crystal, and Ω is the volume of a unit cell. Substituting equation (15.46) into equation (15.44),

$$
\begin{aligned}
\mathbf{P}^{(\lambda)} &= \frac{if q_{el}}{(2\pi)^3} \sum_{n=1}^{M} \frac{1}{N} \int_{BZ} d\mathbf{k} \sum_{\mathbf{R}} \sum_{\mathbf{R}'} [-i(\mathbf{r} - \mathbf{R})] e^{-i\mathbf{k}\cdot(\mathbf{r}-\mathbf{R})} W_n^{(\lambda)}(\mathbf{r} - \mathbf{R}) e^{i\mathbf{k}\cdot(\mathbf{r}-\mathbf{R}')} W_n^{(\lambda)*}(\mathbf{r} - \mathbf{R}') \\
&= \frac{f q_{el}}{N} \sum_{n=1}^{M} \sum_{\mathbf{R}} W_n^{(\lambda)}(\mathbf{r} - \mathbf{R})(\mathbf{r} - \mathbf{R}) W_n^{(\lambda)*}(\mathbf{r} - \mathbf{R}) \\
&= \frac{f q_{el}}{N} \sum_{n=1}^{M} \sum_{\mathbf{R}} \int_{\Omega(\mathbf{R})} \frac{d(\mathbf{r} - \mathbf{R})}{\Omega} W_n^{(\lambda)}(\mathbf{r} - \mathbf{R})(\mathbf{r} - \mathbf{R}) W_n^{(\lambda)*}(\mathbf{r} - \mathbf{R}) \\
&= \frac{f q_{el}}{N} \sum_{n=1}^{M} \frac{N}{\Omega} \int d\mathbf{r} W_n^{(\lambda)}(\mathbf{r}) \mathbf{r} W_n^{(\lambda)*}(\mathbf{r}) = \frac{f q_{el}}{\Omega} \sum_{n=1}^{M} \int d\mathbf{r} |W_n^{(\lambda)}(\mathbf{r})|^2 \mathbf{r},
\end{aligned}
\tag{15.47}
$$

where,

$$
\int_{BZ} d\mathbf{k}\, e^{i\mathbf{k}\cdot(\mathbf{R}-\mathbf{R}')} = (2\pi)^3 \delta(\mathbf{R} - \mathbf{R}').
\tag{15.48}
$$

Equation (15.47) states that the change in polarization of the solid is proportional to the displacement of the center of charge of the Wannier functions induced by the adiabatic change in the Hamiltonian. If the Hamiltonians at $\lambda = 0$ and 1 are identical, $u_{n\mathbf{k}}^{(0)}(\mathbf{r})$ and $u_{n\mathbf{k}}^{(1)}(\mathbf{r})$ can at most differ by a phase factor,

$$
u_{n\mathbf{k}}^{(1)}(\mathbf{r}) = e^{i\theta_{n\mathbf{k}}} u_{n\mathbf{k}}^{(0)}(\mathbf{r}).
\tag{15.49}
$$

Then, equation (15.44) reduces to,

$$
\mathbf{P}^{(\lambda)} = -\frac{f q_{el}}{(2\pi)^3} \sum_{n=1}^{M} \int_{BZ} d\mathbf{k} \frac{\partial \theta_{n\mathbf{k}}}{\partial k_\alpha}.
\tag{15.50}
$$

With a periodic choice of gauge, $e^{i\theta_{n\mathbf{k}}}$ must be periodic in $\mathbf{k}$. The most general form for the phase angle under these circumstances is $\theta_{n\mathbf{k}} = \beta_{n\mathbf{k}} + \mathbf{k} \cdot \mathbf{R}_n$, where $\beta_{n\mathbf{k}}$ is periodic in $\mathbf{k}$. Hence,

$$
\mathbf{P} = \frac{f q_{el}}{\Omega} \sum_{n=1}^{M} \mathbf{R}_n,
\tag{15.51}
$$

and the change in polarization per unit volume for paths where the Hamiltonian returns to itself is quantized in units of $\frac{f q_{el}}{\Omega} \mathbf{R}$. A particularly simple case to consider is the magnitude of $\Delta \mathbf{P}$ for paths of the form $V_{KS}^{(\lambda)}(\mathbf{r}) = V_{KS}^{(0)}(\mathbf{r} - \lambda \mathbf{R})$, which corresponds to a translation of the crystal. It is straightforward to verify by explicit calculation that equation (15.43) yields $\mathbf{P} = \frac{f q_{el}}{\Omega} M \mathbf{R}$, as one would expect on physical grounds.

Equations (15.43) and (15.51) show that for a crystal, $\Delta \mathbf{P}$ can be determined to within a factor of $\frac{fe}{\Omega} \mathbf{R}$ from a knowledge of the valence-band Kohn–Sham wave functions at $\lambda = 0$ and 1. If one is interested in polarization changes for which $|\Delta \mathbf{P}| << \frac{fe}{\Omega} \mathbf{R}_1$, where $\mathbf{R}_1$ is the shortest nonzero real-space lattice vector, the arbitrary factor of $\frac{fe}{\Omega}\mathbf{R}$ can be eliminated by inspection. In more general cases, uncertainties introduced by this factor can be removed by dividing the change in the Hamiltonian into a number of subintervals.

Direct evaluation of $\Delta\mathbf{P}$ via equation (15.44) is cumbersome in numerical calculations, because iwave functions are only computed at a finite number of points in the Brillouin zone, and in general there is no phase relationship between the eigenvectors generated by a diagonalization routine. In actual calculations, this difficulty is circumvented by using the following strategy. First, a direction parallel to a short reciprocal lattice vector of the solid, $\mathbf{G}_{\parallel}$, is chosen. The primitive cell for the k-space integration is chosen to be a prism with its axis aligned along $\mathbf{G}_{\parallel}$. The component of $\Delta\mathbf{P}$ directed along $\mathbf{G}_{\parallel}$ is,

$$\Delta P_{\parallel} = P_{\parallel}^{(1)} - P_{\parallel}^{(0)}, \tag{15.52}$$

where,

$$P_{\parallel}^{(\lambda)} = \frac{ifq_{el}}{(2\pi)^3} \int_A d\mathbf{k}_{\perp} \sum_{n=1}^{M} \int_0^{|\mathbf{G}_{\parallel}|} dk_{\parallel} \langle u_{n\mathbf{k}}^{(\lambda)} | \frac{\partial}{\partial k_{\parallel}} | u_{n\mathbf{k}}^{(\lambda)} \rangle. \tag{15.53}$$

The integration in the perpendicular direction is straightforward and can be performed by sampling over a two-dimensional mesh of k points generated, for example using the Monkhorst–Pack method. To perform the integral over $k_{\parallel}$ at each point in the $k_{\perp}$ mesh, the cell-periodic parts of the wave functions are computed at the string of J k points at $\mathbf{k}_j = \mathbf{k}_{\perp} + j\mathbf{G}_{\parallel}/J$ where j runs from 0 to $J - 1$.

The latter expression can be restated by observing,

$$\log\langle u_{n\mathbf{k}} | u_{n(\mathbf{k}+d\mathbf{k}_{\parallel})} \rangle = \log\langle u_{n\mathbf{k}} | \left(|u_{n\mathbf{k}}\rangle + \frac{\partial|u_{n\mathbf{k}}\rangle}{\partial k_{\parallel}} dk_{\parallel} + \cdots \right)$$

$$= \log\left(1 + \left\langle u_{n\mathbf{k}} | \frac{\partial u_{n\mathbf{k}}}{\partial k_{\parallel}} \right\rangle dk_{\parallel} + \cdots \right) \tag{15.54}$$

$$\approx \left\langle u_{n\mathbf{k}} | \frac{\partial u_{n\mathbf{k}}}{\partial k_{\parallel}} \right\rangle dk_{\parallel} + \cdots.$$

thus,

$$P_{\parallel}^{(\lambda)} = \frac{ifq_{el}}{(2\pi)^3} \int_A d\mathbf{k}_{\perp} \sum_{n=1}^{M} \int_0^{|\mathbf{G}_{\parallel}|} \log\langle u_{n\mathbf{k}} | u_{n(\mathbf{k}+d\mathbf{k}_{\parallel})} \rangle$$

$$= -\frac{fq_{el}}{(2\pi)^3} \int_A d\mathbf{k}_{\perp} \sum_{n=1}^{M} Im\left\{ \int_0^{|\mathbf{G}_{\parallel}|} \log\langle u_{n\mathbf{k}} | u_{n(\mathbf{k}+d\mathbf{k}_{\parallel})} \rangle \right\}, \tag{15.55}$$

where the latter step follows from the fact that $\langle u_{n\mathbf{k}}^{(\lambda)} | \frac{\partial}{\partial k_{\parallel}} | u_{n\mathbf{k}}^{(\lambda)} \rangle$ is purely imaginary according to equation (15.28).

Performing the integration over $k_\parallel$ by summing over j from 0 to $J - 1$ yields,

$$P_\parallel^{(\lambda)} = -\frac{fq_{el}}{(2\pi)^3} \int_A d\mathbf{k}_\perp \sum_{n=1}^M Im\left\{ \log\left[\Pi_{j=0}^{J-1} \langle u_{n\mathbf{k}_j} | u_{n(\mathbf{k}_j + d\mathbf{k}_\parallel)} \rangle \right] \right\}. \tag{15.56}$$

Observing next that,

$$
\begin{aligned}
\sum_{n=1}^M Im\left\{ \log\left[\Pi_{j=0}^{J-1} \langle u_{n\mathbf{k}_j} | u_{n(\mathbf{k}_j + d\mathbf{k}_\parallel)} \rangle \right] \right\} &= Im\left\{ \sum_{n=1}^M \log\left[\Pi_{j=0}^{J-1} \langle u_{n\mathbf{k}_j} | u_{n(\mathbf{k}_j + d\mathbf{k}_\parallel)} \rangle \right] \right\} \\
&= Im\left\{ \log\left[\Pi_{n=1}^M \Pi_{j=0}^{J-1} \langle u_{n\mathbf{k}_j} | u_{n(\mathbf{k}_j + d\mathbf{k}_\parallel)} \rangle \right] \right\} \\
&= Im\left\{ \log\left[\Pi_{j=0}^{J-1} \Pi_{n=1}^M \langle u_{n\mathbf{k}_j} | u_{n(\mathbf{k}_j + d\mathbf{k}_\parallel)} \rangle \right] \right\} \\
&\approx Im\left\{ \log\left[\Pi_{j=0}^{J-1} \det\left\{ \langle u_{m\mathbf{k}_j} | u_{n(\mathbf{k}_j + d\mathbf{k}_\parallel)} \rangle \right\} \right] \right\},
\end{aligned}
\tag{15.57}
$$

where in the last step it was used that since $\mathbf{k}_j$ and $\mathbf{k}_j + d\mathbf{k}_\parallel$ are (infinitesimally) close, a parallel transport representation [5] of the cell-periodic functions $|u_{n\mathbf{k}}\rangle$ shows that the matrix $\bar{M}_{mn}^j = \{\langle u_{m\mathbf{k}_j} | u_{n(\mathbf{k}_j + d\mathbf{k}_\parallel)} \rangle\}$ is close to the identity matrix for which,

$$\Pi_{n=1}^M \langle u_{n\mathbf{k}_j} | u_{n(\mathbf{k}_j + d\mathbf{k}_\parallel)} \rangle \approx \det\left\{ \langle u_{m\mathbf{k}_j} | u_{n(\mathbf{k}_j + d\mathbf{k}_\parallel)} \rangle \right\}. \tag{15.58}$$

Using the latter result in equation (15.56) yields,

$$P_\parallel^{(\lambda)} = -\frac{fq_{el}}{(2\pi)^3} \int_A d\mathbf{k}_\perp Im\left\{ \log\left[\Pi_{j=0}^{J-1} \det\left\{ \langle u_{m\mathbf{k}_j} | u_{n(\mathbf{k}_j + d\mathbf{k}_\parallel)} \rangle \right\} \right] \right\}. \tag{15.59}$$

Evidently, the computation of the right-hand side of equation (15.59) becomes increasingly accurate as $J \to \infty$.

It is easy to confirm that the product over j in equation (15.58) is independent of the phase choice for the wave functions $u_{n\mathbf{k}}$. Other phase choices can only change the value of the integral in equation (15.59) by an integer multiple of 2π. In accordance herewith, the arbitrary constant in the definition of equation (15.58) arises from the fact that the imaginary part of the log of a complex number is only defined up to a constant multiple of 2π. In practice, the arbitrary constant is removed by comparing $P_\parallel^{(1)}$ with $P_\parallel^{(0)}$ as pointed out in the previous paragraph.

Equations (15.52)–(15.59) is a complete set to determine the polarization changes in a practical calculation without the need for supercells or linear-response techniques. Indeed, the method is suited to modern electronic structure methods based on iterative diagonalization techniques in cases where only the valence-band wave functions are sought.

The approach used in reference [1] is illustrated by computing the transverse effective charge tensor and piezoelectric constant of GaAs in a first-principles pseudopotential calculation. The effective charge of GaAs can be determined by computing the change in polarization induced by making a small displacement of one sublattice with the boundary condition $\mathbf{E} = 0$. If the Ga sublattice is moved by a vector $\mathbf{u}$, then the electronic contribution to the polarization difference between the distorted and undistorted structures is,

$$\Delta P = \frac{e}{\Omega} Z_{Ga}^{*(el)} \mathbf{u}, \tag{15.60}$$

where $Z_{Ga}^{*(el)}$ is the electronic contribution to the effective charge. The piezoelectric tensor γ is the strain derivative of the polarization calculated at $\mathbf{E} = 0$. In the zincblende structure there is only one independent component of the piezoelectric tensor, γ_{14}. The piezoelectric tensor can be thought of as the sum of two independent terms The first term, denoted $\gamma_{14}^{(0)}$ following reference [8], arises from the change in polarization when the ions are subjected to a homogeneous strain. The second contribution originates from the relative displacement of the sublattices, and can be expressed in terms of the effective charges and internal strain parameters. It is shown in reference [8] that,

$$(a^2/e)\gamma_{14} \equiv \bar{\gamma}_{14} = (a^2/e)\gamma_{14}^{(0)} + Z_{Ga}^{*(el)}\xi, \tag{15.61}$$

where ξ is the internal strain parameter. The first-principles calculations used norm-conserving nonlocal pseudopotentials. Strictly speaking, a nonlocal potential causes a modification to the momentum operator in equation (15.11). There is, however, a precisely compensating change to the Hamiltonian for the cell-periodic part of the wave function such that results from equation (15.22) remain correct at they stand. The calculation treats exchange and correlation in the local-density approximation using the Wigner form, and wave functions were expanded using a 20-Ry plane-wave cutoff. All calculations of the self-consistent Kohn–Sham potential were performed with a (4, 4, 4) Monkhorst–Pack mesh. Calculations in the cubic structure were performed at the theoretical lattice constant of 5.576 Å using the above parameters. For the calculations of the effective charge, the Ga atom was displaced a distance of $0.01a$ in the (001) direction and the polarization change in the z direction was computed. The integration mesh for computing ΔP used 16 k points in the $\mathbf{k}_\perp$ mesh and a string of 10 k points in the parallel direction. A value for $\gamma_{14}^{(0)}$ was obtained by computing the change in polarization in the z directions induced by applying a 1% xy shear strain to the crystal.

The results in reference [1] are summarized in table 15.1. The total value of Z_{Ga}^* (electronic plus ionic contributions) come out to be 1.984, in excellent agreement with the value of 1.994 obtained from pseudopotential linear response calculations [8]. Both sets of theoretical values of Z_{Ga}^* are about 8% smaller than the experimental value. Calculations on the strained crystal yield a $(a^2/e)\gamma_{14}^{(0)}$ of -1.352. The value agrees to better than 5% with the result obtained from linear-response methods [8]. The overall value for the piezoelectric constant $\bar{\gamma}_{14}$ was -0.28, compared with an experimental value of -0.32. The agreement between the calculated value and experiment is reasonable, given that the two terms in equation (15.61) show a strong tendency to cancel. Convergence with respect to k-point set and plane-wave cutoff was checked and that the polarization response is linear in the applied perturbation. The small differences between the results in reference [1] and those of Gironcoli,

Table 15.1. Piezoelectric response of GaAs. Computed and experimental data adapted from reference [1].

Parameter	reference [1]	Linear response [8]	Experiment
a (Å)	5.576	5.496	5.642
ζ	0.542	0.528	0.55
Z^*_{Ga}	1.984	1.994	2.16
$\frac{a^2}{e}\gamma^{(0)}_{14}$	−1.352	−1.405	
$\bar{\gamma}_{14}$	−0.28	−0.35	−0.32

Baroni, and Resta [8] are attributed to the use of different pseudopotentials and parametrizations of the exchange and correlation potential.

As mentioned in reference [1], it is tempting to physically identify the quantity $\mathbf{P}^{(\lambda)}$ defined in equation (15.44) as the absolute polarization of the perturbed crystal. Clearly, the polarization, defined in this way, can only be well defined modulo $ef\mathbf{R}/\Omega$.

Summarizing, it was shown that adiabatic changes in the Kohn–Sham Hamiltonian lead to polarization changes in the solid which can be computed in terms of the initial and final valence-band wave functions of the system. This is the basis for calculating polarization changes of solids from first-principles total-energy calculations.

15.4 Berry phase: example 1

Consider a path through the four spinor states,

$$|u_0\rangle = \begin{bmatrix} 1 \\ 0 \end{bmatrix}, \; |u_1\rangle = \frac{1}{\sqrt{2}}\begin{bmatrix} 1 \\ 1 \end{bmatrix}, \; |u_2\rangle = \begin{bmatrix} 0 \\ 1 \end{bmatrix}, \; |u_3\rangle = \frac{1}{\sqrt{2}}\begin{bmatrix} 1 \\ i \end{bmatrix}, \tag{15.62}$$

which closes on itself with $|u_4\rangle = |u_0\rangle$. This corresponds to a path in which the spin points along $\hat{\mathbf{z}}$, $\hat{\mathbf{x}}$, $-\hat{\mathbf{z}}$, $\hat{\mathbf{y}}$, and then back to $\hat{\mathbf{z}}$, since from standard quantum mechanics texts, $|\uparrow_{\hat{\mathbf{n}}}\rangle = \begin{pmatrix} \cos(\theta/2) \\ \sin(\theta/2)e^{i\phi} \end{pmatrix}$ where (θ, ϕ) are the polar and azimuthal angles of $\hat{\mathbf{n}}$, respectively.

15.4.1 Compute the discrete Berry phase for the path around this loop

The Berry phase can be approximated by the rough discretization given by the end points in the loop: $\hat{\mathbf{z}}$, $\hat{\mathbf{x}}$, $-\hat{\mathbf{z}}$, $\hat{\mathbf{y}}$, and then back to $\hat{\mathbf{z}}$, i.e., from equations (15.27) and (15.54),

$$\begin{aligned} \phi &= -Im\log\left[\langle \uparrow_{\mathbf{z}} | \uparrow_{\mathbf{x}} \rangle\langle \uparrow_{\mathbf{x}} | \uparrow_{-\mathbf{z}} \rangle\langle \uparrow_{-\mathbf{z}} | \uparrow_{\mathbf{y}} \rangle\langle \uparrow_{\mathbf{y}} | \uparrow_{\mathbf{z}} \rangle\right] \\ &= -Im\log\left[\left(\frac{1}{\sqrt{2}}\right)\left(\frac{1}{\sqrt{2}}\right)\left(\frac{1}{\sqrt{2}}i\right)\left(\frac{1}{\sqrt{2}}\right)\right] = -Im\log\left[\frac{i}{4}\right] = -\frac{\pi}{2}, \end{aligned} \tag{15.63}$$

where, for completeness, we kept real coefficients $\frac{1}{\sqrt{2}}$ in each factor despite the fact that a real multiplication factor (1/4) does not affect the value of the Berry phase.

15.4.2 Construct a parallel transport gauge for this path

Let us construct a parallel transport gauge for this path and check that the Berry phase is the same as found in the preceding subsection.

A parallel transport gauge is a choice of basis where each factor except the last in the computation of the Berry phase is positive and real (in our example, $\langle \uparrow_z | \uparrow_x \rangle$, $\langle \uparrow_x | \uparrow_{-z} \rangle$, $\langle \uparrow_{-z} | \uparrow_y \rangle$). Evidently, such a gauge is,

$$|\bar{u}_0\rangle = \begin{bmatrix} 1 \\ 0 \end{bmatrix}, \ |\bar{u}_1\rangle = \frac{1}{\sqrt{2}}\begin{bmatrix} 1 \\ 1 \end{bmatrix}, \ |\bar{u}_2\rangle = \begin{bmatrix} 0 \\ 1 \end{bmatrix}, \ |\bar{u}_3\rangle = \frac{1}{\sqrt{2}}e^{-i\pi/2}\begin{bmatrix} 1 \\ i \end{bmatrix} = \frac{1}{\sqrt{2}}\begin{bmatrix} -i \\ 1 \end{bmatrix}. \quad (15.64)$$

Thus, in a parallel transport gauge only the last factor (in our example, $\langle \uparrow_y | \uparrow_z \rangle$) can be complex. Then,

$$\begin{aligned}
\phi &= - Im \log[\langle \bar{u}_0|\bar{u}_1\rangle\langle \bar{u}_1|\bar{u}_2\rangle\langle \bar{u}_2|\bar{u}_3\rangle\langle \bar{u}_3|\bar{u}_0\rangle] \\
&= - Im \log\left[\left(\frac{1}{\sqrt{2}}\right)\left(\frac{1}{\sqrt{2}}\right)\left(\frac{1}{\sqrt{2}}\right)\left(\frac{i}{\sqrt{2}}\right)\right] = -Im \log\left[\frac{i}{4}\right] = -\frac{\pi}{2},
\end{aligned} \quad (15.65)$$

which is the same result as found above.

15.4.3 Construct a twisted parallel transport gauge for this path

A twisted parallel transport gauge is a gauge obtained from the parallel transport gauge and applying phase twists,

$$|\tilde{u}_j\rangle = e^{-ij\phi/N}|\bar{u}_j\rangle, \quad (15.66)$$

where ϕ is the Berry phase from above, i.e., $\phi = -\pi/2$. In our case ($N = 4$), a twisted parallel transport gauge is,

$$\begin{aligned}
|\tilde{u}_0\rangle &= |\bar{u}_0\rangle = \begin{bmatrix} 1 \\ 0 \end{bmatrix}, \\
|\tilde{u}_1\rangle &= e^{-i\cdot 1\cdot(-\pi/2)/4}|\bar{u}_1\rangle = e^{i\pi/8}|\bar{u}_1\rangle = e^{i\pi/8}\frac{1}{\sqrt{2}}\begin{bmatrix} 1 \\ 1 \end{bmatrix}, \\
|\tilde{u}_2\rangle &= e^{-i\cdot 2\cdot(-\pi/2)/4}|\bar{u}_2\rangle = e^{i\pi/4}|\bar{u}_2\rangle = e^{i\pi/4}\begin{bmatrix} 0 \\ 1 \end{bmatrix}, \\
|\tilde{u}_3\rangle &= e^{-i\cdot 3\cdot(-\pi/2)/4}|\bar{u}_3\rangle = e^{i3\pi/8}|\bar{u}_3\rangle = e^{-i\pi/8}\frac{1}{\sqrt{2}}\begin{bmatrix} 1 \\ i \end{bmatrix},
\end{aligned} \quad (15.67)$$

Thus, in a twisted parallel transport gauge,

$$\begin{aligned}
\phi &= - Im \log[\langle \tilde{u}_0|\tilde{u}_1\rangle\langle \tilde{u}_1|\tilde{u}_2\rangle\langle \tilde{u}_2|\tilde{u}_3\rangle\langle \tilde{u}_3|\tilde{u}_0\rangle] \\
&= - Im \log\left[\left(\frac{1}{\sqrt{2}}e^{i\pi/8}\right)\left(\frac{1}{\sqrt{2}}e^{i\pi/8}\right)\left(\frac{1}{\sqrt{2}}e^{i\pi/8}\right)\left(\frac{1}{\sqrt{2}}e^{i\pi/8}\right)\right] = -Im \log\left[\frac{1}{4}e^{i\pi/2}\right] = -\frac{\pi}{2},
\end{aligned} \quad (15.68)$$

again, the same result as found above.

15.5 Berry phase of a sequence of N states

For the magnetic field problem the ground state spinor is, according to equation (15.80),

$$| \Uparrow_n \rangle = \begin{bmatrix} \cos(\theta/2) \\ \sin(\theta/2)e^{i\phi} \end{bmatrix}. \tag{15.69}$$

Consider the sequence of N spinor states all with the same θ and with ϕ taking N equally spaced values from 0 to 2π. The Berry phase is,

$$\phi = -Im \log \left[\langle 0 | \frac{2\pi}{N} \rangle \langle \frac{2\pi}{N} | 2 \frac{2\pi}{N} \rangle \cdots \langle \frac{(N-1)2\pi}{N} | 2\pi \rangle \right], \tag{15.70}$$

where $|\frac{2\pi}{N}\rangle$ denotes the spinor state corresponding to θ and $\phi = \frac{2\pi}{N}$, etc. Evidently,

$$\phi = -Im \log \left[\cos^2(\theta/2) + \sin^2(\theta/2)e^{i2\pi/N}\right]^N = -NIm \log \left[\cos^2(\theta/2) + \sin^2(\theta/2)e^{i2\pi/N}\right]$$
$$= -NIm \log \left[\cos^2(\theta/2) + \sin^2(\theta/2)\cos(2\pi/N) + i \sin^2(\theta/2)\sin(2\pi/N)\right]. \tag{15.71}$$

Using,

$$a = \cos^2(\theta/2) + \sin^2(\theta/2)\cos(2\pi/N),$$
$$b = \sin^2(\theta/2)\sin(2\pi/N), \tag{15.72}$$
$$\log[a + ib] = \log\left[\sqrt{a^2 + b^2}\, e^{i \tan^{-1}(b/a)}\right]$$

the Berry phase is evaluated to, from equation (15.71),

$$\phi = -NIm \log\left[\sqrt{a^2 + b^2}\, e^{i \tan^{-1}(b/a)}\right] = -N \tan^{-1}(b/a)$$
$$= -N \tan^{-1}\left(\frac{\sin^2(\theta/2)\sin(2\pi/N)}{\cos^2(\theta/2) + \sin^2(\theta/2)\cos(2\pi/N)}\right). \tag{15.73}$$

15.5.1 Berry phase in the limit $N \rightarrow \infty$

Equation (15.73) can be rewritten as,

$$\tan(-\phi/N) = \frac{\sin^2(\theta/2)\sin(2\pi/N)}{\cos^2(\theta/2) + \sin^2(\theta/2)\cos(2\pi/N)}, \tag{15.74}$$

In the limit $N \rightarrow \infty$,

$$-\phi/N = \frac{\sin^2(\theta/2)2\pi/N}{\cos^2(\theta/2) + \sin^2(\theta/2)} = \sin^2(\theta/2)2\pi/N, \tag{15.75}$$

i.e.,

$$\phi = -2\pi \sin^2(\theta/2). \tag{15.76}$$

A numerical calculation of equation (15.73) shows that a rough discretization (small N) yields a markedly different value for the Berry phase compared to the continuous result ($N \rightarrow \infty$).

15.6 Berry phase: example 2

So far, the discussion above has been entirely mathematical. The unit vector $|u(\lambda)\rangle$ follows a parametrized path in some complex vector space. For physical applications, the case in which $|u(\lambda)\rangle$ is the ground state of some quantum-mechanical Hamiltonian $H(\lambda)$ is usually considered, with the ground state evolving smoothly as a consequence of the smooth evolution of H. For example, it could be relevant to determine the electronic ground state of a molecule as certain atomic structural coordinates are varied or as external electric or magnetic fields are applied. A simple and instructive example is the case of a spin-1/2 particle, such as an electron or a neutron, at rest in free space and subject to a uniform magnetic field $\mathbf{B} = B\hat{\mathbf{n}}$ directed along $\hat{\mathbf{n}}$. Its Hamiltonian is just,

$$H = -\gamma \mathbf{B} \cdot \mathbf{S} = -\frac{\gamma \hbar B}{2}\hat{\mathbf{n}} \cdot \boldsymbol{\sigma}, \tag{15.77}$$

where γ is the gyromagnetic moment, $\mathbf{S} = \hbar\boldsymbol{\sigma}/2$ is the spin, and $\boldsymbol{\sigma} = \{\sigma_1, \sigma_2, \sigma_3\}$ are the Pauli matrices. The ground state $|u_\mathbf{B}\rangle$ is a spin eigenstate of $\hat{\mathbf{n}} \cdot \boldsymbol{\sigma}$ and is therefore completely independent of the magnitude of $\mathbf{B}$. Thus it is natural to write it as $u_{\hat{\mathbf{n}}}$, emphasizing that it depends only on the field direction $\hat{\mathbf{n}}$. The question of interest is now: What is the Berry phase of $u_{\hat{\mathbf{n}}}$ as $\hat{\mathbf{n}}$ is carried around a loop in magnetic-field orientation space as illustrated in figure 15.2? It will be shown that there is an elegant answer to this question, even for a curved loop such as that shown in figure 15.2, but let us first consider a simpler 'triangular' loop in the discretized approximation. Let $\hat{\mathbf{n}}$ start along $\hat{\mathbf{z}}$, then rotate it to $\hat{\mathbf{x}}$, then to $\hat{\mathbf{y}}$, and finally back to $\hat{\mathbf{z}}$, thereby tracing out

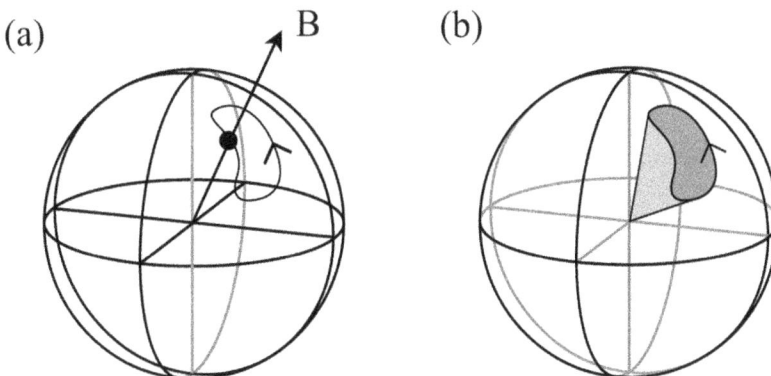

Figure 15.2. (a) Evolution of an applied magnetic field around a closed loop in **B** space. (b) Shaded region shows the solid angle swept out on the unit sphere in **B** space. Adapted from reference [5] with permission from Cambridge University Press. All rights reserved.

one octant of the unit sphere. From equations (15.27) and (15.54), the Berry phase can be written as,

$$\phi = \oint i\langle u(\lambda)| \frac{\partial}{\partial \lambda} |u(\lambda)\rangle d\lambda = -Im \log \Pi_{\lambda=0}^{\lambda=1} \langle u(\lambda)|u(\lambda + d\lambda)\rangle$$

$$= -Im \lim_{N\to\infty} \log(\langle u(0)|u(d\lambda)\rangle\langle u(d\lambda)|u(2d\lambda)\rangle \cdots \langle u((N-1)d\lambda)|u(Nd\lambda)\rangle),$$

(15.78)

where $d\lambda = 1/N$. Thus, for the present example, a rough discretization of the path over the triangular end points yields,

$$\phi = -Im \log[\langle \uparrow_z | \uparrow_x \rangle\langle \uparrow_x | \uparrow_y \rangle\langle \uparrow_y | \uparrow_z \rangle],$$

(15.79)

where $|\uparrow_{\hat{n}}\rangle$ is the spinor that is 'spin up in direction $\hat{n}$'. We can ignore the normalization factors when inserting into the expression for ϕ, obtaining,

$$\phi = -Im \log[(1)(1 + i)(1)] = -\pi/4.$$

(15.80)

As it happens, this result is exact. A more careful treatment using a dense mesh of intermediate points along each great-circle arc does not change this result.

15.6.1 Determining the ground state of an electron or neutron in a magnetic field

It follows from equation (15.77) that the ground state spinor in a magnetic field is found by maximizing the eigenvalue of the operator,

$$\hat{n} \cdot \sigma,$$

(15.81)

equivalent to the minimum of the Hamiltonian. Firstly, let us use the Pauli matrices

$$\sigma_x = \begin{bmatrix} 0 & 1 \\ 1 & 0 \end{bmatrix},$$

(15.82)

$$\sigma_y = \begin{bmatrix} 0 & -i \\ i & 0 \end{bmatrix},$$

(15.83)

$$\sigma_z = \begin{bmatrix} 1 & 0 \\ 0 & -1 \end{bmatrix},$$

(15.84)

$\sigma = (\sigma_x, \sigma_y, \sigma_z)$, and note that a normal vector to a point on the surface of the sphere characterized by (θ, ϕ), where θ and ϕ are the polar and azimuthal angles, is,

$$\hat{n} = \begin{bmatrix} \sin\theta \cos\phi \\ \sin\theta \sin\phi \\ \cos\theta \end{bmatrix},$$

(15.85)

such that,

$$\sigma \cdot \hat{n} = \begin{bmatrix} \cos\theta & \sin\theta(\cos\phi - i\sin\phi) \\ \sin\theta(\cos\phi + i\sin\phi) & -\cos\theta \end{bmatrix}.$$

(15.86)

Eigenvalues λ are found by solving the secular equation,

$$det(\boldsymbol{\sigma} \cdot \hat{\mathbf{n}} - \lambda I_{2\times2}) = 0, \tag{15.87}$$

i.e.,

$$(\cos\theta - \lambda)(-\cos\theta - \lambda) - \sin^2\theta(e^{i\phi}e^{-i\phi})$$
$$= -\cos^2\theta + \lambda^2 - \sin^2\theta = 0, \tag{15.88}$$

giving,

$$\lambda = \pm 1. \tag{15.89}$$

Hence, the ground state has $\lambda = 1$ and eigenvectors (x, y) satisfy,

$$x\cos\theta + y\sin\theta \, e^{-i\phi} = \lambda x = x. \tag{15.90}$$

Thus,

$$y = -\frac{\cos\theta - 1}{\sin\theta}e^{i\phi}x = -\frac{\cos^2(\theta/2) - \sin^2(\theta/2) - (\cos^2(\theta/2) + \sin^2(\theta/2))}{2\sin(\theta/2)\cos(\theta/2)}e^{i\phi}x$$
$$= \frac{\sin(\theta/2)}{\cos(\theta/2)}e^{i\phi}x. \tag{15.91}$$

Choosing $x = \cos(\theta/2)$ gives $y = \sin(\theta/2)e^{i\phi}$, so a normalized ground state eigenvector is,

$$| \uparrow_{\mathbf{n}} \rangle = \begin{bmatrix} x \\ y \end{bmatrix} = \begin{bmatrix} \cos(\theta/2) \\ \sin(\theta/2)e^{i\phi} \end{bmatrix}, \tag{15.92}$$

whereby the expression following equation (15.62) is demonstrated.

15.7 Atomistic approach to strain and elasticity: valence force-field models

Having presented the basic ingredients of the modern theory of polarization, let us now turn our attention to describing a celebrated atomistic model for strain in cubic solids due to Keating [9].

15.7.1 The Keating model

The elastic strain energy of a non-metallic crystal V depends on the position of each nuclei $\mathbf{x}_k$ where k runs over all nuclei. However, it is clear that the strain energy can only depend on the differences in position between any two interacting nuclei to make it independent of a general translation and/or rigid rotation of the nuclei arrangement. Thus the strain energy can be written as,

$$V = V(\mathbf{x}_k - \mathbf{x}_l) = V(\mathbf{x}_{kl}), \tag{15.93}$$

where $\mathbf{x}_{kl} = \mathbf{x}_k - \mathbf{x}_l$. It is convenient to introduce,

$$\lambda_{klmn} = (\mathbf{x}_{kl} \cdot \mathbf{x}_{mn} - \mathbf{X}_{kl} \cdot \mathbf{X}_{mn})/(2a), \tag{15.94}$$

where a is a lattice constant, $\mathbf{X}_{kl}$ is $\mathbf{X}_k - \mathbf{X}_l$, and $\mathbf{X}_k$ is the position vector of the kth nucleus in the undeformed crystal. With this construction, it is evident that λ_{klmn} vanishes in the absence of deformations. It is reasonable to consider the strain energy a function of λ_{klmn}, i.e., $V \equiv V(\lambda_{klmn})$. Since the λ_{klmn}'s are small and the strain energy must be at a minimum in equilibrium, V cannot contain linear terms in the λ_{klmn}'s. With the displacement,

$$\mathbf{u}_{kl} = \mathbf{x}_{kl} - \mathbf{X}_{kl}, \tag{15.95}$$

it is evident that V must be quadratic in the λ_{klmn}'s to leading order (constant terms evidently do not matter), i.e.,

$$V = B^{pqrs}_{klmn} \lambda_{klmn} \lambda_{pqrs} + O(\lambda^3), \tag{15.96}$$

where summation over repeated indices is assumed, and $O(\lambda^3)$ denotes terms of order λ^3 or higher. These higher-order terms are vanishingly small in a small-strain approach and will neglected.

Further simplifications must be made as the number of terms in equation (15.96) is huge and not tractable. First, notice that the number of independent parameters to define any strained configuration of N atoms is $3N - 6$ where the subtraction of 6 accounts for the fact that a general translation in three-dimensional space and 3 rotation degrees-of-freedom of the complete atomic arrangement do not alter the strain configuration.

Consider a non-primitive diatomic (diamond or zincblende) structure with atoms A and B (figure 15.3). Let us write $\mathbf{x}_1(l)$, $\mathbf{x}_2(l)$, and $\mathbf{x}_3(l)$ as the position vectors of the B atoms in neighboring unit cells relative to the A atom of cell (l) and writing $\mathbf{x}_4(l)$ as the position vector of atom B in cell (l) with respect to the A atom there. In this case, we have 5 atoms interacting (atoms 0, 1, 2, 3, 4) and $3N - 6 = 3 \cdot 5 - 6 = 9$ scalar products suffice to define the strain energy. The nine scalar products per unit cell are the 10 scalar products $\mathbf{x}_m(l) \cdot \mathbf{x}_n(l)$ $(m, n = 1, 2, 3, 4)$ less one of the off-diagonal products $(\mathbf{x}_3(l) \cdot \mathbf{x}_4(l)$, for example, is determined for small displacements if the other 9 scalar products are known). The latter follows, e.g., by taking the dot product of $\mathbf{x}_3$ and $\mathbf{x}_4$ written as,

$$\begin{aligned}
\mathbf{x}_3 &= A_1\mathbf{x}_1 + A_2\mathbf{x}_2 + A_3\mathbf{x}_4, \\
\mathbf{x}_4 &= B_1\mathbf{x}_1 + B_2\mathbf{x}_2 + B_3\mathbf{x}_3,
\end{aligned} \tag{15.97}$$

for some constants A_1, A_2, A_3, B_1, B_2, B_3, and the fact that both the vector sets $\{\mathbf{X}_1, \mathbf{X}_2, \mathbf{X}_4\}$ and $\{\mathbf{X}_1, \mathbf{X}_2, \mathbf{X}_3\}$ span the three-dimensional crystal.

The elastic strain energy can be written as,

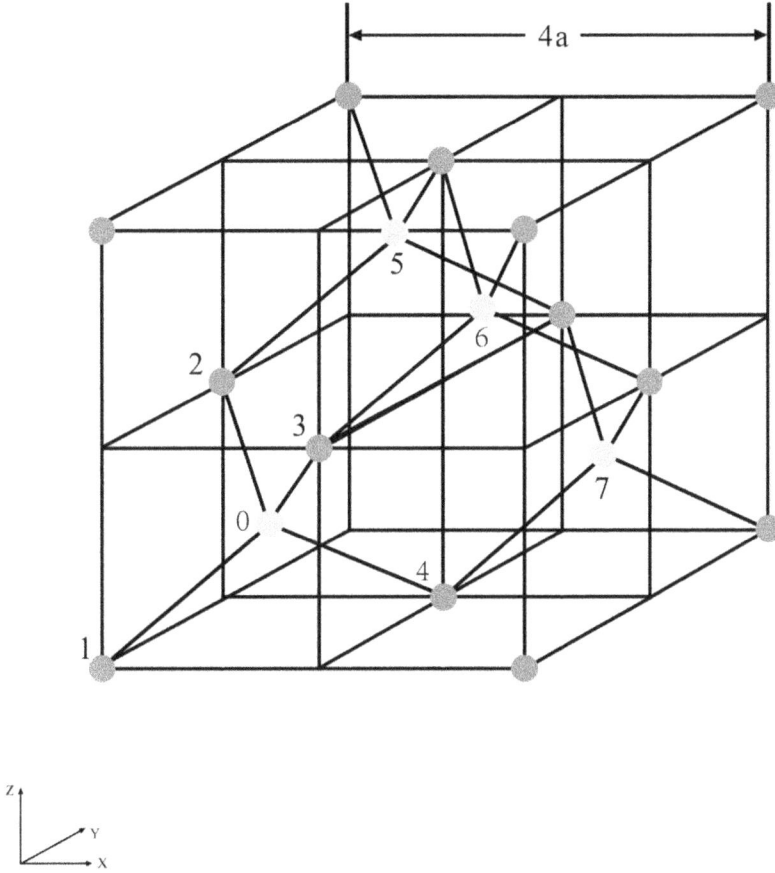

Figure 15.3. The zincblende crystal model. The blue and yellow spheres represent the atoms on the two different sublattices. Adapted from reference [9], copyright (1996) with permission from American Physical Society.

$$V = \frac{1}{2}\sum_{l,l'} \sum_{m,n,m',n'}^{4} {}'B_{m'n'mn}(l - l')\lambda_{m'n'}(l')\lambda_{mn}(l) + \cdots, \qquad (15.98)$$

where the prime denotes that terms λ_{34} (and λ_{43}) are not included (unless they arise in the course of imposing invariance on the B's under symmetry operations). Let us make further simplifications. The basic unit cell of the diamond structure is a rhombohedron with two atoms (1 and 0 of figure 15.3) on its major axis, which is directed along the [111] direction. The three neighboring unit cells of interest contain atoms 2 and 5, 3 and 6, 4 and 7, respectively. Equation (15.98) is used for the strain energy to obtain a two-constant model by including only diagonal products of the λ's. Thus,

$$V = \frac{1}{2}\sum_{l} \sum_{m,n}^{4}{'}B_{mnmn}\lambda_{mn}(l)^2 \qquad (15.99)$$

15-20

$$= \frac{1}{2} \sum_l \left[\frac{\alpha}{4a^2} \sum_{i=1}^{4} \left(x_{0i}^2(l) - 3a^2\right)^2 + \frac{\beta}{2a^2} \sum_{i,j>i}^{4} (\mathbf{x}_{0i}(l) \cdot \mathbf{x}_{0j}(l) + a^2)^2 \right]. \quad (15.100)$$

The above result is obtained by noticing that,

$$\mathbf{X}_{01} = \mathbf{X}_1 - \mathbf{X}_0 = (-a, -a, -a) - (0, 0, 0) = (-a, -a, -a),$$
$$\Longrightarrow \mathbf{X}_{01}^2 = 3a^2, \quad (15.101)$$
$$\lambda_{mm} = \left(\mathbf{x}_{0m}^2 - \mathbf{X}_{0m}^2\right)/(2a) = \left(x_{0m}^2 - 3a^2\right)/(2a),$$

and X_{0i}^2 is the same for all i (evident from figure 15.3). Cubic symmetry reveals that the elastic coefficient $\alpha = B_{nnnn}(0)$ does not depend on n. The number of terms in the summation over m, n is 14 since the prime implies omitting terms $(m, n) = (3, 4)$ and $(m, n) = (4, 3)$. Four of the 14 terms are the diagonal B_{nnnn} terms The remaining 10 terms $B_{nmnm}(n \neq m)$ (and 10 terms $\mathbf{X}_{0m} \cdot \mathbf{X}_{0n}$) are independent of m and n due to cubic symmetry (such that $B_{1212} = B_{3434}$ and $\mathbf{X}_{01} \cdot \mathbf{X}_{02} = \mathbf{X}_{03} \cdot \mathbf{X}_{04}$, etc). Since,

$$\mathbf{X}_{01} \cdot \mathbf{X}_{02} = (-a, -a, -a) \cdot (-a, a, a) = -a^2,$$
$$\lambda_{mn(m \neq n)} = (\mathbf{x}_{0m} \cdot \mathbf{x}_{0n} - \mathbf{X}_{0m} \cdot \mathbf{X}_{0n})/(2a) = (\mathbf{x}_{0m} \cdot \mathbf{x}_{0n} + a^2)/(2a), \quad (15.102)$$

and the summation in the second term $(i, j > i)$ in equation (15.100) covers the six pairs: $(1, 2)$, $(1, 3)$, $(1, 4)$, $(2, 3)$, $(2, 4)$, $(3, 4)$ [note that the term $(m, n) = (3, 4)$ was reintroduced (prime left out from the summation)], i.e.,

$$\sum_{m,n(m \neq n)}^{4} {}'B_{mnmn}\lambda_{mn}(l)^2 = \frac{10}{6} \frac{1}{4a^2} \sum_{i,j>i}^{4} B_{ijij}(\mathbf{x}_{0i}(l) \cdot \mathbf{x}_{0j}(l) + a^2)^2 = \frac{\beta}{2a^2} \sum_{i,j>i}^{4} (\mathbf{x}_{0i}(l) \cdot \mathbf{x}_{0j}(l) + a^2)^2, \quad (15.103)$$

where $\beta = \frac{5}{6}B_{ijij}(i \neq j)$. Note that the ratio 10/6 appears because the sum $\sum_{m,n}^{4}{}'$ in equation (15.99) contains 10 terms while the second sum $\sum_{i,j>i}^{4}$ in equation (15.100) only contains 6 terms.

Further, with reference to figure 15.3,

$$\mathbf{x}_{01} = \mathbf{x}_1 - \mathbf{x}_0 = (u_1, v_1, w_1) - (a + u_0, a + v_0, a + w_0) = (u_1 - u_0 - a, v_1 - v_0 - a, w_1 - w_0 - a)$$
$$= (u_{01} - a, v_{01} - a, w_{01} - a) \Longrightarrow x_{01}^2 - 3a^2 \approx -2au_{01} - 2av_{01} - 2aw_{01} \Longrightarrow \quad (15.104)$$
$$\left(x_{01}^2 - 3a^2\right)^2 \approx 4a^2(u_{01} + v_{01} + w_{01})^2.$$

Similarly,

$$\mathbf{x}_{02} = \mathbf{x}_2 - \mathbf{x}_0 = (u_2, 2a + v_2, 2a + w_2) - (a + u_0, a + v_0, a + w_0) = (u_2 - u_0 - a, v_2 - v_0 + a, w_2 - w_0 + a)$$
$$= (u_{02} - a, v_{02} + a, w_{02} + a) \Longrightarrow x_{02}^2 - 3a^2 \approx -2au_{02} + 2av_{02} + 2aw_{02} \Longrightarrow \quad (15.105)$$
$$\left(x_{02}^2 - 3a^2\right)^2 \approx 4a^2(u_{02} - v_{02} - w_{02})^2,$$

and,

$$\mathbf{x}_{01} \cdot \mathbf{x}_{02} = (u_{01} - a, v_{01} - a, w_{01} - a) \cdot (u_{02} - a, v_{02} + a, w_{02} + a)$$
$$\approx -(u_{01} + u_{02})a + (v_{01} - v_{02})a + (w_{01} - w_{02})a - a^2 = -(u_{01} + u_{02})a + v_{21}a + w_{21}a - a^2 \quad (15.106)$$
$$= -(u_{01} + u_{02} + v_{12} + w_{12})a - a^2 \Longrightarrow (\mathbf{x}_{01} \cdot \mathbf{x}_{02} + a^2)^2 = a^2(u_{01} + u_{02} + v_{12} + w_{12})^2.$$

etc.

Using the above relations in equation (15.100) leads to,

$$V = \frac{1}{2}\alpha \sum_l \left\{ (u_{01} + v_{01} + w_{01})^2 + (u_{02} - v_{02} - w_{02})^2 + (u_{03} + v_{03} - w_{03})^2 + (u_{04} - v_{04} + w_{04})^2 \right\}$$

$$+ \frac{1}{4}\beta \sum_l \left\{ (u_{01} + u_{02} + v_{12} + w_{12})^2 + (u_{32} - v_{32} + w_{02} + w_{03})^2 + (u_{03} + u_{04} + v_{43} - w_{43})^2 + (u_{42} + v_{02} + v_{04} - w_{42})^2 \right. \tag{15.107}$$

$$\left. + (u_{31} - v_{01} - v_{03} - w_{31})^2 + (u_{41} + v_{41} - w_{01} - w_{04})^2 \right\} + O(u^3).$$

An expression for the strain energy density U is also needed. Figure 15.3 shows that four 'blue atoms' belong to a volume $(4a)^3 = 64a^3$. Hence, the primitive cell volume containing one 'blue atom' (and one 'yellow atom') has the volume $16a^3$. The strain energy density U is then obtained from V by removing the summation over all primitive cells and dividing by $16a^3$,

$$U = \frac{\alpha}{32a^3} \left\{ (u_{01} + v_{01} + w_{01})^2 + (u_{02} - v_{02} - w_{02})^2 + (u_{03} + v_{03} - w_{03})^2 + (u_{04} - v_{04} + w_{04})^2 \right\}$$

$$+ \frac{\beta}{64a^3} \left\{ (u_{01} + u_{02} + v_{12} + w_{12})^2 + (u_{32} - v_{32} + w_{02} + w_{03})^2 + (u_{03} + u_{04} + v_{43} - w_{43})^2 + (u_{42} + v_{02} + v_{04} - w_{42})^2 \right. \tag{15.108}$$

$$\left. + (u_{31} - v_{01} - v_{03} - w_{31})^2 + (u_{41} + v_{41} - w_{01} - w_{04})^2 \right\} + O(u^3).$$

15.7.2 Internal strain

In order to proceed, it is necessary to define the displacement *between* sublattices. As we know, strain is defined uniquely by the lattice configuration. Hence, strain does not *a priori* specify the displacement between sublattices (defined by the A and B atoms). The displacement between sublattices is defined as the internal strain. To be precise, for a deformed cubic crystal the displacement of the center atom 0 is defined to be (u', v', w') where u', v', w' are the internal strain components along the x, y, z axes, respectively.

Hence, the differences in atomic positions of atoms 1 and 0 are,

$$u_{01} = u_1 - u_0 = -a\frac{\partial u}{\partial x} - a\frac{\partial u}{\partial y} - a\frac{\partial u}{\partial z} - u',$$

$$v_{01} = v_1 - v_0 = -a\frac{\partial v}{\partial x} - a\frac{\partial v}{\partial y} - a\frac{\partial v}{\partial z} - v', \tag{15.109}$$

$$w_{01} = w_1 - w_0 = -a\frac{\partial w}{\partial x} - a\frac{\partial w}{\partial y} - a\frac{\partial w}{\partial z} - w'.$$

Similarly, the differences in atomic positions of atoms 2 and 0 are,

$$u_{02} = u_2 - u_0 = -a\frac{\partial u}{\partial x} + a\frac{\partial u}{\partial y} + a\frac{\partial u}{\partial z} - u',$$

$$v_{02} = v_2 - v_0 = -a\frac{\partial v}{\partial x} + a\frac{\partial v}{\partial y} + a\frac{\partial v}{\partial z} - v', \tag{15.110}$$

$$w_{02} = w_2 - w_0 = -a\frac{\partial w}{\partial x} + a\frac{\partial w}{\partial y} + a\frac{\partial w}{\partial z} - w'.$$

With these expressions for atomic displacements, the first term in the brackets of the α term in equation (15.108) becomes,

$$u_{01} + v_{01} + w_{01} = -a\left(\frac{\partial u}{\partial x} + \frac{\partial v}{\partial y} + \frac{\partial w}{\partial z}\right) - a\left(\frac{\partial v}{\partial z} + \frac{\partial w}{\partial y}\right)$$

$$-a\left(\frac{\partial w}{\partial x} + \frac{\partial u}{\partial z}\right) - a\left(\frac{\partial u}{\partial y} + \frac{\partial v}{\partial x}\right) - u' - v' - w' \tag{15.111}$$

$$= -ae_d - ae_{yz} - ae_{xz} - ae_{xy} - u' - v' - w',$$

where the strain coefficients are,

$$e_d = \frac{\partial u}{\partial x} + \frac{\partial v}{\partial y} + \frac{\partial w}{\partial z},$$

$$e_{xx} = \frac{\partial u}{\partial x},$$

$$e_{yy} = \frac{\partial v}{\partial y},$$

$$e_{zz} = \frac{\partial w}{\partial z}, \tag{15.112}$$

$$e_{yz} = \frac{\partial v}{\partial z} + \frac{\partial w}{\partial y},$$

$$e_{xz} = \frac{\partial u}{\partial z} + \frac{\partial w}{\partial x},$$

$$e_{xy} = \frac{\partial u}{\partial y} + \frac{\partial v}{\partial x}.$$

Similarly, the second term in the brackets of the α term in equation (15.108) becomes,

$$u_{02} - v_{02} - w_{02} = -a\frac{\partial u}{\partial x} + a\frac{\partial u}{\partial y} + a\frac{\partial u}{\partial z} + a\frac{\partial v}{\partial x} - a\frac{\partial v}{\partial y} - a\frac{\partial v}{\partial z}$$

$$+ a\frac{\partial w}{\partial x} - a\frac{\partial w}{\partial y} - a\frac{\partial w}{\partial z} - u' + v' + w' \tag{15.113}$$

$$= -ae_d + ae_{xz} + ae_{xy} - ae_{yz} - u' + v' + w'.$$

Continuing, note first that,

$$\mathbf{X}_{12} = \mathbf{X}_2 - \mathbf{X}_1 = (0, 2a, 2a) - (0, 0, 0) = (0, 2a, 2a), \tag{15.114}$$

and,

$$u_{12} = u_2 - u_1 = 2a\frac{\partial u}{\partial y} + 2a\frac{\partial v}{\partial z},$$

$$v_{12} = v_2 - v_1 = 2a\frac{\partial v}{\partial y} + 2a\frac{\partial v}{\partial z}, \tag{15.115}$$

$$w_{12} = w_2 - w_1 = 2a\frac{\partial w}{\partial y} + 2a\frac{\partial w}{\partial z}.$$

The internal strain components disappear in the expressions of displacement differences between blue (or yellow) atoms as they sit on the same sublattice.

The first term in the brackets of the β term in equation (15.108) can then be written as,

$$
\begin{aligned}
u_{01} + u_{02} + v_{12} + w_{12} = & -a\frac{\partial u}{\partial x} - a\frac{\partial u}{\partial y} - a\frac{\partial u}{\partial z} \\
& - a\frac{\partial u}{\partial x} + a\frac{\partial u}{\partial y} + a\frac{\partial u}{\partial z} + 2a\frac{\partial v}{\partial y} + 2a\frac{\partial v}{\partial z} + 2a\frac{\partial w}{\partial y} + 2a\frac{\partial w}{\partial z} - 2u' \\
= & -2a\left(\left[\frac{\partial u}{\partial x} - \frac{\partial v}{\partial y} - \frac{\partial w}{\partial z}\right] - \left[\frac{\partial v}{\partial z} + \frac{\partial w}{\partial y}\right] + \frac{u'}{a}\right) \\
= & -2a\left(e_{xx} - e_{yy} - e_{zz} - e_{yz} + \frac{u'}{a}\right).
\end{aligned}
\tag{15.116}
$$

Combining the above equations and similar expressions for the other terms in the equation for U allows us to write,

$$
\begin{aligned}
U = \frac{\alpha}{32a}\Bigg\{ & \left(e_d + e_{yz} + e_{xz} + e_{xy} + \frac{u' + v' + w'}{a}\right)^2 + \left(e_d + e_{yz} - e_{xz} - e_{xy} + \frac{u' - v' - w'}{a}\right)^2 \\
& \left(e_d - e_{yz} - e_{xz} + e_{xy} - \frac{u' + v' - w'}{a}\right)^2 + \left(e_d - e_{yz} + e_{xz} - e_{xy} - \frac{u' - v' + w'}{a}\right)^2 \Bigg\} \\
+ \frac{\beta}{16a}\Bigg\{ & \left(e_{xx} - e_{yy} - e_{zz} - e_{yz} + \frac{u'}{a}\right)^2 + \left(e_{xx} + e_{yy} - e_{zz} - e_{xy} + \frac{w'}{a}\right)^2 + \left(e_{xx} - e_{yy} - e_{zz} + e_{yz} - \frac{u'}{a}\right)^2 \\
& + \left(e_{xx} - e_{yy} + e_{zz} - e_{xz} + \frac{v'}{a}\right)^2 + \left(e_{xx} - e_{yy} + e_{zz} + e_{xz} - \frac{v'}{a}\right)^2 + \left(e_{xx} + e_{yy} - e_{zz} + e_{xy} - \frac{w'}{a}\right)^2 \Bigg\}.
\end{aligned}
\tag{15.117}
$$

Since equilibrium corresponds to a minimum in the elastic energy as a function of the internal strain (relaxation of internal strain), the conditions,

$$
\frac{\partial U}{\partial u'} = \frac{\partial U}{\partial v'} = \frac{\partial U}{\partial w'} = 0,
\tag{15.118}
$$

must be imposed. Performing the calculation $\frac{\partial U}{\partial u'} = 0$ explicitly yields,

$$
\frac{\alpha}{32a}\left(8e_{yz} + 8\frac{u'}{a}\right)\frac{1}{a} + \frac{\beta}{16a}\left(-4e_{yz} + 4\frac{u'}{a}\right)\frac{1}{a} = 0 \Longrightarrow
$$

$$
u' = -a\frac{\alpha - \beta}{\alpha + \beta}e_{yz} = -a\zeta e_{yz},
\tag{15.119}
$$

where,

$$
\zeta = \frac{\alpha - \beta}{\alpha + \beta}.
\tag{15.120}
$$

Similarly, the two other equilibrium conditions (in v', w') give,

$$
\begin{aligned}
v' &= -a\zeta e_{xz}, \\
w' &= -a\zeta e_{xy}.
\end{aligned}
\tag{15.121}
$$

Hence, the internal strains u', v', w' are entirely fixed by the shear strain components e_{yz}, e_{xz}, e_{xy}, and the total elastic energy density U in equilibrium is a function of the strains alone.

Inserting expressions for the internal strains u', v', w' in equation (15.108) and comparing with the familiar expression of the elastic energy density in terms of elastic stiffnesses,

$$U = \frac{1}{2}c_{11}\left(e_{xx}^2 + e_{yy}^2 + e_{zz}^2\right) + c_{12}(e_{yy}e_{zz} + e_{xx}e_{zz} + e_{xx}e_{yy}) + \frac{1}{2}c_{44}\left(e_{yz}^2 + e_{xz}^2 + e_{xy}^2\right), \quad (15.122)$$

yields,

$$c_{11} = \frac{\alpha + 3\beta}{4a},$$

$$c_{12} = \frac{\alpha - \beta}{4a}, \quad (15.123)$$

$$c_{44} = \frac{\alpha\beta}{a(\alpha + \beta)}.$$

Since a cubic crystal from symmetry has three *independent* elastic coefficients this latter result can only be approximative. An 'artificial coupling' among the elastic constants is obtained by augmenting α and β from two of the three equations in equation (15.123) and inserting in the third equation,

$$2c_{44}(c_{11} + c_{12}) = (c_{11} - c_{12})(c_{11} + 3c_{12}). \quad (15.124)$$

While the latter coupling is not exact the relation is expected to be quite accurate. Deviations must be due to elastic couplings between second-nearest, third-nearest, etc atoms that were neglected in writing down the approximative strain energy in equation (15.100).

In table 15.2, a quantitative comparison is listed of how accurate the Keating model is for three diamond-like crystals: Diamond, silicon, and germanium. The accuracy is indeed satisfactory.

15.8 equation-of-motion and dynamical matrix

In this section, the equation-of-motion is determined using the dynamical matrix formulation for the motion of crystal atoms. Consider a perfect crystal with periodic boundary conditions. We define $\mathbf{R}_\alpha$ as the position vectors to the αth atom within an arbitrarily chosen unit cell (designated the zeroth unit cell), and the position vector to the nth unit cell is denoted by $\mathbf{R}_n$. Then, the equilibrium position of atom α in unit cell n can be written as $\mathbf{R}_{n\alpha} = \mathbf{R}_n + \mathbf{R}_\alpha$. Denoting the Cartesian components of the displacement vector of atom α in unit cell n by $s_{n\alpha i}$, where i takes the three values x, y, z, the kinetic energy becomes,

$$\mathcal{T} = \sum_{n=1}^{N} \sum_{\alpha=1}^{r} \sum_{i=1}^{3} \frac{M_\alpha}{2}\left(\frac{ds_{n\alpha i}(t)}{dt}\right)^2. \quad (15.125)$$

Table 15.2. Accuracy of the approximative relation, equation (15.124), between elastic coefficients in the Keating model for three diamond-like cubic crystals, diamond, silicon, and germanium. Units for elastic stiffnesses c_{11}, c_{12}, c_{44} are in GPa. Force constants α, β are in N m^{-1}. Values are taken from reference [9].

	Diamond	Silicon	Germanium
$\dfrac{2c_{44}(c_{11}+c_{12})}{(c_{11}-c_{12})(c_{11}+3c_{12})}$	0.99	0.99	1.07
α	129	48.5	38
β	85	13.8	12
c_{11} (expt)	107.6	16.6	12.9
c_{11} (theoret)	107.6	16.6	13.1
c_{12} (expt)	12.5	6.4	4.8
c_{12} (theoret)	12.5	6.4	4.6
c_{44} (expt)	57.6	7.9	6.7
c_{44} (theoret)	57.5	7.9	6.5

Here, M_α is the mass of atom α, and N is the number of unit cells, r the number of atoms per unit cell, and i stands for the three Cartesian coordinates x, y, z. Hence the total number of degrees of freedom is $3rN$. The potential energy $\mathcal{W}$ can be expanded into a Taylor series in the displacement vectors,

$$\mathcal{W}(\mathbf{x}) = \mathcal{W}(\mathbf{R}_{n\alpha}) + \sum_{n\alpha i}\left(\frac{\partial \mathcal{W}(\mathbf{x})}{\partial s_{n\alpha i}}\right)_{\mathbf{x}=\mathbf{R}_{n\alpha}} s_{n\alpha i} + \frac{1}{2}\sum_{n\alpha i}\sum_{n'\alpha'i'}\left(\frac{\partial^2 \mathcal{W}(\mathbf{x})}{\partial s_{n\alpha i}\partial s_{n'\alpha'i'}}\right)_{\mathbf{x}=\mathbf{R}_{n\alpha}} s_{n\alpha i} s_{n'\alpha'i'} + O(s^3). \quad (15.126)$$

The linear terms in s vanish if expansion is made around the equilibrium configuration (vanishing forces). Introducing the force constants,

$$\Phi_{n\alpha i}^{n'\alpha'i'} = \sum_{n\alpha i}\sum_{n'\alpha'i'}\left(\frac{\partial^2 \mathcal{W}(\mathbf{x})}{\partial s_{n\alpha i}\partial s_{n'\alpha'i'}}\right)_{\mathbf{x}=\mathbf{R}_{n\alpha}}. \quad (15.127)$$

Note that higher terms than the harmonic ones are neglected. In order to obtain the equations-of-motion, the Lagrange function $\mathcal{L} = \mathcal{T} - \mathcal{W}$ is defined, and the Euler–Lagrange equations become,

$$\frac{d}{dt}\frac{\partial \mathcal{L}}{\partial \dot{s}_{n\alpha i}} - \frac{\partial \mathcal{L}}{\partial s_{n\alpha i}} = 0. \quad (15.128)$$

Inserting the expressions for the kinetic energy $\mathcal{T}$ and the potential expansion $\mathcal{W}$ into the Euler–Lagrange equation, and taking account of the symmetry of the force constants, $\Phi_{n\alpha i}^{n'\alpha'i'} = \Phi_{n'\alpha'i'}^{n\alpha i}$, leads to,

$$M_\alpha \frac{d^2 s_{n\alpha i}}{dt^2} = -\sum_{n'\alpha'i'}\Phi_{n\alpha i}^{n'\alpha'i'} s_{n'\alpha'i'}. \quad (15.129)$$

With the ansatz,

$$s_{n\alpha i} = \frac{1}{\sqrt{M_\alpha}} u_{n\alpha i} e^{-i\omega t}, \tag{15.130}$$

the equations-of-motion is,

$$\omega^2 u_{n\alpha i} = \sum_{n'\alpha'i'} \frac{\Phi_{n\alpha i}^{n'\alpha'i'}}{\sqrt{M_\alpha M_{\alpha'}}} u_{n'\alpha'i'}, \tag{15.131}$$

which is an eigenvalue equation for the $3rN$ normal frequencies ω. From translational symmetry, the force constants $\Phi_{n\alpha i}^{n'\alpha'i'}$, which can be identified as the ith component of the force on atom α in unit cell n when atom α' in unit cell n' is displaced a unit distance in the i' direction, only depend on the difference $n - n'$, i.e., $\Phi_{n\alpha i}^{n'\alpha'i'} = \Phi_{\alpha i}^{\alpha'i'}(n - n')$. In order to take account of the translational symmetry, the ansatz,

$$u_{n\alpha i} = c_{\alpha i} e^{-i\mathbf{q}\cdot\mathbf{R}_n}, \tag{15.132}$$

is made, and,

$$
\begin{aligned}
\omega^2 c_{\alpha i} e^{-i\mathbf{q}\cdot\mathbf{R}_n} &= \sum_{n'\alpha'i'} \frac{\Phi_{\alpha i}^{\alpha'i'}(n - n')}{\sqrt{M_\alpha M_{\alpha'}}} c_{\alpha'i'} e^{-i\mathbf{q}\cdot\mathbf{R}_n'} \\
&= \sum_{n'\alpha'i'} \frac{\Phi_{\alpha i}^{\alpha'i'}(n - n')}{\sqrt{M_\alpha M_{\alpha'}}} c_{\alpha'i'} e^{-i\mathbf{q}\cdot\mathbf{R}_n} e^{i\mathbf{q}\cdot\left(\mathbf{R}_n - \mathbf{R}_n'\right)} \Longrightarrow \\
\omega^2(\mathbf{q}) c_{\alpha i} &= \sum_{n'\alpha'i'} \frac{\Phi_{\alpha i}^{\alpha'i'}(n - n')}{\sqrt{M_\alpha M_{\alpha'}}} c_{\alpha'i'} e^{i\mathbf{q}\cdot\left(\mathbf{R}_n - \mathbf{R}_n'\right)} \\
&= \sum_{n\alpha'i'} \frac{\Phi_{\alpha i}^{\alpha'i'}(n)}{\sqrt{M_\alpha M_{\alpha'}}} c_{\alpha'i'} e^{i\mathbf{q}\cdot\mathbf{R}_n} \\
&= \sum_{\alpha'i'} \left[\sum_n \frac{\Phi_{\alpha i}^{\alpha'i'}(n)}{\sqrt{M_\alpha M_{\alpha'}}} c_{\alpha'i'} e^{i\mathbf{q}\cdot\mathbf{R}_n} \right] \\
&= \sum_{\alpha'i'} \left[\sum_n \frac{\Phi_{\alpha i}^{\alpha'i'}(n)}{\sqrt{M_\alpha M_{\alpha'}}} e^{i\mathbf{q}\cdot\mathbf{R}_n} \right] c_{\alpha'i'} \\
&= \sum_{\alpha'i'} D_{\alpha i}^{\alpha'i'}(\mathbf{q}) c_{\alpha'i'},
\end{aligned}
\tag{15.133}
$$

where the dynamical matrix is introduced as,

$$D_{\alpha i}^{\alpha'i'}(\mathbf{q}) = \sum_n \frac{\Phi_{\alpha i}^{\alpha'i'}(n)}{\sqrt{M_\alpha M_{\alpha'}}} e^{i\mathbf{q}\cdot\mathbf{R}_n}. \tag{15.134}$$

Thus, an eigenvalue equation [equation (15.133)] is obtained, which is only three-dimensional, but now the eigenfrequencies ω are functions of $\mathbf{q}$, which is a wave vector within the first Brillouin zone.

Note that a static (dc) strain configuration (eigenvector $c_{\alpha i}$) of the atomic positions corresponds to $\omega = 0$. Phonon properties are found by searching for ω_n solutions as a function of $\mathbf{q}$. The different phonon bands are assigned different n values.

15.8.1 Potential energy of the wurtzite crystal

With the above preliminaries, the short-range part of the potential energy of a wurtzite crystal is determined following Nusimovici and Birman [10],

$$
\begin{aligned}
V^{SR} &= \frac{1}{2} \sum_{1 \, n.n.} \lambda (\delta r_{lm})^2 + \frac{1}{2} \sum_{2 \, n.n.cat.} \mu (\delta r_{mo})^2 + \frac{1}{2} \sum_{2 \, n.n.an.} \nu (\delta r_{ln})^2 \\
&+ \frac{1}{2} \sum_{3 \, n.n.} \delta (\delta r_{lo})^2 + \frac{1}{2} \sum_{an.-cat.-an.} k_\theta r_0^2 (\delta \theta_{lmn})^2 \\
&+ \frac{1}{2} \sum_{cat.-an.-cat.} k_{\theta'} r_0^2 (\delta \theta_{mno})^2 + \frac{1}{2} \sum_{cat.-an.-cat.} k_{r\theta'} r_0^2 (\delta \theta_{mno})(\delta r_{mn}) \\
&+ \frac{1}{2} \sum_{an.-cat.-an.} k_{r\theta} r_0^2 (\delta \theta_{lmn})(\delta r_{lm}).
\end{aligned}
\tag{15.135}
$$

The spring constants in equation (15.135) are λ, μ, ν, δ, k_θ, $k_{\theta'}$, $k_{r\theta'}$, $k_{r\theta}$. The spring constants λ, μ, ν, δ are two-body interactions which oppose bond extensions. These represent springs opposing first-neighbor cation–anion extension (λ), second-neighbor cation–cation bond extension (μ), second-neighbor anion–anion bond extension (ν), and third-neighbor cation–anion bond extension (δ). The springs k_θ and $k_{\theta'}$ represent three-body interactions opposing changes in bond angles: k_θ refers to changes in first-neighbor angles anion-cation–anion while k_θ' to refers to changes in first-neighbor angles cation-anion-cation. The constants $k_{r\theta}$ and $k_{r\theta'}$ are cross coupling constants.

The potential-energy function V^{SR} includes interactions of a given ion with its neighbors in wurtzite up to and including third neighbors, as indicated under the summation indices in equation (15.135).

To utilize equation (15.135) in the analysis, the derivatives in equation (15.127) must be evaluated, or equivalently, transformed from the set of independent valence force-field variables to the usual Cartesian displacement variables $s_{n\alpha i}$. The short-range force constants are given by [11, 12],

$$
\begin{aligned}
\frac{\partial (\delta r_{lm})}{\partial s_i(l)} &= \frac{s_i(l) - s_i(m)}{r_0^2}, \\
\frac{\partial (\theta_{lmn})}{\partial s_i(l)} &= \frac{1}{2\sqrt{2}\, r_0^2} [3(s_i(n) - s_i(m)) + (s_i(l) - s_i(m))], \\
\frac{\partial (\theta_{lmn})}{\partial s_{i'}(m)} &= \frac{2}{\sqrt{2}\, r_0^2} [s_{i'}(l) + s_{i'}(n) - 2s_{i'}(m)].
\end{aligned}
\tag{15.136}
$$

In equation (15.136) the atom index pair n, α were replaced by a single index l, m, n, o and r_0 is the initial distance between the first neighbors. Using equation

(15.136) the elementary short-range force matrices are easily obtained from equation (15.135). Similarly, the corresponding short-range part of the dynamical matrix $D_{ii'}^{SR}(\mathbf{q}, \alpha\alpha')$ are found from equation (15.134). In constructing the latter, it is assumed that the force matrix connecting atom pairs n, α and n', α', which are farther apart than third neighbors, vanishes identically. The evaluation of the short-range force matrices is elementar, and we refer to examples given by Merten [13]. Note also that use is made of the ideal tetrahedral arrangement of the first neighbors about each ion, so that the last two equations in equation (15.136) must be modified for non-ideal wurtzite structures.

15.8.2 Coulomb contribution

To determine the Coulomb part of the force matrix, the cation and anion are taken to be point ions of (unknown) charge $\pm q_k$, respectively. When the ions are at instantaneous positions $\mathbf{r}(n, i)$, see (3.1), the Coulomb contribution to the potential energy in wurtzite is,

$$\mathcal{W}^C = \frac{1}{4\pi\epsilon} \sum_{n\alpha;n'\alpha'}{}' \frac{q_\alpha q_{\alpha'}}{|\mathbf{r}(n, \alpha) - \mathbf{r}(n', \alpha')|}, \tag{15.137}$$

where ϵ is the permittivity of the material. In equation (15.137) the prime signifies that the divergent term due to the Coulomb self-interaction of a point charge must be omitted. The force-constant matrix element corresponding to equation (15.137) is,

$$\Phi_{ii'}^C(nn', \alpha\alpha') = \left[\frac{\partial^2 \mathcal{W}^C}{\partial s_i(n, \alpha)\partial s_{i'}(n', \alpha')} \right]_0. \tag{15.138}$$

The corresponding Coulomb contribution to the dynamical matrix is,

$$\begin{aligned} D_{ii'}^C(\mathbf{q}, \alpha\alpha') &= \sum_{n-n'} \frac{\Phi_{ii'}^C(n - n', \alpha\alpha')}{\sqrt{M_\alpha M_{\alpha'}}} e^{-i\mathbf{q}\cdot\mathbf{R}_{n-n'}} \\ &= \frac{q_\alpha q_{\alpha'}}{4\pi\epsilon\sqrt{M_\alpha M_{\alpha'}}} \sum_{n-n'}{}' e^{-i\mathbf{q}\cdot\mathbf{R}_{n-n'}} \left[\frac{\partial^2}{\partial s_i(n, \alpha)\partial s_{i'}(n', \alpha')} \left(\frac{1}{|\mathbf{r}(n, \alpha) - \mathbf{r}(n', \alpha')|} \right) \right], \end{aligned} \tag{15.139}$$

where $\mathbf{R}_{n-n'} = \mathbf{R}_n - \mathbf{R}_{n'}$. The element $D_{ii'}^C(\mathbf{q}, \alpha\alpha')$ is interpreted as the nth component of the electric field at site $\mathbf{r}(\alpha\alpha') = \mathbf{r}(\alpha) - \mathbf{r}(\alpha')$ when a fictitious dipole wave of polarization i', wave vector $\mathbf{q}$, magnitude $\frac{q_\alpha q_{\alpha'}}{4\pi\epsilon\sqrt{M_\alpha M_{\alpha'}}}$ exists at each lattice site of the hexagonal Bravais lattice. Thus, the equivalent electric field is defined as the dot product of a dyadic with a vector representing the magnitude and polarization of a fictitious dipole. Writing,

$$\mathbf{p} = \rho\hat{\mathbf{p}}, \tag{15.140}$$

where $\rho = \frac{q_\alpha q_{\alpha'}}{4\pi\epsilon\sqrt{M_\alpha M_{\alpha'}}} = -\frac{q^2}{4\pi\epsilon\sqrt{M_\alpha M_{\alpha'}}}$ and $\hat{\mathbf{p}}$ is a unit polarization vector, we can write,

$$D_{ii'}^C(\mathbf{q}, \alpha\alpha') = (\mathbf{B}(\mathbf{q}, \alpha\alpha') \cdot \rho\hat{p}_{i'})_i, \tag{15.141}$$

where the basic quantity to compute is,

$$\mathbf{B}(\mathbf{q}, \alpha\alpha') = \sum_{n-n'}{}' e^{-i\mathbf{q}\cdot\mathbf{R}_{n-n'}}\left[\nabla\nabla\frac{1}{|\mathbf{r} - \mathbf{R}_{n-n'}|}\right]_0. \tag{15.142}$$

The gradient in equation (15.142) is to be taken with respect to $\mathbf{r}$; the sum is over a hexagonal lattice. The evaluation of equation (15.142) requires special care [14]. The most exact and general procedure utilizes the Ewald transformation [15, 16]. Once the parameter ρ is determined, and the quantities in equation (15.142) computed, the contribution of the Coulomb field to the dynamical matrix element is completely determined.

15.9 The dynamical matrix

Owing to the four-atom basis (cation I, cation II, anion I, anion II) of wurtzite, there are 12 dynamical degrees of freedom per cell. Hence equation (15.132) is a 12×12 matrix. In sphalerite with a two-atom basis, the secular problem, equation (15.133), is 6×6. To compose the complete dynamical matrix from its constituents we write

$$\begin{aligned} D_{ii'}(\mathbf{q}, \alpha\alpha') &= D_{ii'}^{SR}(\mathbf{q}, \alpha\alpha') + D_{ii'}^C(\mathbf{q}, \alpha\alpha'), \\ i, i' &= 1, 2, 3, \\ \alpha, \alpha' &= 1, 2, 3, 4, \end{aligned} \tag{15.143}$$

for wurtzite. For sphalerite (also known as zincblende), $\alpha, \alpha' = 1, 2$ only (figure 15.4).

In order to proceed with actual calculations, the numerical values of the nine parameters must be specified: eight short-range, plus one Coulomb parameter.

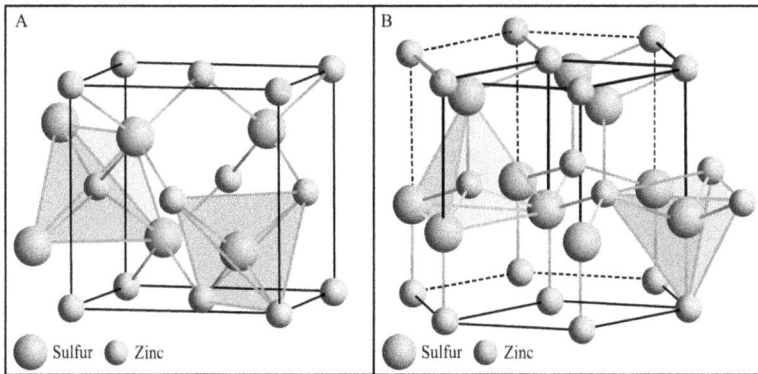

Figure 15.4. Crystal structures of the two main forms of ZnS, showing the tetrahedral relationship between both Zn and S atoms. (A) Cubic structure of sphalerite. (B) Hexagonal structure of wurtzite.

15.10 CdS wurtzite case

The values of the nine parameters were obtained by fitting the expressions for the $\mathbf{q} = 0$ solutions of the equation-of-motion, equation (15.133), to the experimentally observed one-phonon eigenfrequencies in CdS wurtzite [10]. In this way, the following parameter values are found,

$$
\begin{aligned}
\lambda &= 106.1 \text{ Nm}^{-1}, \\
\mu &= 20.3 \text{ Nm}^{-1}, \\
\nu &= -8.6 \text{ Nm}^{-1}, \\
\delta &= -8.5 \text{ Nm}^{-1}, \\
k_\theta &= 17.0 \text{ Nm}^{-1}, \\
k_{\theta'} &= 15.4 \text{ Nm}^{-1}, \\
K_{r\theta} &= k_{r\theta} + k_{r\theta'} = -2.16 \text{ Nm}^{-1}, \\
\rho &= 3.2 \text{ Nm}^{-1}.
\end{aligned}
\tag{15.144}
$$

These values were used in all calculations for wurtzite and sphalerite. For the sphalerite calculation, $\delta = 0$ was taken. Observe that the $\mathbf{q} = 0$ work only permits a determination of the sum of the cross-force constants $K_{r\theta} = k_{r\theta} + k_{r\theta'}$. Fortunately, it turns out that even for $\mathbf{q} \neq 0$ only the sum of these constants appears in the equations.

15.11 Theory of the local electric field

In this section we will derive an expression for the frequency-dependent dielectric constant of a cubic solid. In order to accomplish this we need a theory relating the polarization density $\mathbf{P}$ to the macroscopic electric field $\mathbf{E}$. Since each ion has microscopic dimensions, its displacement and distortion will be determined by the force due to the microscopic electric field at its position in addition to the microscopic field from the ion itself. This electric field is called the local electric field $\mathbf{E}^{\text{loc}}\mathbf{r}$. A macroscopic theory of the electric field can be used to evaluate the local electric field by dividing space into regions near to and far from $\mathbf{r}$. The far region contains all external sources, all points outside the crystal, and only points inside the crystal that are far from $\mathbf{r}$ compared to the dimensions r_0 of the averaging region defined by a weight function f (refer to figure 15.5),

$$
\mathbf{E}(\mathbf{r}) = \int d\mathbf{r}' \mathbf{E}(\mathbf{r} - \mathbf{r}') f(\mathbf{r}'),
$$

$$
\int d\mathbf{r} f(\mathbf{r}) = 1,
\tag{15.145}
$$

$$
f(\mathbf{r}) = 0, \quad r > r_0.
$$

All other points are said to be in the near region. The reason for this division is that the contribution to $\mathbf{E}^{\text{loc}}(\mathbf{r})$ of charges in the far region will vary negligibly over a distance r_0 about $\mathbf{r}$. Therefore the contribution to $\mathbf{E}^{\text{loc}}(\mathbf{r})$ of charges in the far region

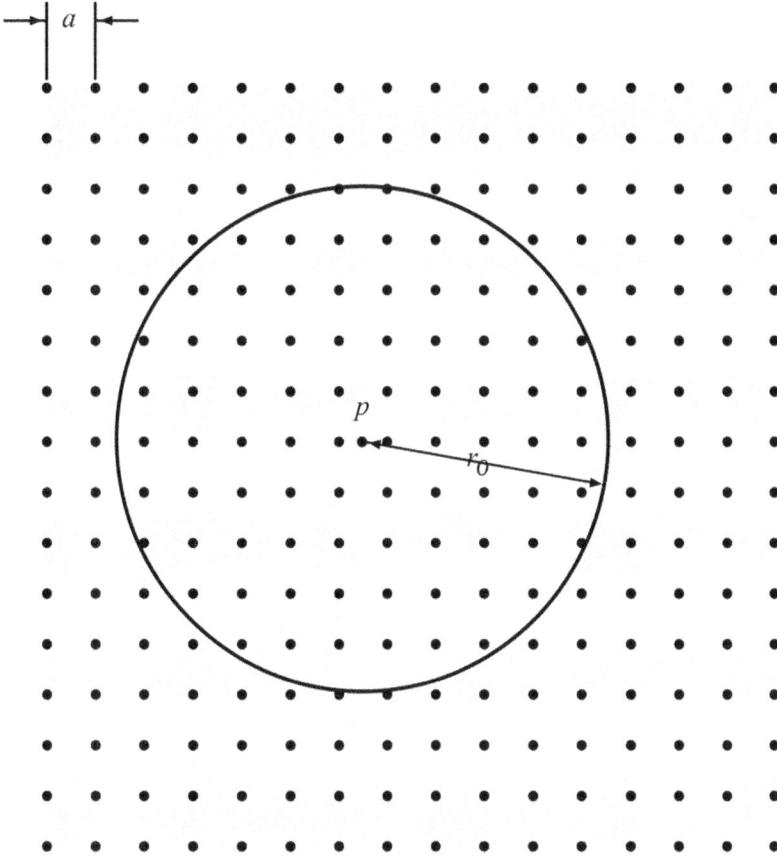

Figure 15.5. The value of a macroscopic quantity at a point P is an average of the microscopic quantity over a region of dimensions r_0 in the neighborhood of P where r_0 is large compared to the lattice spacing.

is just the macroscopic electric field, $\mathbf{E}_{far}^{macro}(\mathbf{r})$, that would exist at $\mathbf{r}$ if only charges in the far region were present,

$$\mathbf{E}^{loc}(\mathbf{r}) = \mathbf{E}_{near}^{loc}(\mathbf{r}) + \mathbf{E}_{far}^{micro}(\mathbf{r}) = \mathbf{E}_{near}^{loc}(\mathbf{r}) + \mathbf{E}_{far}^{macro}(\mathbf{r}). \tag{15.146}$$

Now, $\mathbf{E}$, the full macroscopic electric field at $\mathbf{r}$, is constructed by averaging the microscopic electric field within r_0 of $\mathbf{r}$ due to all charges, in both the near and the far regions, i.e.,

$$\mathbf{E}(\mathbf{r}) = \mathbf{E}_{far}^{macro}(\mathbf{r}) + \mathbf{E}_{near}^{macro}(\mathbf{r}), \tag{15.147}$$

where $\mathbf{E}_{near}^{macro}(\mathbf{r})$ is the macroscopic electric field that would exist at $\mathbf{r}$ if only the charges in the near region were present.

Combining equations (15.146) and (15.147), the expression for the local electric field can be written as,

$$\mathbf{E}^{loc}(\mathbf{r}) = \mathbf{E}(\mathbf{r}) + \mathbf{E}_{near}^{loc}(\mathbf{r}) - \mathbf{E}_{near}^{macro}(\mathbf{r}). \tag{15.148}$$

equation (15.148) will be used only for nonequilibrium configurations of the crystal with negligible spatial variation from cell to cell over distances of the order of the size of the near region. In this case, $E_{near}^{macro}(r)$ will be the macroscopic electric field due to a uniformly polarized medium whose shape is that of the near region. If the near region is chosen to be a sphere, then this field is given by the elementary result of electrostatics (the proof is given in the following section: 'Electric field from a polarized medium'),

$$E_{near}^{macro}(r) = -\frac{P(r)}{3\epsilon_0},$$ (15.149)

where P is the polarization density. Therefore, if the near region is a sphere over which P has negligible spatial variation, then equation (15.148) becomes

$$E^{loc}(r) = E(r) + E_{near}^{loc}(r) + \frac{P(r)}{3\epsilon_0}.$$ (15.150)

In order to proceed from here, additional simplifying assumptions will be made:
- The spatial dimensions and the displacement from equilibrium of each ion are considered to be so small that the polarizing field acting on it can be taken to be uniform over the whole ion and equal to the value of E^{loc} at the equilibrium position of the ion.
- The spatial dimensions and the displacement from equilibrium of each ion are considered to be so small that the contribution to the local electric field at the equilibrium position of the given ion, from the ion whose equilibrium position $R + d$, where R and d are the lattice point of the ion's cell and the position of the ion with respect to R, respectively, is accurately given by the electric field of a dipole moment $e(R + d)u(R + d) + p(R + d)$ where u is the displacement. Here, $p(R + d)$ is the dipole moment of the ion at $R + d$ in the absence of displacements.

Now, since the dipole moments of ions at equivalent sites are identical within the near region, the calculation of $E_{near}^{loc}(r)$ at equilibrium sites reduces to a lattice sum. Furthermore, in the special case in every equilibrium site in the crystal is a center of *cubic* symmetry, this lattice sum must vanish, i.e., $E_{near}^{loc}(r) = 0$ at every equilibrium site. Then, equation (15.150) reduces to

$$E^{loc}(r) = E(r) + \frac{P(r)}{3\epsilon_0}.$$ (15.151)

From the definition of the electric displacement,

$$P(r) = D(r) - \epsilon_0 E(r) = (\epsilon - \epsilon_0)E(r),$$ (15.152)

we have

$$E^{loc}(r) = E(r) + \frac{(\epsilon - \epsilon_0)E^{loc}(r)}{3\epsilon_0} = \frac{(\epsilon + 2\epsilon_0)E^{loc}(r)}{3\epsilon_0}.$$ (15.153)

Further, from the definition of the polarizability α,

$$\mathbf{P}(\mathbf{r}) = \frac{\alpha}{V_{\text{prim}}} \mathbf{E}^{\text{loc}}(\mathbf{r}), \qquad (15.154)$$

where V_{prim} is the volume of a primitive cell in the neighborhood of $\mathbf{r}$, we get

$$\frac{\alpha}{V_{\text{prim}}} = \frac{\mathbf{P}(\mathbf{r})}{\mathbf{E}^{\text{loc}}(\mathbf{r})} = \frac{(\epsilon - \epsilon_0)\mathbf{E}(\mathbf{r})}{\dfrac{\epsilon + 2\epsilon_0}{3\epsilon_0}\mathbf{E}(\mathbf{r})} \Longrightarrow$$

$$\frac{1}{3\epsilon_0} \frac{\alpha}{V_{\text{prim}}} = \frac{\epsilon - \epsilon_0}{\epsilon + 2\epsilon_0}, \qquad (15.155)$$

which is known as the Clausius–Mossotti relation.

15.12 Electric field from a polarized medium

We will give the full derivation of the electric field inside a uniformly polarized sphere. With reference to figure 15.6, the electric potential φ from an electric dipole moment $\mathbf{p} = q\mathbf{d}$ is,

$$\phi(\mathbf{x}) = \frac{1}{4\pi\epsilon_0}\left(\frac{q}{r_+} - \frac{q}{r_-}\right) = \frac{1}{4\pi\epsilon_0}\left(\frac{1}{\mathbf{x} - \mathbf{d}/2} - \frac{1}{\mathbf{x} + \mathbf{d}/2}\right). \qquad (15.156)$$

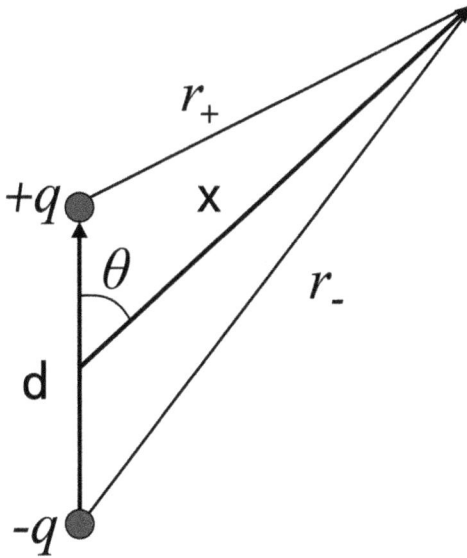

Figure 15.6. An electric dipole consists of two equal and opposite charges $+q$ and $-q$ separated by a displacement d.

In the far-field region for which $r = |\mathbf{x}| > > d$,

$$\phi(\mathbf{x}) \approx \frac{1}{4\pi\epsilon_0 r}\left[\left(1 + \frac{d}{2r}\cos\theta\right) - \left(1 - \frac{d}{2r}\cos\theta\right)\right] = \frac{1}{4\pi\epsilon_0 r}\frac{qd\cos\theta}{r^2} = \frac{1}{4\pi\epsilon_0}\frac{\mathbf{p}\cdot\mathbf{x}}{r^3}, \quad (15.157)$$

where θ is the angle between $\mathbf{x}$ and $\mathbf{d}$.

If an electric field is applied to a medium composed of many atoms or molecules, each atom or molecule forms a dipole $\mathbf{p}_i$ due to the field-induced displacement of the bound charges. The small dipoles are aligned along the direction of the field and the material becomes polarized. The polarization of a small volume element dV which contains many atoms (or molecules) is then (figure 15.7),

$$\mathbf{P} = \frac{1}{dV}\sum_i \mathbf{p}_i, \quad (15.158)$$

i.e.,

$$\mathbf{P}dV = \mathbf{P}(\mathbf{x}')d\mathbf{x}', \quad (15.159)$$

and the electric potential from a volume V becomes,

$$\phi(\mathbf{x}) = \frac{1}{4\pi\epsilon_0}\int\frac{\mathbf{P}(\mathbf{x}')\cdot(\mathbf{x} - \mathbf{x}')}{|\mathbf{x} - \mathbf{x}'|^3}d\mathbf{x}'. \quad (15.160)$$

The latter expression can be rewritten as,

$$\phi(\mathbf{x}) = \frac{1}{4\pi\epsilon_0}\int\mathbf{P}(\mathbf{x}')\cdot\nabla'\left(\frac{1}{|\mathbf{x} - \mathbf{x}'|}\right)d\mathbf{x}'. \quad (15.161)$$

Integration by parts yields,

$$\phi(\mathbf{x}) = \frac{1}{4\pi\epsilon_0}\left\{\int\nabla'\cdot\left(\frac{\mathbf{P}(\mathbf{x}')}{|\mathbf{x} - \mathbf{x}'|}\right)d\mathbf{x}' - \int\frac{\nabla'\cdot\mathbf{P}(\mathbf{x}')}{|\mathbf{x} - \mathbf{x}'|}d\mathbf{x}'\right\}, \quad (15.162)$$

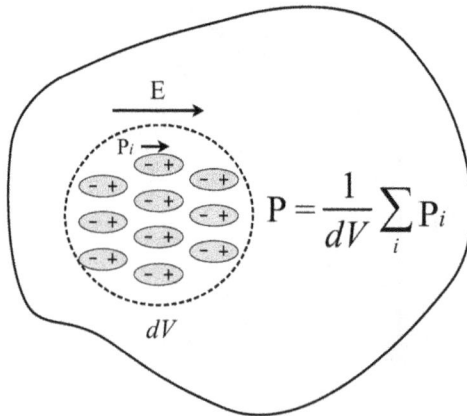

Figure 15.7. An external electric field induces electric polarization in a dielectric medium.

and using the divergence theorem,

$$\phi(\mathbf{x}) = \frac{1}{4\pi\epsilon_0}\left\{\int\frac{\mathbf{P}(\mathbf{x}')\cdot\mathbf{n}'}{|\mathbf{x}-\mathbf{x}'|}ds' - \int\frac{\nabla'\cdot\mathbf{P}(\mathbf{x}')}{|\mathbf{x}-\mathbf{x}'|}d\mathbf{x}'\right\}, \tag{15.163}$$

where ds' is a surface area element. Defining surface and volume charge densities,

$$\sigma_b = \mathbf{P}\cdot\mathbf{n}, \\ \rho_b = -\nabla\cdot\mathbf{P}, \tag{15.164}$$

the electric potential becomes,

$$\phi(\mathbf{x}) = \frac{1}{4\pi\epsilon_0}\left\{\int\frac{\sigma_b(\mathbf{x}')}{|\mathbf{x}-\mathbf{x}'|}ds' + \int\frac{\rho_b(\mathbf{x}')}{|\mathbf{x}-\mathbf{x}'|}d\mathbf{x}'\right\}. \tag{15.165}$$

15.12.1 Uniformly polarized sphere

In the case of a uniformly polarized sphere,

$$\rho_b = -\nabla\cdot\mathbf{P} = 0, \\ \sigma_b = \mathbf{P}\cdot\mathbf{n} = P\cos\theta', \\ ds' = Rd\theta'R\sin\theta'd\phi, \tag{15.166}$$

and equation (15.165) becomes,

$$\phi(\mathbf{x}) = \frac{1}{4\pi\epsilon_0}\int_0^{2\pi}d\phi\int_0^{\pi}\frac{P\cos\theta'}{|\mathbf{x}-\mathbf{x}'|}R^2\sin\theta'd\theta'd\phi. \tag{15.167}$$

By using the expansion,

$$\frac{1}{|\mathbf{x}-\mathbf{x}'|} = \sum_{i=0}^{\infty}\frac{x_<^i}{x_>^{i+1}}P_i(\cos\theta), \tag{15.168}$$

where θ is the angle between $\mathbf{x}$ and $\mathbf{x}'$ and P_i is the Legendre polynomial,

$$P_i(x) = \frac{1}{2^i i!}\frac{d^i}{dx^i}(x^2-1)^i, \tag{15.169}$$

and,

$$x_< = x, \ \text{if } x < x'; \ \ x_< = x', \ \text{if } x > x', \\ x_> = x, \ \text{if } x > x'; \ \ x_> = x', \ \text{if } x < x'. \tag{15.170}$$

If $\mathbf{x}$ is chosen to be along the z axis, then,

$$\theta = \theta', \tag{15.171}$$

and equation (15.167) for φ becomes, for $|\mathbf{x}| = r < |\mathbf{x}'| = R$,

$$\phi(\mathbf{x}) = \frac{1}{4\pi\epsilon_0} 2\pi \int_0^\pi \frac{1}{R} \sum_{i=0}^\infty \left(\frac{r}{R}\right)^i P_i(\cos\theta') P \cos\theta' R^2 \sin\theta' d\theta'$$

$$= \frac{1}{2\epsilon_0} \int_{-1}^1 \frac{1}{R}\left(\frac{r}{R}\right) P_1(\cos\theta') P \cos\theta' R^2 d\cos\theta' \qquad (15.172)$$

$$= \frac{P}{3\epsilon_0} r,$$

where, in the second step, orthogonality of the Legendre polynomials ensures that only the term $i = 1$ in the sum contributes. Hence, the electric field component along the axis of polarization (z axis) is minus the r derivative of the potential just found in equation (15.172),

$$E_z = -\frac{\partial\phi}{\partial z} = -\frac{\partial\phi}{\partial r} = -\frac{P}{3\epsilon_0}. \qquad (15.173)$$

Note that for a polarization along the z axis, spherical symmetry dictates that the two other spherical electric field components E_θ, E_ϕ must vanish along the z axis. Hence, along the z axis the electric field $\mathbf{E}$ only has a z component: $-\frac{P}{3\epsilon_0}\hat{\mathbf{z}}$.

Next, it is used that since,

$$\mathbf{P} = (\epsilon - \epsilon_0)\mathbf{E}, \qquad (15.174)$$

and the sphere polarization $\mathbf{P}$ is uniform *everywhere* inside the sphere, then the electric field $\mathbf{E}$ must also be uniform *everywhere* inside the sphere and given by $-\frac{P}{3\epsilon_0}\hat{\mathbf{z}}$. This concludes the proof.

15.12.2 Linear-chain model

The linear-chain model of a one-dimensional diatomic crystal is based upon a system of two atoms with masses, m and M, placed along a one-dimensional chain as depicted in figure 15.8. As for a diatomic lattice, the masses are situated alternately along the chain and their separation is a. On such a chain the displacement of one atom from its equilibrium position will perturb the positions of its neighboring atoms.

In the simple linear-chain model considered in this section, it is assumed that only nearest neighbors are coupled and that the interaction between these atoms is described by Hooke's law; the spring constant K is taken to be that of a harmonic oscillator. This model describes many of the basic properties of a diatomic lattice.

Let us write for harmonic oscillations of the two ions,

$$\begin{aligned} u_{2r} &= A_1 e^{i(2rqa-\omega t)}, \\ u_{2r+1} &= A_2 e^{i((2r+1)qa-\omega t)}, \end{aligned} \qquad (15.175)$$

Figure 15.8. One-dimensional linear-chain representation of a diatomic lattice.

where q is the phonon wave vector, m and M are the masses of the two ions forming the diatomic crystal, and ω is the angular frequency. In the nearest-neighbor approximation, these longitudinal displacements satisfy,

$$m\frac{d^2 u_{2r}}{dt^2} = -K(u_{2r} - u_{2r-1}) - K(u_{2r} - u_{2r+1}) = K(u_{2r+1} + u_{2r-1} - 2u_{2r}),$$

$$M\frac{d^2 u_{2r+1}}{dt^2} = -K(u_{2r+1} - u_{2r}) - K(u_{2r+1} - u_{2r+2}) = K(u_{2r+2} + u_{2r} - 2u_{2r+1}).$$

$$(15.176)$$

The signs in the four terms on the right-hand sides of these equations are determined by considering the relative displacements of neighboring atoms. For example, if the positive displacement of u_{2r} is greater than that of u_{2r-1} there is a restoring force $-K(u_{2r+1} - u_{2r})$. Hence,

$$-m\omega^2 A_1 = KA_2(e^{iqa} + e^{-iqa}) - 2KA_1,$$
$$-M\omega^2 A_2 = KA_1(e^{iqa} + e^{-iqa}) - 2KA_2.$$

$$(15.177)$$

Eliminating A_1 and A_2,

$$\omega^2 = K\left(\frac{1}{m} + \frac{1}{M}\right) \pm K\left[\left(\frac{1}{m} + \frac{1}{M}\right)^2 - \frac{4\sin^2(qa)}{mM}\right]^{1/2}.$$

$$(15.178)$$

This relationship between frequency and wave vector is commonly called a dispersion relation. The higher-frequency solution is known as the optical mode since, for many semiconductors, its frequency is in the terahertz range, which happens to coincide with the infrared portion of the electromagnetic spectrum. The lower-frequency solution is known as the acoustic mode. More precisely, since only longitudinal displacements have been modeled, these two solutions correspond to the longitudinal optical (LO) and longitudinal acoustic (LA) modes of the linear-chain lattice. Clearly, the displacements along this chain can be described in terms of wavevectors q in the range from $-\frac{\pi}{2a}$ to $\frac{\pi}{2a}$. From the solution for ω, it is evident that over this Brillouin zone the LO modes have a maximum frequency $[2K(\frac{1}{m} + \frac{1}{M})]^{1/2}$ at the center of the Brillouin zone and a minimum frequency $(\frac{2K}{m})^{1/2}$ at the edge of the Brillouin zone. Likewise, the LA modes have a maximum frequency $(\frac{2K}{M})^{1/2}$ at the edge of the Brillouin zone and a minimum frequency equal to zero at the center of the Brillouin zone.

In polar semiconductors, the masses m and M carry opposite charges, e^* and $-e^*$, respectively, as a result of the redistribution of the charge associated with polar bonding. In polar materials such ionic bonding is characterized by values of e^* equal to 1, to an order-of-magnitude. When there is an electric field E present in the semiconductor, it is necessary to augment the previous force equation with terms describing the interaction with the charge. In the long-wavelength limit of the electric field E, the force equations then become,

$$\begin{aligned}
- m\omega^2 u_{2r} &= K(u_{2r+1} + u_{2r-1} - 2u_{2r}) + e^*E \\
&= K(e^{i2qa} + 1)u_{2r-1} - 2Ku_{2r} + e^*E, \\
- M\omega^2 u_{2r+1} &= K(u_{2r+2} + u_{2r} - 2u_{2r+1}) - e^*E \\
&= K(1 + e^{-i2qa})u_{2r+2} - 2Ku_{2r+1} - e^*E.
\end{aligned} \tag{15.179}$$

Regarding the phonon displacements, in the long-wavelength limit there is no need to distinguish between the different sites for a given mass type since all atoms of the same mass are displaced by the same amount. In this limit, $q \to 0$. Denoting the displacements on even-numbered sites by u_1 and those on odd-numbered sites by u_2, in the long-wavelength limit the force equations reduce to,

$$\begin{aligned}
- m\omega^2 u_1 &= 2K(u_2 - u_1) + e^*E, \\
- M\omega^2 u_2 &= 2K(u_1 - u_2) - e^*E.
\end{aligned} \tag{15.180}$$

Simple manipulations then yield,

$$\begin{aligned}
u_1 &= \frac{e^*E}{\left(\omega_0^2 - \omega^2\right)m}, \\
u_2 &= \frac{e^*E}{\left(\omega^2 - \omega_0^2\right)M},
\end{aligned} \tag{15.181}$$

where,

$$\omega_0^2 = 2K\left(\frac{1}{m} + \frac{1}{M}\right), \tag{15.182}$$

is the resonant frequency squared, in the absence of Coulomb effects; that is, for $e^* = 0$.

The induced dipole moment is now,

$$p = e^*(u_1 - u_2) = \frac{e^{*2}}{\omega_0^2 - \omega^2}\left(\frac{1}{m} + \frac{1}{M}\right)E. \tag{15.183}$$

and the associated (displacement) polarization becomes,

$$\alpha^{dis} = \frac{p}{E} = \frac{e^{*2}}{\omega_0^2 - \omega^2}\left(\frac{1}{m} + \frac{1}{M}\right) \tag{15.184}$$

Let us take a naive approach and write the full polarization as,

$$\alpha = (\alpha^+ + \alpha^-) + \frac{e^{*2}}{\omega_0^2 - \omega^2}\left(\frac{1}{m} + \frac{1}{M}\right), \tag{15.185}$$

where α^+ and α^- are the polarizations of the positive and negative ions, respectively. There is no real justification for this expression since the first term in equation (15.185) is calculated on the assumption that the ions were immobile but polarizable while the second term is calculated for ions that could be moved but not deformed.

A better theory for polarization is to calculate the full polarization is to include both movement and deformation at once. Such theories are known as shell-model theories [17].

Combining equation (15.185) with the Clausius–Mossotti relation yields,

$$\frac{\epsilon(\omega) - \epsilon_0}{\epsilon(\omega) + 2\epsilon_0} = \frac{1}{3\epsilon_0} \frac{\alpha}{V_{\text{prim}}} = \frac{1}{3\epsilon_0 V_{\text{prim}}} \left[(\alpha^+ + \alpha^-) + \frac{e^{*2}}{\omega_0^2 - \omega^2} \left(\frac{1}{m} + \frac{1}{M} \right) \right], \quad (15.186)$$

where the dependence of ϵ on frequency was highlighted.

In particular, the static (low-frequency) permittivity is given by,

$$\frac{\epsilon(0) - \epsilon_0}{\epsilon(0) + 2\epsilon_0} = \frac{1}{3\epsilon_0 V_{\text{prim}}} \left[(\alpha^+ + \alpha^-) + \frac{e^{*2}}{\omega_0^2} \left(\frac{1}{m} + \frac{1}{M} \right) \right], \quad \omega < <\omega_0, \quad (15.187)$$

while the high-frequency permittivity satisfies,

$$\frac{\epsilon(\infty) - \epsilon_0}{\epsilon(\infty) + 2\epsilon_0} = \frac{1}{3\epsilon_0 V_{\text{prim}}} (\alpha^+ + \alpha^-), \quad \omega > >\omega_0. \quad (15.188)$$

It is convenient to write $\epsilon(\omega)$ in terms of $\epsilon(0)$ and $\epsilon(\infty)$ since the two limiting forms are readily measured. $\epsilon(0)$ is the static permittivity of the crystal, while $\epsilon(\infty)$ is the permittivity at optical frequencies, and is therefore related to the index of refraction, n by $n^2 = \epsilon(\infty)$. Then,

$$\frac{\epsilon(\omega) - \epsilon_0}{\epsilon(\omega) + 2\epsilon_0} = \frac{\epsilon(\infty) - \epsilon_0}{\epsilon(\infty) + 2\epsilon_0} + \frac{1}{1 - \dfrac{\omega^2}{\omega_0^2}} \left(\frac{\epsilon(0) - \epsilon_0}{\epsilon(0) + 2\epsilon_0} - \frac{\epsilon(\infty) - \epsilon_0}{\epsilon(\infty) + 2\epsilon_0} \right), \quad (15.189)$$

which can be solved for $\epsilon(\omega)$,

$$\epsilon(\omega) = \epsilon(\infty) + \frac{\epsilon(\infty) - \epsilon(0)}{\dfrac{\omega^2}{\omega_T^2} - 1}, \quad (15.190)$$

where,

$$\omega_T^2 = \omega_0^2 \frac{\epsilon(\infty) + 2\epsilon_0}{\epsilon(0) + 2\epsilon_0}. \quad (15.191)$$

From the Maxwell equation,

$$\nabla \cdot \mathbf{D} = 0, \quad (15.192)$$

it follows for plane-wave propagation $\propto e^{i\mathbf{k} \cdot \mathbf{r}}$ that,

$$\mathbf{k} \cdot \mathbf{D} = 0, \quad (15.193)$$

which requires,

$$\mathbf{D} = 0, \quad \text{or } \mathbf{D}, \mathbf{E} \text{ and } \mathbf{P} \text{ are orthogonal to } \mathbf{k}. \quad (15.194)$$

Furthermore, if the electric field is the gradient of a scalar potential, it follows that,

$$\nabla \times \mathbf{E} = 0, \qquad (15.195)$$

and,

$$\mathbf{k} \times \mathbf{E} = 0, \qquad (15.196)$$

which requires,

$$\mathbf{E} = 0, \quad \text{or } \mathbf{E}, \mathbf{D} \text{ and } \mathbf{P} \text{ are parallel to } \mathbf{k}. \qquad (15.197)$$

In a longitudinal optical mode the nonzero polarization density $\mathbf{P}$ is parallel to $\mathbf{k}$, and therefore equation (15.194) requires that $\mathbf{D}$ must vanish such that,

$$\epsilon = 0,$$
$$\implies \quad \epsilon(\infty) + \frac{\epsilon(\infty) - \epsilon(0)}{\dfrac{\omega_L^2}{\omega_T^2} - 1} = 0, \qquad (15.198)$$

which determines the longitudinal optical-mode frequency ω_L. On the other hand, in a transverse optical mode the nonzero polarization density $\mathbf{P}$ is perpendicular to $\mathbf{k}$, which is consistent with equation (15.197) only if $\mathbf{E}$ vanishes. From,

$$\mathbf{D} = \epsilon_0 \mathbf{E} + \mathbf{P} = \epsilon \mathbf{E}, \qquad (15.199)$$

it follows that,

$$\epsilon = \infty. \qquad (15.200)$$

According to equation (15.190), $\epsilon = \infty$ when $\omega^2 = \omega_T^2$, and therefore this result identifies ω_T as the frequency of the long-wavelength transverse optical mode.

Observe also that equation (15.198) yields,

$$\frac{\omega_L^2}{\omega_T^2} = \frac{\epsilon(0)}{\epsilon(\infty)}, \qquad (15.201)$$

which is known as the Lyddane–Sachs–Teller relation.

15.13 Born–Huang theory

Huang [18] in 1951, and Born and Huang [19] took the most general form of the microscopic theory of diatomic polar crystals to be described by a pair of phenomenological equations relating $\mathbf{w}$, $\mathbf{E}$, and $\mathbf{P}$,

$$\ddot{\mathbf{w}} = a\mathbf{w} + b\mathbf{E},$$
$$\mathbf{P} = d\mathbf{w} + c\mathbf{E}, \qquad (15.202)$$

where,

$$\mathbf{w} = \sqrt{\frac{mM}{m+M}\frac{N}{V}}\,\mathbf{u}, \tag{15.203}$$

and $\mathbf{u} = \mathbf{u}_1 - \mathbf{u}_2$ is the relative displacement of the two ions, N is the number of unit cells in the crystal, and V is the volume of the crystal.

Assuming a time dependence,

$$\begin{aligned}
\mathbf{E} &= \mathbf{E}_0 e^{-i\omega t}, \\
\mathbf{w} &= \mathbf{w}_0 e^{-i\omega t}, \\
\mathbf{P} &= \mathbf{P}_0 e^{-i\omega t},
\end{aligned} \tag{15.204}$$

yields,

$$\begin{aligned}
\mathbf{w}_0 &= -\frac{b}{a+\omega^2}\mathbf{E}_0, \\
\mathbf{P}_0 &= \left(-\frac{db}{a+\omega^2} + c\right)\mathbf{E}_0 = \left(\frac{b^2}{-a-\omega^2} + c\right)\mathbf{E}_0,
\end{aligned} \tag{15.205}$$

where it is used that $d = b$; a result that follows from energy conservation [18, 19].

From,

$$\mathbf{D} = \epsilon_0 \mathbf{E} + \mathbf{P} = \epsilon(\omega)\mathbf{E}, \tag{15.206}$$

it follows that,

$$\epsilon(\omega) = \epsilon_0 + c + \frac{b^2}{-a-\omega^2}. \tag{15.207}$$

Comparison with equation (15.190) allows us to identify,

$$\begin{aligned}
a &= -\omega_T^2, \\
b &= \sqrt{\epsilon(0) - \epsilon(\infty)}\,\omega_T, \\
c &= \epsilon(\infty) - \epsilon_0.
\end{aligned} \tag{15.208}$$

Inserting the latter expressions in the Born–Huang relations,

$$\begin{aligned}
\left(\omega_T^2 - \omega^2\right)\mathbf{w} &= \sqrt{\epsilon(0) - \epsilon(\infty)}\,\omega_T \mathbf{E}, \\
\mathbf{P} &= \sqrt{\epsilon(0) - \epsilon(\infty)}\,\omega_T \mathbf{w} + (\epsilon(\infty) - \epsilon_0)\mathbf{E},
\end{aligned} \tag{15.209}$$

and upon re-introducing the relative displacement $\mathbf{u}$,

$$\begin{aligned}
\left(\omega_T^2 - \omega^2\right)\mathbf{u} &= \sqrt{\left(\frac{1}{m} + \frac{1}{M}\right)\frac{V}{N}}\sqrt{\epsilon(0) - \epsilon(\infty)}\,\omega_T \mathbf{E}, \\
\mathbf{P} &= \sqrt{\epsilon(0) - \epsilon(\infty)}\,\omega_T\sqrt{\frac{mM}{m+M}\frac{N}{V}}\,\mathbf{u} + (\epsilon(\infty) - \epsilon_0)\mathbf{E}.
\end{aligned} \tag{15.210}$$

15.14 Piezoelectric vibrations at optical frequencies

In this section, we will derive a frequency-dependent description of piezoelectricity related to the dielectric constant. The idea follows a simple and elegant approach by Tan [20] applied to α-quartz. Under the influence of an incident electric field $\mathbf{E}$, the forced oscillation of the charged ions is described by the forced oscillator system,

$$m_j \frac{d^2 x_j}{dt^2} + \gamma_j \frac{dx_j}{dt} + q_j x_j = p_j e E_j, \quad j = 1, 2, 3, \tag{15.211}$$

where $\mathbf{x} = (x_1, x_2, x_3)$ is the displacement vector of the charged ions, m_j is the reduced mass, γ_j is the damping constant, p_j is the valence number of the ions in one of the three mutually orthogonal directions, and e is the charge of the electron. Note that in anisotropic systems, m_j, γ_j, p_j, and q_j depend on the direction x_j. For $\mathbf{E} = \mathbf{E}_0 e^{i\omega t}$, equation (15.211) gives,

$$x_j = \frac{p_j e}{m_j \left(\omega_{0j}^2 - \omega^2 + i\gamma_j \omega \right)} E_j, \tag{15.212}$$

where,

$$\omega_{0j}^2 = q_j / m_j, \tag{15.213}$$

denotes the resonance frequency in the infrared range. Tan generalizes phenomenologically to the case of many resonances and tacitly assumes,

$$x_j = \frac{p_j e}{m_j} \sum_k \frac{f_{kj}}{\left(\omega_{kj}^2 - \omega^2 + i\gamma_{kj}\omega \right)} E_j, \tag{15.214}$$

where f_{kj} are the oscillations strengths at the resonance frequency ω_{kj},

$$\omega_{kj}^2 = q_{kj} / m_j. \tag{15.215}$$

The real part of the displacement, $Re(x_j)$, is expressed by,

$$Re(x_j) = \frac{p_j e}{m_j} \sum_k \frac{f_{kj} \left(\omega_{kj}^2 - \omega^2 \right)}{\left(\omega_{kj}^2 - \omega \right)^2 + \gamma_{kj}^2 \omega^2} E_j. \tag{15.216}$$

An uniaxial strain induced by the inverse piezoelectric effect in the direction E_j produces longitudinal strain. Let a be a piezoelectric cell parameter in the direction E_j. The longitudinal strain S_{jj} in this direction is then given by $Re(x_j)/a$,

$$S_{jj} = \frac{p_j e}{m_j a} \sum_k \frac{f_{kj} \left(\omega_{kj}^2 - \omega^2 \right)}{\left(\omega_{kj}^2 - \omega^2 \right)^2 + \gamma_{kj}^2 \omega^2} E_j, \tag{15.217}$$

and the piezoelectric $\mathbf{d}$ tensor becomes,

$$\mathbf{d}_{ijl} = \frac{\partial S_{jl}}{\partial E_i}. \tag{15.218}$$

This implies,

$$d_{jjj} = \frac{p_j e}{m_j a} \sum_k \frac{f_{kj}\left(\omega_{kj}^2 - \omega^2\right)}{\left(\omega_{kj}^2 - \omega^2\right)^2 + \gamma_{kj}^2 \omega^2}, \tag{15.219}$$

Note that the cell parameter a and the quantities f_{kj}, p_j, γ_{kj}, ω_{kj} all depend on the direction of E_j in piezoelectric crystals. Observe that, in the visible frequency range where $\omega > \omega_{kj}$ that the piezoelectric constant is negative. For a static electric field, $\omega = 0$,

$$d_{jjj}(0) = \frac{p_j e}{m_j a} \sum_k \frac{f_{kj}}{\omega_{kj}^2} = \frac{p_j e}{a} \sum_k \frac{f_{kj}}{q_{kj}}, \tag{15.220}$$

where $m_j \omega_{kj}^2 = q_{kj}$ was used. In the visible and infrared frequency region, there are two contributions to the polarization, electronic and ionic polarizations. In a harmonic field of the incident light, ionic polarization causes piezoelectric vibrations. Further, the ionic displacement is related to the optical dispersion of the permittivity [19]. For the real part of the dielectric constant,

$$Re(\epsilon_{jj}(\omega)) = \epsilon_{jj}(\infty) + \frac{N(p_j e)^2}{m_j \epsilon_0} \sum_k \frac{f_{kj}\left(\omega_{kj}^2 - \omega^2\right)}{\left(\omega_{kj}^2 - \omega^2\right)^2 + \gamma_{kj}^2 \omega^2}, \tag{15.221}$$

where $\epsilon_{jj}(\infty)$ is the high-frequency dielectric constant, N is the number of charged particles per unit volume,

$$N = N_A \frac{\rho}{W}, \tag{15.222}$$

and N_A, ρ, and W are Avogadro's number, the density, and the molecular weight, respectively. Combining equations (15.219) and (15.221) yields a relationship between the piezoelectric coefficient and the dielectric constant,

$$d_{jjj}(\omega) = \frac{\epsilon_0(Re(\epsilon_{jj}(\omega)) - \epsilon_{jj}(\infty))}{N p_j e a}. \tag{15.223}$$

From the latter relation, observe that the piezoelectric constant approaches zero at high frequencies since $\lim_{\omega \to \infty} Re(\epsilon_{jj}(\omega)) = \epsilon_{jj}(\infty)$. At low frequencies, the values of d_{jjj} are comparable with those determined by conventional direct measurements.

For α-quartz, which is an important and widely used piezoelectric crystal, the piezoelectric coefficients have been accurately determined by direct longitudinal and transverse methods [23] and by x-ray spectrometry [24, 25]. The value of d_{111} is measured to 2.31×10^{-12} m/V [24, 25]. The calculated piezoelectric coefficient is found to be 2.199×10^{-12} m/V [20] which is in good agreement.

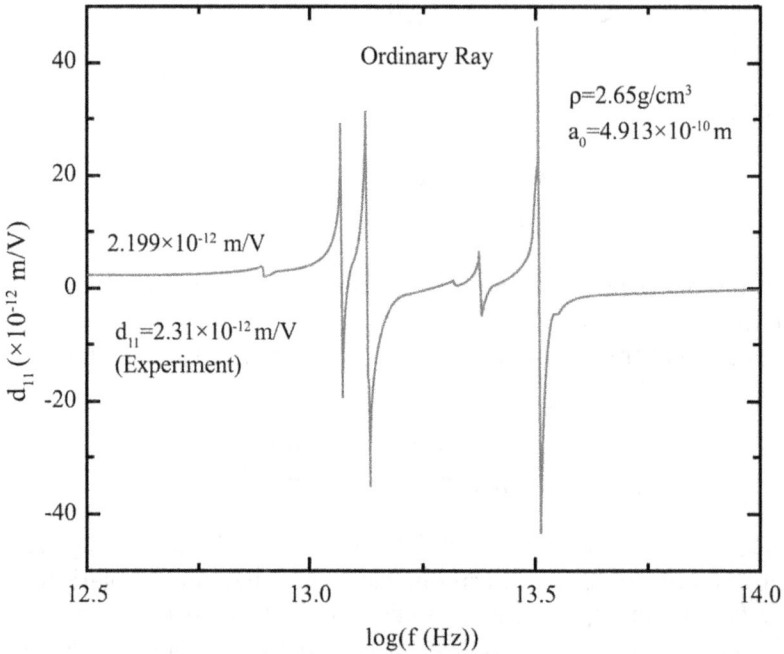

Figure 15.9. Piezoelectric coefficient calculated by equation (15.223) at different frequencies, $f = \omega/(2\pi)$ of the incident light. Optical constants of α-quartz are taken from references [21, 22], obtained by a Kramers–Kronig analysis of the infrared reflection spectra of α-quartz. In the case of propagation of the ordinary ray, the vibration direction of the incident light is perpendicular to the optic axis of α-quartz. The piezoelectric cell parameter in this case is equal to a_0; and the calculated piezoelectric coefficient corresponds to d_{111} (note that $d_{11} = d_{111}$). Adapted from reference [20], copyright (2004) with permission from Elsevier.

The full spectrum of the piezoelectric constant and the dielectric constant of α-quartz can be evaluated from knowledge of the optical constants of α-quartz, the density $\rho = 2650$ kg m^{-3}, and the valence number of the ions is four, $p_j = 4$, and the cell parameter a is chosen to be the lattice constant $a_0 = 4.913$ Å. In figure 15.9, the spectrum of $d_{111}(\omega)$ for α-quartz is shown.

References

[1] King-Smith R D and Vanderbilt D 1993 Theory of polarization of crystalline solids *Phys. Rev.* **B47** 1651

[2] Resta R 1992 Theory of the electric polarization in crystals *Ferroelectrics* (Taylor & Francis) vol 136 pp 51–5

[3] Resta R and Vanderbilt D 2007 Theory of polarization: a modern approach *Physics of Ferroelectrics: A Modern Perspective, Topics Applied Physics* ed K Rabe, C H Ahn and J-M Triscone (Berlin: Springer) vol 105

[4] Lundqvist S and March N H (ed) 1983 *Theory of the Inhomogeneous Electron Gas* (New York: Plenum)

[5] Vanderbilt D 2018 *Berry Phases in Electronic Structure Theory* (Cambridge: Cambridge University Press)

[6] Zak J 1989 Berry's phase for energy bands in solids *Phys. Rev. Lett.* **62** 2747

[7] Michel L and Zak J 1992 Physical equivalence of energy bands in solids *Europhys. Lett.* **18** 239

[8] de Gironcoli S, Baroni S and Resta R 1998 Piezoelectric properties of III-V semiconductors from first-principles linear-response theory *Phys. Rev. Lett.* **62** 2853

[9] Keating P N 1966 Effect of invariance requirements on the elastic strain energy of crystals with application to the diamond structure *Phys. Rev.* **B145** 637

[10] Nusimovici M A and Birman J L 1967 Lattice dynamics of wurtzite: CdS *Phys. Rev.* **B156** 925

[11] Musgrave M J P and Pople J A 1962 A general valence force field for diamond *Proc. R. Soc.* **268** 474

[12] Wilson E B, Decius J C, Cross P C and Sundheim B R 1955 Molecular vibrations–the theory of infrared and Raman vibration spectra *J. Electrochem. Soc.* **102** 235Ca

[13] Merten L 1962 Über die Gitterschwingungen in Kristallen mit Wurtzitstruktur *Z. Naturforsch.* **17a** 65

[14] Kellerman E W 1940 Theory of the vibrations of the sodium chloride lattice *Phil. Trans. Roy. Soc. (London)* **A238** 513

[15] Frenkel D and Smit B 2001 *Understanding molecular simulation: From algorithms to applications* (Amsterdam: Elsevier)

[16] Deserno M and Holm C 1998 How to mesh up Ewald sums. A theoretical and numerical comparison of various particle mesh routines *J. Chem. Phys.* **109** 7678–93

[17] Ashcroft N and Mermin N D 1976 *Solid State Physics* (Philadelphia, PA: Saunders)

[18] Huang K 1951 On the interaction between the radiation field and ionic crystals *Proc. R. Soc.* **A208** 352

[19] Born M and Huang K 1954 *Dynamical Theory of Crystal Lattices* (Oxford: Oxford University Press)

[20] Tan C Z 2004 Piezoelectric lattice vibrations at optical frequencies *Solid State Commun.* **131** 405

[21] Spitzer W G and Kleinman D A 1961 Infrared lattice bands of quartz *Phys. Rev.* **121** 1324

[22] Russell E E and Bell E E 1967 Optical constants of sapphire in the far infrared *J. Opt. Soc. Am.* **57** 341

[23] Cady W G 1964 *Piezoleectricity I* (New York: Dover)

[24] Bhalla A S, Bose D N, White E W and Cross L E 1971 Precise x-ray determination of small homogeneous strains applied to the direct measurement of piezoelectric constants *Phys. Stat. Solidi* A **7** 335

[25] Davaasambuu J, Pucher A, Cochin V and Pietsch U 2002 Atomistic origin of the inverse piezoelectric effect in α-SiO$_2$ and α-GaPO$_4$ *Europhys. Lett.* **62** 834

IOP Publishing

Piezoelectricity in Classical and Modern Systems

Morten Willatzen

Chapter 16

Optical properties of piezoelectric materials

In chapter 16, the general theory of optical absorption in materials is presented. Two-level systems are first considered before generalizing the analysis to semiconductors. In the second part of the chapter, the influence of piezoelectricity and spontaneous polarization on electronic band structure and wave functions is discussed for zinc blende and wurtzite quantum-confined structures. The final part of the chapter highlights the influence of piezoelectricity on optical properties.

16.1 Optical absorption in a semiconductor

In the following, we discuss the absorption coefficient for photons in a semiconductor. The presentation follows to a large extent the analysis outlined in reference [1]. Firstly, the one-electron description of electrons coupled to an electromagnetic vector potential in a periodic crystal is introduced. This coupling leads to interband transitions and can be treated in a perturbative way using Fermi's Golden Rule for calculation of the photon emission/absorption rate [2].

16.1.1 A non-relativistic description of the electron–photon interaction

A simple, and to a large extent adequate, description of electrons in a crystal lattice is based on the one-particle Schrödinger equation,

$$H\psi(\mathbf{r}) = E\psi(\mathbf{r})$$
$$H = \frac{\mathbf{p}^2}{2m_0} + V(\mathbf{r}) = -\frac{\hbar^2\mathbf{\nabla}^2}{2m_0} + V(\mathbf{r}), \tag{16.1}$$

where ψ is the electron wave function, E is the energy eigenvalue, V is the crystalline potential, $\mathbf{p}$ is the momentum operator, and m_0 is the free-electron mass.

We know from classical mechanics that the presence of an electromagnetic field modifies the Hamiltonian in equation (16.1) according to the replacement,

$$\mathbf{p} \rightarrow \mathbf{p} - e\mathbf{A}(\mathbf{r}, t), \tag{16.2}$$

doi:10.1088/978-0-7503-5557-5ch16

where e is the (negative) electron charge and $\mathbf{A}$ is the electromagnetic vector potential satisfying,

$$\mathcal{E}(\mathbf{r},\, t) = -\frac{\partial \mathbf{A}(\mathbf{r},\, t)}{\partial t}. \tag{16.3}$$

In equation (16.3), $\mathcal{E}$ is the electric field. Inserting equation (16.3) in equation (16.1) leads to the following expression for the interaction Hamiltonian H_{int},

$$H = H_0 + H_{\text{int}},$$
$$H_{\text{int}} = -\frac{e}{m_0}\mathbf{A}(\mathbf{r},\, t) \cdot \mathbf{p}, \tag{16.4}$$

where H_0 is the unperturbed Hamiltonian equal to the Hamiltonian in equation (16.1). In deriving equation (16.4) it has been used that terms of second order in $\mathbf{A}$ are neglected (lowest order approximation) and that $\mathbf{A}$ commutes with the momentum operator $\mathbf{p}$ due to the transversality condition $\nabla \cdot \mathbf{A} = 0$.

The electron–photon interaction given by equation (16.4) induces optical interband transitions between the conduction and valence band because momentum matrix elements between conduction band states and valence band states in general are non-vanishing. We can obtain information about the optical absorption, spontaneous emission, etc in a semiconductor by use of time-dependent perturbation theory by summing over all available electron and hole states.

16.1.2 Absorption and spontaneous emission in a two-level system

For a simple two-energy level system, the emission rate W between an initial state of energy E_i and a final state E_f is given by Fermi's Golden Rule,

$$W = \frac{2\pi}{\hbar}|\langle f|H_{\text{int}}^e|i\rangle|^2 \rho_f(E)\delta(E - E_i + E_f), \tag{16.5}$$

where H_{int}^e is the time-independent part of the interaction Hamiltonian H_{int} responsible for emission of photons,

$$H_{\text{int}} = H_{\text{int}}^e e^{i\frac{E}{\hbar}t} + H_{\text{int}}^a e^{-i\frac{E}{\hbar}t}. \tag{16.6}$$

Here, superscripts a and e refer to emission and absorption, respectively. Further, $|i\rangle$ and $|f\rangle$ are the initial and final states, respectively, $\rho_f(E)$ is the density of final states, and E is the energy of the emitted photon, i.e., $E = E_i - E_f$.

For a plane wave with the angular frequency ω, the electric field is written as,

$$\mathcal{E} = \varepsilon\left(\frac{\mathcal{E}_0}{2}e^{i\omega t} + c.\, c.\right), \tag{16.7}$$

where ε is the unit polarization vector. An expression for the magnitude of the coefficient $\mathcal{E}_0$ associated with one photon can be found by evaluating the energy flux S using Maxwell's equations,

$$S = \frac{1}{2}|\mathcal{E}_0|^2 n \epsilon_0 c. \tag{16.8}$$

In equation (16.8), n is the refractive index of the medium. The energy flux is also given by the product of the photon energy density $\hbar\omega/V$ and the group velocity c/n_g so that,

$$|\mathcal{E}_0| = \sqrt{\frac{2\hbar\omega}{nn_g \epsilon_0 V}}, \tag{16.9}$$

where V is the volume of the enclosure confining the electromagnetic field. The interaction Hamiltonian in equation (16.4) can now be written as,

$$H_{\text{int}} = -\frac{e}{m_0} i \sqrt{\frac{\hbar}{2nn_g \epsilon_0 \omega V}} [e^{i(\delta+\omega t)} - e^{-i(\delta+\omega t)}] \varepsilon \cdot \mathbf{p}, \tag{16.10}$$

where δ is defined as $\mathcal{E}_0 = |\mathcal{E}_0|e^{i\delta}$, and equations (16.4), (16.7), and (16.9) are used. The emission rate defined in equation (16.5) is given by,

$$W = \frac{\pi e^2 \hbar}{m_0^2 nn_g \epsilon_0 EV}|\langle f|\varepsilon \cdot \mathbf{p}|i\rangle|^2 \rho_f(E)\delta(E - E_i + E_f), \tag{16.11}$$

using equation (16.10). In the case of stimulated emission or absorption for a two-level system $\rho_f(E) = 1$. In a similar way, we can obtain the absorption coefficient from the (stimulated) absorption rate. The absorption coefficient $\alpha(E)$ for photons of energy $E = \hbar\omega$ becomes,

$$\alpha(E) = \frac{\pi e^2 \hbar}{m_0^2 nc\epsilon_0 EV}|\langle f|\varepsilon \cdot \mathbf{p}|i\rangle|^2 \delta(E + E_i - E_f). \tag{16.12}$$

Here, $\alpha(E)$ is the number of photons absorbed per unit distance, $\alpha = \frac{n_g W}{c}$ where W now refers to the absorption rate. Note that the absorption coefficient is proportional to $1/V$ because we consider absorption of a photon within a box of volume V by a single two-level system.

For spontaneous emission of ε-polarized photons, on the other hand, the density of final states in a solid angle element $d\Omega$, $\rho_{f,d\Omega}$, equals the number of photon states of energy E per unit volume per unit energy with wave vector pointing into $d\Omega$,

$$\frac{d\rho_f}{d\Omega} = \frac{k^2 dk d\Omega}{(2\pi)^3 dE}. \tag{16.13}$$

Since $E = \hbar\omega$ and $k = n\omega/c$ we may write for the spontaneous emission rate into $d\Omega$ per unit volume at the photon energy E for ε-polarized photons, $r_{sp,d\Omega}(E)$,

$$r_{sp,d\Omega}(E) = \frac{ne^2 Ed\Omega}{8\pi^2 m_0^2 c^3 \epsilon_0 \hbar^2 V}|\langle f|\varepsilon \cdot \mathbf{p}|i\rangle|^2 \delta(E - E_i + E_f). \tag{16.14}$$

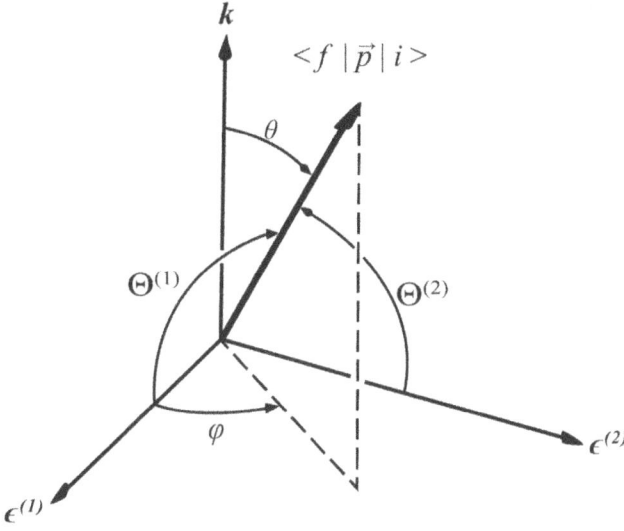

Figure 16.1. Orientation of $\langle f|\mathbf{p}|i\rangle$.

So far, we have been concerned with a radiative transition in which a photon with definite propagation direction $\mathbf{k}$ and polarization ε is emitted. To get the total spontaneous emission rate per volume at the photon energy E, r_{sp}, we must sum over the two independent polarization directions for a given $\mathbf{k}$ and integrate over all possible propagation orientations. From figure 16.1 it is evident that

$$|\langle f|\varepsilon^{(1)}\cdot\mathbf{p}|i\rangle| = |\langle f|\mathbf{p}|i\rangle|\sin\theta\cos\phi,$$
$$|\langle f|\varepsilon^{(2)}\cdot\mathbf{p}|i\rangle| = |\langle f|\mathbf{p}|i\rangle|\sin\theta\sin\phi. \tag{16.15}$$

The sum over two polarization states gives $\sin^2\theta$. Performing the integration over all possible angles, with $\langle f|\mathbf{p}|i\rangle$ fixed in space, gives $8\pi/3$, and we obtain,

$$r_{sp} = \int r_{sp,d\Omega}d\Omega = \frac{e^2 nE}{3\pi m_0^2 c^3 \epsilon_0 \hbar^2 V}|\langle f|\mathbf{p}|i\rangle|^2 \delta(E - E_i + E_f). \tag{16.16}$$

Equations (16.12) and (16.16) are the final results for the absorption coefficient and the spontaneous emission rate per unit volume for a two-level system, respectively. Note that stimulated emission (photons present in the initial state) and spontaneous emission are discussed on equal footing, the only difference appears in the expressions for the density of final states.

16.1.3 Absorption and spontaneous emission in a semiconductor

The generalization of transition rates for discrete levels to the case of a semiconductor is straightforward. The expressions for the absorption and spontaneous emission rates in equations (16.12) and (16.16) are integrated over the occupied electron and hole states which for an unexcited semiconductor is given by the Fermi distribution and determined by the position of the Fermi level E_f.

The crystalline potential $\mathbf{V}$ is periodic in the Bravais lattice, i.e., the translation operator $T_{\mathbf{R}}$ defined by,

$$T_{\mathbf{R}}\psi(\mathbf{r}) = \psi(\mathbf{r} + \mathbf{R}), \tag{16.17}$$

where $\mathbf{R}$ is a Bravais lattice vector, commutes with the total Hamiltonian. Therefore, the eigenstates of H can be chosen in Bloch form,

$$\psi_{\mathbf{k}}(\mathbf{r}) = e^{i\mathbf{k}\cdot\mathbf{r}}u_{\mathbf{k}}(\mathbf{r}), \tag{16.18}$$

where $u_{\mathbf{k}}$ is periodic in the Bravais lattice. More precisely, a conduction band state $|co\rangle$ and a valence band state $|vm\rangle$ for the bulk semiconductor, are normalized according to,

$$\langle co|co\rangle = \langle vm|vm\rangle = 1, \tag{16.19}$$

are written as,

$$|co\rangle = \frac{1}{\sqrt{V}}|So\rangle e^{i\mathbf{k}_c\cdot\mathbf{r}},$$
$$|vm\rangle = \frac{1}{\sqrt{V}}\sum_{\nu=1}^{6}g_{\nu}^{m}(\mathbf{k}_v)|U_{\nu}\rangle e^{i\mathbf{k}_v\cdot\mathbf{r}}. \tag{16.20}$$

In the following the lowest conduction band is assumed to be spanned by $|S\rangle$ cell states with spin σ ($|S\uparrow\rangle$ and $|S\downarrow\rangle$) and the highest valence band states are mixtures of $|P\rangle$ cell states with spin σ, i.e., $|U_{\nu}\rangle$ is an orthonormal linear combination of six states,

$$|X\uparrow\rangle, |Y\uparrow\rangle, |Z\uparrow\rangle, X\downarrow\rangle, |Y\downarrow\rangle, |Z\downarrow\rangle. \tag{16.21}$$

This assumption is a good approximation for many zincblende and wurtzite semiconductors, especially, wide-bandgap zincblende and wurtzite semiconductors. Here, $\mathbf{k}_c$ ($\mathbf{k}_v$) is the conduction (valence) band wave vector, $e^{i\mathbf{k}_c\cdot\mathbf{r}}$ ($e^{i\mathbf{k}_v\cdot\mathbf{r}}$) is the conduction (valence) band envelope part of the total wave function, and V is the crystal volume. Let us now evaluate the matrix element $|\langle vm|\varepsilon\cdot\mathbf{p}|co\rangle|^2$ appearing in the transition rate,

$$|\langle vm|\varepsilon\cdot\mathbf{p}|co\rangle|^2 = \left|\frac{1}{V}\int e^{-i\mathbf{k}_v\cdot\mathbf{r}}\langle\sum_{\nu=1}^{6}g_{\nu}^{m}(\mathbf{k}_v)U_{\nu}|\varepsilon\cdot\mathbf{p}|So\rangle e^{i\mathbf{k}_c\cdot\mathbf{r}}d\mathbf{r}\right|^2$$
$$= \frac{(2\pi)^3}{V}\delta(\mathbf{k}_c - \mathbf{k}_v)\left|\langle\sum_{\nu=1}^{6}g_{\nu}^{m}(\mathbf{k}_v)U_{\nu}|\varepsilon\cdot\mathbf{p}|So\rangle\right|^2. \tag{16.22}$$

In deriving the second step in equation (16.22), it was used that the envelope part of the wave function is slowly varying and therefore approximately constant over a unit cell. The Bloch form in equation (16.20) evidently leads to the $\mathbf{k}$-selection rule $\mathbf{k}_c = \mathbf{k}_v$.

The absorption coefficient can now be obtained by integrating over the available electron and hole states,

$$\alpha(E) = \frac{\pi e^2 \hbar}{\epsilon_0 m_0^2 cnEV} \sum_\sigma \sum_m \sum_{\mathbf{k}_c, \mathbf{k}_v} (f(E^m(\mathbf{k}_v))(1 - f(E^\sigma(\mathbf{k}_c))) - f(E^\sigma(\mathbf{k}_x))(1 - f(E^m(\mathbf{k}_v))))$$
$$\times |\langle vm|\varepsilon \cdot \mathbf{p}|c\sigma\rangle|^2 \delta(E^\sigma(\mathbf{k}_c) - E^m(\mathbf{k}_v) - E)$$
$$= \frac{e^2 \hbar}{8\pi^2 \epsilon_0 m_0^2 cnE} \sum_\sigma \sum_m \int_{1.BZ} d\mathbf{k}_c \int_{1.BZ} d\mathbf{k}_v \left| \langle \sum_{\nu=1}^{6} g_\nu^m(\mathbf{k}_v) U_\nu |\varepsilon \cdot \mathbf{p}|S\sigma\rangle \right|^2 \qquad (16.23)$$
$$\times \delta(\mathbf{k}_c - \mathbf{k}_v)(f(E^m(\mathbf{k}_v)) - f(E^\sigma(\mathbf{k}_c)))\delta(E^\sigma(\mathbf{k}_c) - E^m(\mathbf{k}_v) - E)$$
$$= \frac{e^2 \hbar}{8\pi^2 \epsilon_0 m_0^2 cnE} \sum_\sigma \sum_m \int_{1.BZ} d\mathbf{k} \left| \langle \sum_{\nu=1}^{6} g_\nu^m(\mathbf{k}) U_\nu |\varepsilon \cdot \mathbf{p}|S\sigma\rangle \right|^2$$
$$\times (f(E^m(\mathbf{k})) - f(E^\sigma(\mathbf{k})))\delta(E^\sigma(\mathbf{k}) - E^m(\mathbf{k}) - E),$$

where equation (16.22) is used in obtaining the second equality. As always, the integration in $\mathbf{k}$ space is confined to the first Brillouin zone (1. BZ). Also, the Fermi function,

$$f(E) = \frac{1}{1 + e^{(E - E_F)/(k_B T)}}, \qquad (16.24)$$

was introduced where E_F is the Fermi level, k_B is Boltzmann's constant, and T is the absolute temperature. Note that the Fermi level is defined as an electron energy, and usually zero is chosen at the bottom of the conduction band. For an unexcited semiconductor where the Fermi level is in the bandgap, it is a good approximation to use $f(E^m(\mathbf{k})) \approx 1$ and $f(E^\sigma(\mathbf{k})) \approx 0$ such that

$$\alpha(E) \approx \frac{e^2 \hbar}{8\pi^2 \epsilon_0 m_0^2 cnE} \sum_\sigma \sum_m \int_{1.BZ} d\mathbf{k} \left| \langle \sum_{\nu=1}^{6} g_\nu^m(\mathbf{k}) U_\nu |\varepsilon \cdot \mathbf{p}|S\sigma\rangle \right|^2 \qquad (16.25)$$
$$\times \delta(E^\sigma(\mathbf{k}) - E^m(\mathbf{k}) - E).$$

A similar calculation of the total spontaneous emission rate per volume, R_{sp}, in a semiconductor using equation (16.16) gives,

$$R_{sp} = \int r_{sp}(E)dE = \int dE \frac{e^2 nE}{24\pi^4 m_0^2 c^3 \epsilon_0 \hbar^2} \sum_\sigma \sum_m \int_{1.BZ} d\mathbf{k} \left| \langle \sum_{\nu=1}^{6} g_\nu^m(\mathbf{k}) U_\nu |\mathbf{p}|S\sigma\rangle \right|^2$$
$$\times f(E^\sigma(\mathbf{k}))(1 - f(E^m(\mathbf{k})))\delta(E^\sigma(\mathbf{k}) - E^m(\mathbf{k}) - E)$$
$$= \frac{e^2 n}{24\pi^4 m_0^2 c^3 \epsilon_0 \hbar^2} \sum_\sigma \sum_m \int_{1.BZ} d\mathbf{k}(E^\sigma(\mathbf{k}) - E^m(\mathbf{k})) \left| \langle \sum_{\nu=1}^{6} g_\nu^m(\mathbf{k}) U_\nu |\mathbf{p}|S\sigma\rangle \right|^2 \qquad (16.26)$$
$$\times f(E^\sigma(\mathbf{k}))(1 - f(E^m(\mathbf{k}))).$$

Note, as expected, that for an unexcited semiconductor where $E^\sigma(\mathbf{k}) = 0$ and $f(E^m(\mathbf{k})) = 1$, the spontaneous emission rate vanishes since the two factors appearing in the integrand of R_{sp}, $f(E^\sigma(\mathbf{k}))$ and $(1 - f(E^m(\mathbf{k})))$, both vanish.

The expressions equations (16.23) and (16.26) for the absorption coefficient and the spontaneous emission, respectively, are of central importance in all discussions of semiconductor optical properties. We shall discuss how these optical key parameters depend on the electronic band structure, strain, spontaneous polarization, and piezoelectric properties in the following sections.

16.2 The k · p method

We will in the following derive some general equations governing the electronic band structures of semiconductors. The theory we use is the k · p method dating back to the 1950s, where a number of key papers were published most notably by Dresselhaus *et al* [3], Luttinger and Kohn [4], and Kane [5]. The k · p method is a perturbative theory. We shall give a brief introduction to the k · p method before discussing some of the essential k · p Hamiltonians for important semiconductor crystal structures such as zincblende and wurtzite.

16.2.1 The k · p equation

The basis for a relativistic quantum mechanical analysis of electrons is the Dirac equation [2]. In the case of a bulk crystal the one-electron Schrödinger equation derived from the Dirac equation can be written as,

$$\left[\frac{\mathbf{p}^2}{2m_0} + \frac{\hbar}{4m_0^2 c^2}(\nabla V \times \mathbf{p}) \cdot \boldsymbol{\sigma} - \frac{\mathbf{p}^4}{8m_0^3 c^2} + \frac{1}{8}e\hbar^2 m_0^2 c^2 \rho(\mathbf{r}) + V(\mathbf{r}) \right]\psi_{n\mathbf{k}}(\mathbf{r}) = E_n(\mathbf{k})\psi_{n\mathbf{k}}(\mathbf{r}), \quad (16.27)$$

where $\boldsymbol{\sigma}$ are the Pauli spin matrices, ρ is the charge density, V is the crystalline potential, n is the band index, and k is the wave vector. The second, third, and fourth terms in equation (16.27) are relativistic corrections. The second term represents the spin interaction of the moving electron with the electric field, a term known as the spin–orbit interaction. The third and the fourth terms are the relativistic correction to the kinetic energy and the Darwin term accounting for electron coordinate fluctuations known as Zitterbewegung, respectively. In our description, however, we will neglect the relativistic correction terms except for the spin–orbit term. This is not because this term is numerically larger than the remaining corrections (in fact, its magnitude is smaller). The reason is that the spin–orbit term lifts the symmetry, splitting states otherwise degenerate in energy, thereby modifying the band structure and eigenstates significantly with important implications for, e.g., optical properties. Thus,

$$H\psi_{n\mathbf{k}}(\mathbf{r}) = E_n(\mathbf{k})\psi_{n\mathbf{k}}(\mathbf{r}),$$
$$H = \frac{\mathbf{p}^2}{2m_0} + \frac{\hbar}{4m_0^2 c^2}(\nabla V \times \mathbf{p}) \cdot \boldsymbol{\sigma} + V(\mathbf{r}). \quad (16.28)$$

Since crystal eigenstates must be Bloch states [6],

$$\psi_{n\mathbf{k}}(\mathbf{r}) = e^{i\mathbf{k}\cdot\mathbf{r}}u_{n\mathbf{k}}(\mathbf{r}), \quad (16.29)$$

where the cellular part $u_{n\mathbf{k}}$ obeys the crystal translation symmetry,

$$u_{nk}(\mathbf{r} + \mathbf{R}) = u_{nk}(\mathbf{r}). \tag{16.30}$$

The Bloch and cellular functions satisfy the following set of properties,

$$\langle \psi_{nk} | \psi_{n'k'} \rangle = \int dV \; \psi_{nk}^*(\mathbf{r}) \psi_{n'k'}(\mathbf{r}') = \delta_{nn'} \delta(\mathbf{k} - \mathbf{k}'), \tag{16.31}$$

$$\langle u_{nk} | u_{n'k'} \rangle = \int d\Omega \; u_{nk}^*(\mathbf{r}) u_{n'k'}(\mathbf{r}') = \delta_{nn'} \frac{\Omega}{(2\pi)^3}, \tag{16.32}$$

where V (Ω) is the crystal (unit-cell) volume.

Inserting equation (16.29) in equation (16.28) the Schrödinger equation written in terms of the cellular functions becomes,

$$H(\mathbf{k}) u_{nk}(\mathbf{r}) = E_n(\mathbf{k}) u_{nk}(\mathbf{r}), \tag{16.33}$$

where

$$H(\mathbf{k}) = H + H_{k \cdot p} + \frac{\hbar^2 k^2}{2m_0},$$

$$H_{k \cdot p} = \frac{\hbar}{m_0} \mathbf{k} \cdot \boldsymbol{\pi}, \tag{16.34}$$

$$\boldsymbol{\pi} = \mathbf{p} + \frac{\hbar}{4m_0 c^2} (\boldsymbol{\sigma} \times \boldsymbol{\nabla} V).$$

16.2.2 Effective masses in k · p

Let us proceed to determine effective masses using the **k · p** method. We will use a transformation due to Luttinger and Kohn [4, 6]. Starting out by expanding the cellular functions in terms of a complete set of periodic functions,

$$u_{nk}(\mathbf{r}) = \sum_{n'} A_{nn'}(\mathbf{k}) u_{n'0}(\mathbf{r}), \tag{16.35}$$

the **k · p** equation, equation (16.33), becomes,

$$\sum_{nn'} A_{nn'}(\mathbf{k}) \left(H + H_{k \cdot p} + \frac{\hbar^2 k^2}{2m_0} \right) u_{n'0}(\mathbf{r}) = \sum_{nn'} A_{nn'}(\mathbf{k}) \left(E_{n'}(0) + H_{k \cdot p} + \frac{\hbar^2 k^2}{2m_0} \right) u_{n'0}(\mathbf{r})$$
$$= E_{n'}(\mathbf{k}) u_{n'0}(\mathbf{r}). \tag{16.36}$$

Multiplying by $(2\pi)^3/\Omega \int_\Omega d\mathbf{r} \; u_{n0}^*(\mathbf{r})$ gives,

$$\left(E_n(0) + \frac{\hbar^2 k^2}{2m_0} \right) A_{nn}(\mathbf{k}) + \sum_{n'} \frac{\hbar k}{m_0} \cdot \boldsymbol{\pi}_{nn'} A_{nn'}(\mathbf{k}) = E_n(\mathbf{k}) A_{nn}(\mathbf{k}), \tag{16.37}$$

where

$$\boldsymbol{\pi}_{nn'} = \boldsymbol{\pi}_{nn'}(0) = \frac{(2\pi)^3}{\Omega} \int d\Omega \; u_{n0}^*(\mathbf{r}) \boldsymbol{\pi} u_{n'0}(\mathbf{r}), \tag{16.38}$$

where the orthogonal relations, equation (16.32) were used. Now one can write the preceding equation, dropping one band index,

$$H(\mathbf{k})A = E_n(\mathbf{k})A, \quad A = \begin{pmatrix} \vdots \\ \vdots \\ A_n \\ \vdots \\ \vdots \end{pmatrix}. \tag{16.39}$$

This is a set of linear equations. The solution involves uncoupling them which can be achieved by a canonical transformation,

$$A = TB, \tag{16.40}$$

where T is unitary in order to preserve normalization. Then,

$$\bar{H}(\mathbf{k})B = E_n(\mathbf{k})B, \tag{16.41}$$

where

$$\bar{H}(\mathbf{k}) = T^{-1}H(\mathbf{k})T. \tag{16.42}$$

Writing, $T = e^S$, $T^{-1} = e^{-S} = T^\dagger$,

$$\begin{aligned}
\bar{H}(\mathbf{k}) &= \left(1 - S + \frac{1}{2!}S^2 - \cdots\right)H(\mathbf{k})\left(1 + S + \frac{1}{2!}S^2 + \cdots\right) \\
&= H(\mathbf{k}) + [H(\mathbf{k}), S] + \frac{1}{2!}[[H(\mathbf{k}), S], S] + \cdots \\
&= H + H_{k\cdot p} + \frac{\hbar^2 k^2}{2m_0} + [H, S] + [H_{k\cdot p}, S] \\
&\quad + \frac{1}{2!}[[H, S], S] + \frac{1}{2!}[[H_{k\cdot p}, S], S] + \cdots.
\end{aligned} \tag{16.43}$$

Since $H_{k\cdot p}$ induces the coupling, one would like to remove it to order S by,

$$H_{k\cdot p} + [H, S] = 0, \tag{16.44}$$

so, with $|n\rangle = |u_{n0}\rangle$,

$$\begin{aligned}
\langle n|H_{k\cdot p}|n'\rangle + \sum_{n''}[\langle n|H|n''\rangle\langle n''|S|n'\rangle - \langle n|S|n''\rangle\langle n''|H|n'\rangle] &= 0, \\
\frac{\hbar}{m_0}\mathbf{k}\cdot\boldsymbol{\pi}_{nn'} + E_n(0)\langle n|S|n'\rangle - \langle n|S|n'\rangle E_{n'}(0) &= 0,
\end{aligned} \tag{16.45}$$

thus, assuming $E_n(0) \neq E_{n'}(0)$,

$$\langle n|S|n'\rangle = -\frac{\hbar}{m_0}\frac{\mathbf{k}\cdot\boldsymbol{\pi}_{nn'}}{[E_n(0) - E_{n'}(0)]}. \tag{16.46}$$

Now, equation (16.43) becomes,

$$\bar{H}(\mathbf{k}) = H + \frac{\hbar^2 k^2}{2m_0} + \frac{1}{2}[H_{k \cdot p}, S] + \frac{1}{2}[[H_{k \cdot p}, S], S] + \cdots, \qquad (16.47)$$

and, to second order in $\mathbf{k}$,

$$\langle n | \bar{H}(\mathbf{k}) | n' \rangle \approx \langle n | H | n' \rangle + \frac{\hbar^2 k^2}{2m_0} \delta_{nn'} + \frac{1}{2} \sum_{n''} \left[\langle n | H_{k \cdot p} | n'' \rangle \langle n'' | S | n' \rangle - \langle n | S | n'' \rangle \langle n'' | H_{k \cdot p} | n' \rangle \right]$$

$$= \left(E_n(0) + \frac{\hbar^2 k^2}{2m_0} \right) \delta_{nn'} + \frac{\hbar^2}{2m_0^2} \sum_{n''} \left[\frac{\mathbf{k} \cdot \boldsymbol{\pi}_{nn''} \mathbf{k} \cdot \boldsymbol{\pi}_{n''n'}}{[E_{n'}(0) - E_{n''}(0)]} + \frac{\mathbf{k} \cdot \boldsymbol{\pi}_{nn''} \mathbf{k} \cdot \boldsymbol{\pi}_{n''n'}}{[E_n(0) - E_{n''}(0)]} \right] \qquad (16.48)$$

$$= \left[E_n(0) + \frac{\hbar^2}{2} \sum_{\alpha\beta} k_\alpha \left(\frac{1}{m_n} \right)_{\alpha\beta} k_\beta \right] \delta_{nn'} + \text{interband terms of order } k^2,$$

which is the effective mass equation and $(\frac{1}{m_n})_{\alpha\beta}$ is the inverse effective mass tensor of band n defined by,

$$\left(\frac{1}{m_n} \right)_{\alpha\beta} \approx \frac{1}{m_0} \delta_{\alpha\beta} + \frac{2}{m_0^2} {\sum_l}' \frac{\pi_{nl}^\alpha \pi_{ln}^\beta}{E_n(0) - E_l(0)}, \qquad (16.49)$$

where $\sum_l'$ indicates that the sum is over states l for which the energy $E_l \neq E_n$.

16.3 Piezoelectric potential

In order to determine the influence of strain, spontaneous polarization, and piezoelectric properties on optical properties we first need to determine how the electronic band structure is affected by strain, spontaneous polarization, and piezoelectric properties.

Let us recapitulate how strain and electric fields are determined in solids. In general, the static strain and electric fields in a semiconductor in the absence of free charges are found by solving the coupled set of equations,

$$\frac{\partial T_{ij}}{\partial x_j} = 0, \qquad (16.50)$$

$$\frac{\partial D_i}{\partial x_i} = 0, \qquad (16.51)$$

and a complete set of boundary conditions, where $\mathbf{T}, \mathbf{D}$ are the stress tensor and the electric displacement, respectively, and summation over repeated indices $(i, j = 1, 2, 3)$ is assumed. Choosing, e.g., Cartesian coordinates, $x_1 = x$, $x_2 = y$, $x_3 = z$. The constitutive equations for stress and electric displacement are,

$$\mathbf{T} = \mathbf{cS} - \mathbf{eE}, \qquad (16.52)$$

$$\mathbf{D} = \epsilon \mathbf{E} + \mathbf{P} = \epsilon \mathbf{E} + \mathbf{P}_{\text{spon}} + \mathbf{eS}, \qquad (16.53)$$

$$P = P_{spon} + eS, \tag{16.54}$$

where S, E, P, and P_{spon} are the strain tensor, the electric field, the total polarization, and the spontaneous polarization, respectively, and c, e, and ϵ are the stiffness tensor, the piezoelectric e tensor, and the permittivity tensor, respectively. In the case of static fields, the electric field can be calculated as minus the gradient of a scalar (electric) potential φ,

$$E = -\nabla\phi,$$
$$\phi = \tilde{\phi} + \phi_{spon} + \phi_{piezo} + \phi_{spon-piezo}, \tag{16.55}$$

where the different contributions to the electric potential ϕ can be computed as follows: $\tilde{\phi}$ is the electric potential in the absence of both spontaneous polarization and piezoelectricity, ϕ_{spon} is the change in ϕ due to a non-vanishing spontaneous polarization and fixing the piezoelectric tensor to zero ($P_{spon} \neq 0$, $e = 0$), ϕ_{piezo} is the change in the scalar potential ϕ in the presence of piezoelectricity and fixing the spontaneous polarization to zero ($P_{spon} = 0$, $e \neq 0$). The electric potential contributions from coupling terms between spontaneous polarization and piezoelectricity, $\phi_{spon-piezo}$, are non-zero only if both e and P_{spon} are non-vanishing. When piezoelectricity is present ($e \neq 0$), the potential ϕ is often named the *piezoelectric potential*.

In zincblende semiconductors where the spontaneous polarization is zero but piezoelectricity is present,

$$E = -\nabla\phi,$$
$$\phi = \tilde{\phi} + \phi_{piezo}. \tag{16.56}$$

In wurtzite semiconductors where both spontaneous polarization and piezoelectricity are present,

$$E = -\nabla\phi,$$
$$\phi = \tilde{\phi} + \phi_{spon} + \phi_{piezo} + \phi_{spon-piezo}. \tag{16.57}$$

16.3.1 Zincblende structures

In zincblende crystals, point group $\bar{4}3m$ symmetry leads to the following relations for the (total) polarization P,

$$P_x = e_{14}S_{yz},$$
$$P_y = e_{14}S_{xz}, \tag{16.58}$$
$$P_z = e_{14}S_{xy}.$$

The stiffness and permittivity tensors are given by,

$$
\mathbf{c} = \begin{pmatrix}
c_{11} & c_{12} & c_{12} & 0 & 0 & 0 \\
c_{12} & c_{11} & c_{12} & 0 & 0 & 0 \\
c_{12} & c_{12} & c_{11} & 0 & 0 & 0 \\
0 & 0 & 0 & c_{44} & 0 & 0 \\
0 & 0 & 0 & 0 & c_{44} & 0 \\
0 & 0 & 0 & 0 & 0 & c_{44}
\end{pmatrix}, \quad
\boldsymbol{\epsilon} = \begin{pmatrix}
\epsilon_{11} & 0 & 0 \\
0 & \epsilon_{11} & 0 \\
0 & 0 & \epsilon_{11}
\end{pmatrix}.
\tag{16.59}
$$

16.3.2 Wurtzite structures

In wurtzite crystals, point group $6mm$ symmetry leads to the following relations for the polarization $\mathbf{P}$,

$$
\begin{aligned}
P_x &= e_{15} S_{xz}, \\
P_y &= e_{15} S_{yz}, \\
P_z &= e_{31}(S_{xx} + S_{yy}) + e_{33} S_{zz} + P_{\text{spon}},
\end{aligned}
\tag{16.60}
$$

where the z axis is chosen to correspond to the hexagonal axis (c axis) of the wurtzite crystal, i.e., $\mathbf{P}_{\text{spon}} = P_{\text{spon}}\hat{\mathbf{z}}$.

The stiffness and permittivity tensors are given by,

$$
\mathbf{c} = \begin{pmatrix}
c_{11} & c_{12} & c_{13} & 0 & 0 & 0 \\
c_{12} & c_{11} & c_{13} & 0 & 0 & 0 \\
c_{13} & c_{13} & c_{33} & 0 & 0 & 0 \\
0 & 0 & 0 & c_{44} & 0 & 0 \\
0 & 0 & 0 & 0 & c_{44} & 0 \\
0 & 0 & 0 & 0 & 0 & \frac{1}{2}(c_{11} - c_{12})
\end{pmatrix}, \quad
\boldsymbol{\epsilon} = \begin{pmatrix}
\epsilon_{11} & 0 & 0 \\
0 & \epsilon_{11} & 0 \\
0 & 0 & \epsilon_{33}
\end{pmatrix}.
\tag{16.61}
$$

16.4 Electron Hamiltonian

In accordance with the previous description of optical properties, we shall assume the conduction and valence bands be sufficiently separated in energy such that electron states can be treated independent of hole states. That is, we assume electron eigenstates are spin-degenerate $|S\rangle$ states with envelope functions Ψ_e that obey the one-particle effective-mass equation (following reference [7]),

$$
H_e \Psi_e(\mathbf{r}_e) = E_e \Psi_e(\mathbf{r}_e),
\tag{16.62}
$$

$$
H_e = H_e^K(\mathbf{r}_e) + H_e^S(\mathbf{r}_e) + E_c(\mathbf{r}_e) + e\phi(\mathbf{r}_e),
\tag{16.63}
$$

where H_e^K is the kinetic energy part of the electron Hamiltonian, H_e^S is the strain-dependent part of the electron Hamiltonian, E_c is the conduction-band edge, e is the positive elementary charge, and φ is the electric potential. The electric field is obtained from the electric potential as $\mathbf{E} = -\nabla\phi$. The total electron wave function ψ_e can be written as spin-degenerate S states,

$$\begin{aligned}
\psi_e^\uparrow(\mathbf{r}_e) &= \Psi_e(\mathbf{r}_e)|S\uparrow\rangle, \\
\psi_e^\downarrow(\mathbf{r}_e) &= \Psi_e(\mathbf{r}_e)|S\downarrow\rangle.
\end{aligned} \tag{16.64}$$

16.5 Hole Hamiltonian

Hole envelope functions are eigenstates of a six-band Hamiltonian,

$$H_h\Psi_h(\mathbf{r}_h) = E_h\Psi_h(\mathbf{r}_h), \tag{16.65}$$

where H_h is the 6×6 matrix of the hole Hamiltonian, Ψ_h is the six-component hole envelope function, and E_h is the hole energy. The total hole wave function can be written as a linear combination of P spin states, written as a matrix product,

$$(|X\uparrow\rangle, |Y\uparrow\rangle, |Z\uparrow\rangle, |X\downarrow\rangle, |Y\downarrow\rangle, |Z\downarrow\rangle) \cdot \Psi_h. \tag{16.66}$$

The hole Hamiltonian H_h can be written as

$$H_h = \begin{pmatrix} H_h^K(\mathbf{r}_h) + H_h^S(\mathbf{r}_h) & 0 \\ 0 & H_h^K(\mathbf{r}_h) + H_h^S(\mathbf{r}_h) \end{pmatrix} + (E_v(\mathbf{r}_h) + e\phi(\mathbf{r}_h))\mathbf{I}_{6\times6} + H_{so}(\mathbf{r}_h), \tag{16.67}$$

where H_h^K is a 3×3 matrix of the kinetic part of the hole Hamiltonian including the crystal field splitting for wurtzite structures. H_h^S is a 3×3 matrix of the strain-dependent part of the hole Hamiltonian, E_v is the valence band edge in the absence of strain, e is the positive elementary charge, φ is the electric potential, $\mathbf{I}_{6\times6}$ is the 6×6 identity matrix, and H_{so} is the spin–orbit Hamiltonian defined by,

$$H_{so}(\mathbf{r}) = \frac{\Delta_{so}}{3} \begin{pmatrix} -1 & -i & 0 & 0 & 0 & 1 \\ i & -1 & 0 & 0 & 0 & -i \\ 0 & 0 & -1 & -1 & i & 0 \\ 0 & 0 & -1 & -1 & i & 0 \\ 0 & 0 & -i & -i & -1 & 0 \\ 1 & i & 0 & 0 & 0 & -1 \end{pmatrix}, \tag{16.68}$$

where Δ_{so} is the spin–orbit splitting energy.

16.6 Zincblende

In this section the electron and hole Hamiltonian contributions that apply to the zincblende structure are given.

16.6.1 Electrons

The kinetic part of the electron Hamiltonian for zincblende structures is,

$$H_e^K = \frac{\hbar^2\mathbf{k}^2}{2m_0} = \frac{\hbar^2}{2m_0}(k_x^2 + k_y^2 + k_z^2). \tag{16.69}$$

The strain contribution to the electron Hamiltonian for zincblende structures is,

$$H_e^S = a_c(S_{xx} + S_{yy} + S_{zz}),$$ (16.70)

where a_c is the conduction-band deformation potential.

16.6.2 Holes

The kinetic part of the hole Hamiltonian for zincblende structures in the X, Y, Z states with spin is,

$$H_h^K = -\frac{\hbar^2}{2m_0}\begin{pmatrix} \tilde{\gamma}_1 k_x^2 + \tilde{\gamma}_2\left(k_y^2 + k_z^2\right) & 6\gamma_3 k_x k_y & 6\gamma_3 k_x k_z \\ 6\gamma_3 k_x k_y & \tilde{\gamma}_1 k_y^2 + \tilde{\gamma}_2\left(k_x^2 + k_z^2\right) & 6\gamma_3 k_y k_z \\ 6\gamma_3 k_x k_z & 6\gamma_3 k_y k_z & \tilde{\gamma}_1 k_z^2 + \tilde{\gamma}_2\left(k_x^2 + k_y^2\right) \end{pmatrix},$$ (16.71)

where $\tilde{\gamma}_1 = \gamma_1 + 4\gamma_2$, $\tilde{\gamma}_2 = \gamma_1 - 2\gamma_2$, and γ_1, γ_2, γ_3 are the Luttinger parameters of the valence band. The strain-dependent part of the Hamiltonian can be written as,

$$H_h^S = -a_v(S_{xx} + S_{yy} + S_{zz})\mathbf{I}_{3\times 3}$$
$$+ \begin{pmatrix} b(2S_{xx} - S_{yy} - S_{zz}) & \sqrt{3}\,dS_{xy} & \sqrt{3}\,dS_{xz} \\ \sqrt{3}\,dS_{xy} & b(2S_{yy} - S_{xx} - S_{zz}) & \sqrt{3}\,dS_{yz} \\ \sqrt{3}\,dS_{xz} & \sqrt{3}\,dS_{yz} & b(2S_{zz} - S_{xx} - S_{yy}) \end{pmatrix},$$ (16.72)

where $\mathbf{I}_{3\times 3}$ is the 3×3 identity matrix, and a_v, b, and d are the hydrostatic and two shear valence band deformation potential parameters, respectively.

16.7 Wurtzite

In this section the electron and hole Hamiltonian contributions that apply to the wurtzite structure are given.

16.7.1 Electrons

The kinetic part of the electron Hamiltonian for wurtzite structures is,

$$H_e^K = \frac{\hbar^2}{2m_e^\|}k_z^2 + \frac{\hbar^2}{2m_e^\perp}\left(k_x^2 + k_y^2\right),$$ (16.73)

where $m_e^\|$ and $m_e^\perp$ are electron effective masses in the absence of strain. The strain contribution to the electron Hamiltonian for wurtzite structures is,

$$H_e^S = a_c^\| S_{zz} + a_c^\perp(S_{xx} + S_{yy}),$$ (16.74)

where $a_c^\|$ and $a_c^\perp$ are the conduction-band deformation potentials.

16.7.2 Holes

The kinetic part of the hole Hamiltonian for wurtzite structures in the X, Y, Z states with spin is,

$$H_h^K = -\frac{\hbar^2}{2m_0} \begin{pmatrix} L_1 k_x^2 + M_1 k_y^2 + M_2 k_z^2 & 2A_5 k_x k_y & \sqrt{2} A_6 k_x k_z \\ 2A_5 k_x k_y & M_1 k_x^2 + L_1 k_y^2 + M_2 k_z^2 & \sqrt{2} A_6 k_y k_z \\ \sqrt{2} A_6 k_x k_z & \sqrt{2} A_6 k_y k_z & M_3 k_x^2 + M_3 k_y^2 + L_2 k_z^2 - \delta_{cr} \end{pmatrix}, \quad (16.75)$$

where

$$\begin{aligned} L_1 &= A_2 + A_4 + A_5, \quad L_2 = A_1, \\ M_1 &= A_2 + A_4 - A_5, \quad M_2 = A_1 + A_3, \quad M_3 = A_2, \\ \delta_{cr} &= 2m_0 \Delta_{cr}/\hbar^2, \end{aligned} \quad (16.76)$$

where $A_k (k = 1,\dots, 6)$ are the Rashba–Sheka–Pikus parameters of the valence band and Δ_{cr} is the crystal-field splitting energy.

The strain-dependent part H_h^S of the wurtzite Hamiltonian is,

$$H_h^S = \begin{pmatrix} l_1 S_{xx} + m_1 S_{yy} + m_2 S_{zz} & n_1 S_{xy} & n_2 S_{xz} \\ n_1 S_{xy} & m_1 S_{xx} + l_1 S_{yy} + m_2 S_{zz} & n_2 S_{yz} \\ n_2 S_{xz} & n_2 S_{yz} & m_3 (S_{xx} + S_{yy}) + l_2 S_{zz} \end{pmatrix}, \quad (16.77)$$

where

$$\begin{aligned} l_1 &= D_2 + D_4 + D_5, \quad l_2 = D_1, \\ m_1 &= D_2 + D_4 - D_5, \quad m_2 = D_1 + D_3, \quad m_3 = D_2, \\ n_1 &= 2D_5, \quad n_2 = \sqrt{2} D_6, \end{aligned} \quad (16.78)$$

and $D_k (k = 1,\dots, 6)$ are valence-band deformation potentials.

16.7.3 Exciton effects in strongly confined systems

Later, the electronic band structures and radiative decay times of quantum-dot heterostructure materials for which strain and piezoelectric effects play a significant role will be addressed. For that purpose, we now discuss exciton effects in heterostructures as well as bulk semiconductors.

For a quantum-dot heterostructure, the Coulomb potential energy of the electron–hole system is [8],

$$U(\mathbf{r}_e, \mathbf{r}_h) = U_{int}(\mathbf{r}_e, \mathbf{r}_h) + U_{self}(\mathbf{r}_e) + U_{self}(\mathbf{r}_h), \quad (16.79)$$

where U_{int} is the electron–hole interaction energy determined from the Maxwell–Poisson equation,

$$\nabla_{\mathbf{r}_h} \cdot [\epsilon_r(\mathbf{r}_h) \nabla_{\mathbf{r}_h} U_{int}(\mathbf{r}_e, \mathbf{r}_h)] = \frac{e^2}{\epsilon_0 \delta(\mathbf{r}_e - \mathbf{r}_h)}, \quad (16.80)$$

and $\epsilon_r(\mathbf{r}_h)$ is the optical dielectric constant of the semiconductor. The electron and hole self-interaction energies are obtained as,

$$U_{\text{self}}(\mathbf{r}) = -\frac{1}{2} \lim_{\mathbf{r}' \to \mathbf{r}} \left[U_{\text{int}}(\mathbf{r}, \mathbf{r}') - U_{\text{int}}^{\text{bulk}}(\mathbf{r}, \mathbf{r}') \right], \tag{16.81}$$

where $U_{\text{int}}^{\text{bulk}}(\mathbf{r}, \mathbf{r}')$ is the local bulk Coulomb potential energy of an electron–hole system,

$$U_{\text{int}}^{\text{bulk}}(\mathbf{r}, \mathbf{r}') = -\frac{e^2}{4\pi \epsilon_0 \epsilon_r(\mathbf{r})|\mathbf{r} - \mathbf{r}'|}. \tag{16.82}$$

For a strongly confined system, the exciton wave function can be written as the product of the single electron and hole wave functions in the absence of the Coulomb interaction,

$$\psi_{\text{exc}}(\mathbf{r}_e, \mathbf{r}_h) = \psi_e^*(\mathbf{r}_e)\psi_e(\mathbf{r}_h), \tag{16.83}$$

and the exciton energy E_{exc} can be calculated by considering the Coulomb interaction as a perturbation,

$$E_{\text{exc}} = E_e - E_h + \int_V d\mathbf{r}_e \int_V d\mathbf{r}_h \, U(\mathbf{r}_e, \mathbf{r}_h)|\psi_{\text{exc}}(\mathbf{r}_e, \mathbf{r}_h)|^2, \tag{16.84}$$

where V is the system volume and E_e and E_h are the single-particle electron and hole energies, respectively, found by solving separately the $\mathbf{k} \cdot \mathbf{p}$ electron and hole problems in accordance with the description above. Note that the exciton energy is less than the energy difference $E_e - E_h$ due to the negativity of the Coulomb potential energy U. In bulk systems a perturbative treatment fails since the exciton energy shifts are larger than or comparable to differences between the unperturbed single-particle energies. Instead, for bulk systems, one can resort to variational methods or solving the full exciton band structure problem in a truncated set of basis states where usually the latter are chosen as eigenstates of the electron–hole problem in the absence of Coulomb interaction [9–11].

The oscillator strength f is determined as,

$$f = \frac{2\hbar^2}{m_0 E_{\text{exc}}} \sum_\alpha \left| \int_V d\mathbf{r} \, \psi_e^*(\mathbf{r})(\varepsilon \cdot \mathbf{k})\psi_h^\alpha(\mathbf{r}) \right|^2, \tag{16.85}$$

where ε is the polarization of incident light, $\mathbf{k} = -i\nabla$ is the wave number operator, and the sum over α is over all degenerate hole states with energy E_h.

Having determined the oscillator strength, the radiative decay time τ is obtained as,

$$\tau = \frac{2\pi \epsilon_0 m_0 c^3 \hbar^2}{ne^2 E_{\text{exc}}^2 f}. \tag{16.86}$$

16.7.4 Bulk wurtzite GaN band structure

We now discuss electron and valence band structures of bulk wurtzite GaN following Chuang and Chang [12]. They used two sets of band parameters to compute the solid and dashed lines shown in figure 16.2 for the A, B, and C transition lines (corresponding to transitions between the lowest conduction band

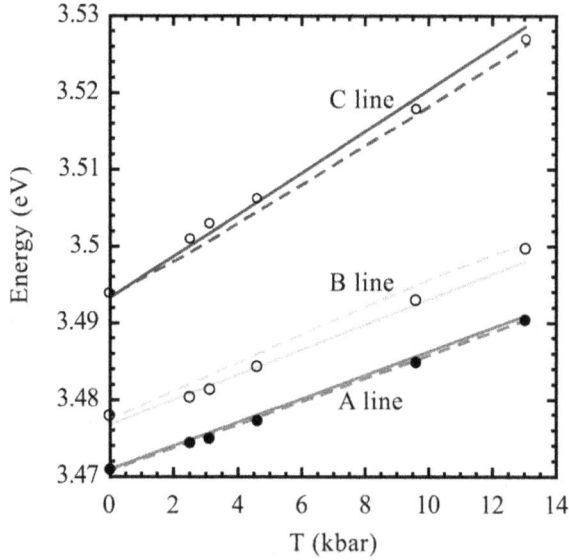

Figure 16.2. Interband energies of bulk wurtzite GaN calculated for the transitions between the conduction band and the three valence bands, also called the A, B, and C lines. The symbols are experimental data from GaN samples at temperatures below 10 K collected from reference [13]. The solid and dashed lines are calculated from two sets of parameter data. Calculated transition energies are corrected by an exciton energy E_{ex} of approximately 0.028 eV. Adapted from reference [12], copyright (1996) with permission from American Physical Society.

and the first three valence bands). The anisotropy of the conduction band deformation potentials is neglected, i.e., $a_c^{\parallel} = a_c^{\perp} = a_c$ and the energy gap at $T = 300$ K is $E_g = 3.44$eV. For the dashed lines, spin–orbit parameters are fixed to $\Delta_2 = 6.2$ meV and $\Delta_3 = 5.5$ meV, respectively [12]. The crystal field splitting is $\Delta_{cr} = 16$ meV. Similarly, for the dashed lines, the conduction-band deformation potential is $a_c = 0$ and the valence-band deformation potentials are set to: $D_1 = D_2 = 8.16$eV, $D_3 = D_4 = -3.71$ eV. The solid lines in figure 16.2 correspond to a different set of parameter values: $a_c = -4.08$ eV, $\Delta_2 = \Delta_3 = \Delta_{so}/3 = 4$ meV, $D_1 = 0.7$ eV, $D_2 = 2.1$ eV, $D_3 = 1.4$ eV, and $D_4 = -0.7$ eV. Experimental data points from reference [13] are shown as symbols. We note in passing that the spin–orbit Hamiltonian in reference [7], our equation (16.68), assumes $\Delta_2 = \Delta_3 = \Delta_{so}/3$. For both the solid and dashed lines, transition energies are corrected by subtracting the exciton energy $E_{ex} \approx 0.028$eV. The agreement between model results and experimental data is evident for a range of applied compressive stresses T [12].

In figure 16.3, the conduction- and valence-band energy shifts versus in-plane compressive strain ϵ_{xx} are plotted. It can be seen that the bandgap energy increases with increasing compressive strain. If $\Delta_{cr} > \Delta_2 = \Delta_{so}/3$, which is the case for wurtzite GaN, the topmost valence band is a heavy-hole band (HH), the middle valence band is a light-hole band (LH), and the bottom valence band is the crystal-field split-off hole band (CH).

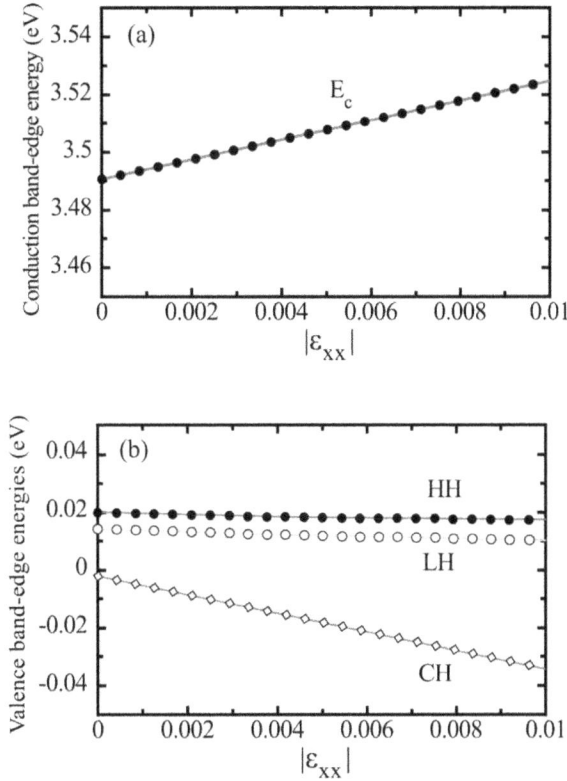

Figure 16.3. Compressive strain effects on the conduction and valence bands of bulk wurtzite GaN. The conduction-band edge increases while the three valence-band edges decrease with increasing in-plane compressive strain. The energy shifts are linear in the strain. Adapted from reference [12], copyright (1996) with permission from American Physical Society.

In addition to the last set of parameters above (used to obtain the solid lines in figure 16.2) we calculate the band structure of bulk wurtzite GaN using the conduction-band effective masses $m_e^\parallel = 0.18m_0$, $m_e^\perp = 0.20m_0$. The valence-band effective mass parameters are: $A_1 = -6.56$, $A_2 = -0.91$, $A_3 = 5.65$, $A_4 = -2.83$, $A_5 = -3.13$, and $A_6 = -4.86$.

Figure 16.4 shows the valence band structures of bulk wurtzite GaN as function of wavenumbers $k^\parallel = k_x$ and $k_z = k^\perp$. In the upper panel, the *HH*, *LH*, and *CH* bands are plotted in the absence of strain. In the lower panel, the same bands are shown for a compressive negative strain $\epsilon_{xx} = -1\%$. Note that the crystal-field split-off band separates from the heavy-hole and light-hole bands in the presence of compressive strain. The band curvatures (effective masses) are clearly sensitive to strain.

16.8 Band structure of heterostructures

With the possibility to grow atomically-sharp heterostructures in the 1970s using molecular beam epitaxy and metal-organo chemical vapor deposition techniques,

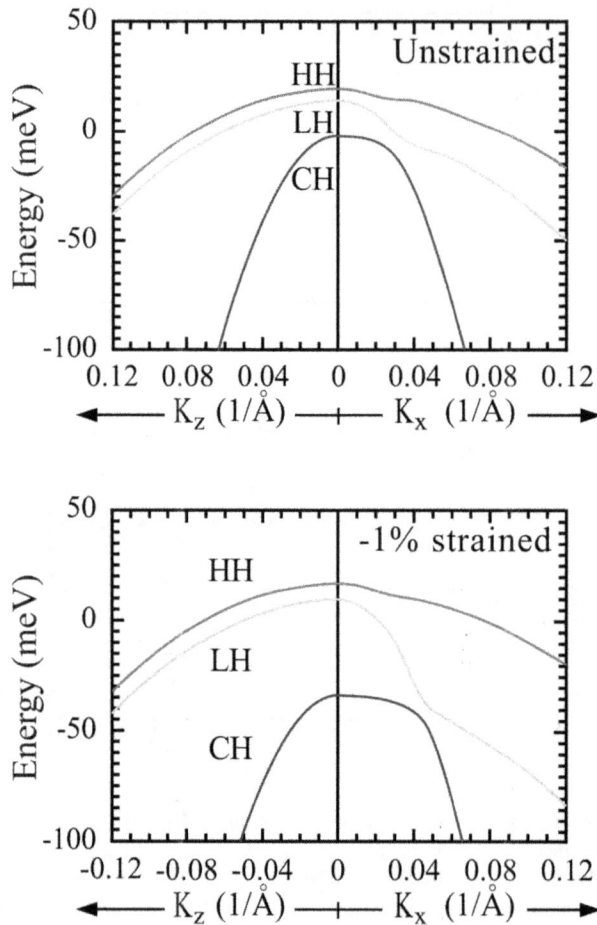

Figure 16.4. Valence bands of bulk wurtzite GaN plotted as function of wavenumbers k_x and k_z for (upper panel) unstrained and (lower panel) 1% negative compressive strain Adapted from reference [12], copyright (1996) with permission from American Physical Society.

new methods were needed to compute band structure effects due to varying material properties [6]. In essence, three continuous methods have been applied to attack heterostructure (or nanostructure) band structures: (1) a one-band or particle-in-a-box model [14]. Another is the envelope-function theory of Bastard [15]. Both are ad hoc theories in the sense that Hamiltonian hermiticity is restored by invoking symmetrization of all operator terms The final is the so-called first-principles envelope function theory due to Burt [16] where operator term ordering is dictated by imposing envelope functions to satisfy differentiability to infinite order. Burt's theory, however, requires a number of approximations and leads to a non-local term in the differential equations of the envelope functions which is important only near interfaces and therefore can be dropped in a first attempt. Foreman later used Burt's method to derive the band structure of the valence band of zincblende crystals [17].

In the following, a brief discussion of Bastard's approach to determine envelope functions of heterostructures is presented.

Consider a heterostructure composed of materials A and B. The simple envelope function model assumes the wave function in each material can be expanded in the periodic (cellular) parts $u_{n\mathbf{k}_0}^{(A,B)}$ at a given wave vector $\mathbf{k}_0$ (for example the Γ point of cubic crystals),

$$\psi(\mathbf{r}) = \sum_n f_n^{(A,B)}(\mathbf{r}) u_{n\mathbf{k}_0}^{(A,B)}(\mathbf{r}), \tag{16.87}$$

where the sum over n runs over the number of bands in consideration. The idea is then to match solutions across material interfaces to uniquely determine the envelope functions $f_n^{(A,B)}$.

It is then assumed that the cellular parts are the same in the two materials, i.e., $u_{n\mathbf{k}_0}^A = u_{n\mathbf{k}_0}^B = u_n$ which is a good approximation for chemically and structurally similar materials such as the GaAs–AlAs material system.

16.8.1 Application to a planar interface, e.g., a quantum well

For a planar interface between two materials A and B with the growth direction along the z direction, Bloch periodicity in the plane allows us to write,

$$f_n^{(A,B)}(\mathbf{r}_\parallel, z) = \frac{1}{\sqrt{A}} e^{i\mathbf{k}_\parallel \cdot \mathbf{r}_\parallel} \chi_n^{(A,B)}(z), \tag{16.88}$$

where $\mathbf{r}_\parallel = (x, y)$ is the in-plane position vector, A is the surface area in the same plane, and $\mathbf{k}_\parallel$ is the in-plane wave vector. The Schrödinger equation can be written as,

$$H\psi(\mathbf{r}) = E\psi(\mathbf{r}), \tag{16.89}$$

where

$$H = \frac{p^2}{2m_0} + V_A \theta_A(\mathbf{r}) + V_B \theta_B(\mathbf{r}), \tag{16.90}$$

and θ is a step function.

In order to obtain a set of differential equations in the unknown envelope functions, $f_n^{(A,B)}$, it is now used that the envelope functions are slowly varying and approximately constant over a unit cell. Hence, when integrating over a unit cell V_0, the following approximation applies,

$$\int_{V_0} d\mathbf{r} \, \chi_n(\mathbf{r}) u_n(\mathbf{r}) \approx \chi_n(\mathbf{r}) \int_{V_0} d\mathbf{r} \, u_n(\mathbf{r}). \tag{16.91}$$

Inserting equations (16.87) and (16.88) in equation (16.89), premultiplying by $e^{-i\mathbf{k}_\parallel \cdot \mathbf{r}_\parallel} u_n^*(\mathbf{r})$, and integrating over a unit cell gives [18], gives a set of differential equations in the envelope-function parts χ_n,

$$D\left(z, k_z = -i\frac{\partial}{\partial z}\right)\chi = E\chi, \tag{16.92}$$

where $\chi = \{\chi_n\}$ is a $N \times 1$ vector of the z-part envelope functions, and N is the number of bands in the multiband model. Note that the breaking of translational symmetry along the z coordinate implies that k_z is not a good quantum number so $k_z = -i\frac{\partial}{\partial z}$. The operator D is a $N \times N$ matrix operator and becomes, for the first-order Kane model [18],

$$\begin{aligned}
D_{mn}^{(1)}\left(z, -i\frac{\partial}{\partial z}\right) = &\left[E_m^{(A)}(0)\theta_A(z) + E_m^{(B)}(0)\theta_B(z) + \frac{\hbar^2 k_\parallel^2}{2m_0} - \frac{\hbar^2}{2m_0}\frac{\partial^2}{\partial z^2}\right]\delta_{mn} \\
&+ \frac{\hbar \mathbf{k}_\parallel}{m_0} \cdot \langle m|\mathbf{p}_\parallel|n\rangle - i\frac{\hbar}{m_0} \cdot \langle m|p_z|n\rangle\frac{\partial}{\partial z},
\end{aligned} \tag{16.93}$$

where $E_m^{(A)}(0)$ denotes the energy of the mth band at the Γ point in material A, etc.

For a second-order Kane model where coupling to remote bands ν is included, the **D** matrix is,

$$\begin{aligned}
D_{mn}^{(2)}\left(z, -i\frac{\partial}{\partial z}\right) = &\left[E_m^{(A)}(0)\theta_A(z) + E_m^{(B)}(0)\theta_B(z) + \frac{\hbar^2 k_\parallel^2}{2m_0} - \frac{\hbar^2}{2m_0}\frac{\partial^2}{\partial z^2}\right]\delta_{mn} \\
&+ \frac{\hbar \mathbf{k}_\parallel}{m_0} \cdot \langle m|\mathbf{p}_\parallel|n\rangle - i\frac{\hbar}{m_0} \cdot \langle m|p_z|n\rangle\frac{\partial}{\partial z} - \frac{\hbar^2}{2}\frac{\partial}{\partial z}\frac{1}{M_{mn}^{zz}}\frac{\partial}{\partial z} \\
&- \frac{i\hbar^2}{2}\sum_{i=x,y}\left[k_i\frac{1}{M_{mn}^{iz}}\frac{\partial}{\partial z} + \frac{\partial}{\partial z}\frac{1}{M_{mn}^{zi}}k_i\right] + \frac{\hbar^2}{2}\sum_{i,j=x,y}k_i\frac{1}{M_{mn}^{ij}}k_j,
\end{aligned} \tag{16.94}$$

where

$$\frac{m_0}{M_{mn}^{ij}} = \frac{2}{m_0}\sum_\nu \langle m|p_i|\nu\rangle \frac{1}{\bar{E} - E_\nu^{(A)}(0) - V_\nu(z)}\langle \nu|p_j|n\rangle, \tag{16.95}$$

and $V_\nu(z) = (E_\nu^{(B)}(0) - E_\nu^{(A)}(0))\theta_B(z)$. It should be pointed out that a symmetrization scheme is imposed on equation (16.94). In noting this, observe that momentum matrix elements are assumed to be the same in materials A and B which is a good approximation for III–V heterostructure materials.

16.8.2 Strain and piezoelectricity effects in wurtzite and zincblende quantum-dot heterostructures

Using Burt's envelope function theory, applied to the bulk wurtzite and zincblende electron and hole Hamiltonians above, Balandin *et al* [7] calculated excitonic properties of strained wurtzite and zincblende GaN quantum dots embedded in a AlN matrix. The analysis of Balandin *et al* will now be presented in detail [7]. With reference to figure 16.5 they considered truncated pyramidal wurtzite quantum-dot structures of variable heights H, wetting-layer thickness $w = 0.5$ nm, quantum-dot

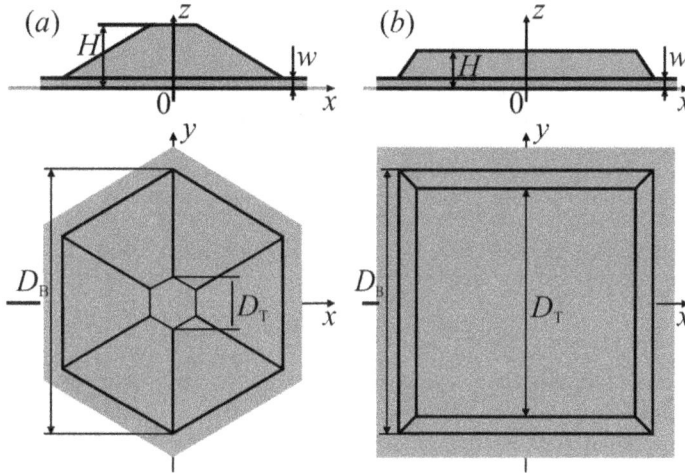

Figure 16.5. Shapes and dimensions of (a) wurtzite GaN/AlN and (b) zincblende GaN/AlN quantum dot heterostructures. Adapted from reference [7], copyright (2003) with permission from American Institute of Physics.

bottom diameter $D_B = 5(H - w)$, and top diameter $D_T = H - w$. Similarly, for truncated pyramidal zincblende quantum dots, parameter values $w = 0.5$ nm, quantum-dot bottom base length $D_B = 10(H - w)$, and quantum-dot top base length $D_T = 8.6(H - w)$ were used. For material parameters of wurtzite and zincblende GaN/AlN they used parameters from Vurgaftman *et al* [19] and other references in reference [7].

From the calculated strain tensor of the GaN/AlN quantum-dot structure, the total electric potential φ in the wurtzite and zincblende quantum-dot structures, from which the electric field is obtained as $\mathbf{E} = -\boldsymbol{\nabla}\phi$, was calculated by solving the Maxwell–Poisson equation using the finite-difference method [7]. Note that the piezoelectric potential V_p (equation (13) in reference [7]) corresponds to φ in the present work.

Figure 16.6 shows the piezoelectric potentials in wurtzite and zincblende GaN/AlN QDs with height 3 nm. Surprisingly, it is seen that the magnitude of the piezoelectric potential in the wurtzite GaN/AlN quantum dot is about ten times its magnitude in the zincblende GaN/AlN quantum dot. Moreover, the piezoelectric potential in the wurtzite quantum dot has maxima near the quantum dot top and bottom, while the maxima of the piezoelectric potential in the zincblende quantum dot lie outside the quantum dot. The above facts explain why the piezoelectric field has a strong effect on the excitonic properties of wurtzite GaN/AlN quantum dots while it has very little effect on those in zincblende GaN/AlN quantum dots.

In figure 16.7, the conduction and valence band edges of wurtzite GaN/AlN quantum-dot structures (height 3 nm) in the presence (solid) and absence (dashed–dotted) of strain are plotted as function of (a) z and (b) x. The gray curves are the electron and hole groundstate energies. Note that the piezo potential shifts electron

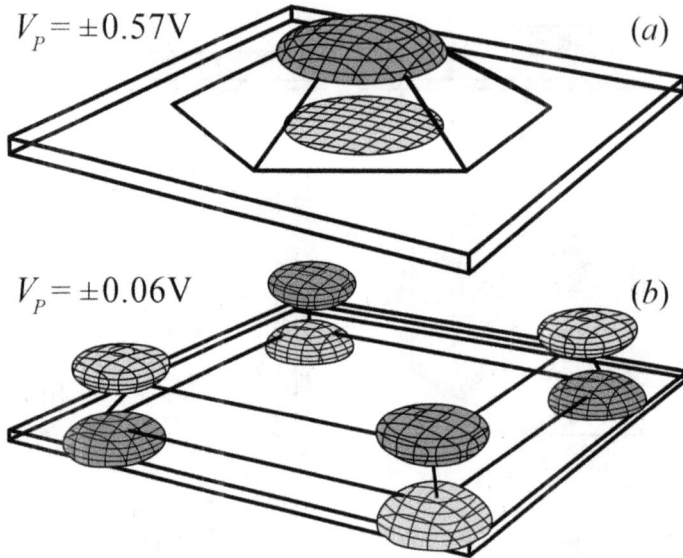

Figure 16.6. Piezoelectric potential in wurtzite GaN/AlN and zincblende GaN/AlN quantum dots with height 3 nm. Light and dark surfaces represent positive and negative values of the piezoelectric potential, correspondingly. Adapted from reference [7], copyright (2003) with permission from American Institute of Physics.

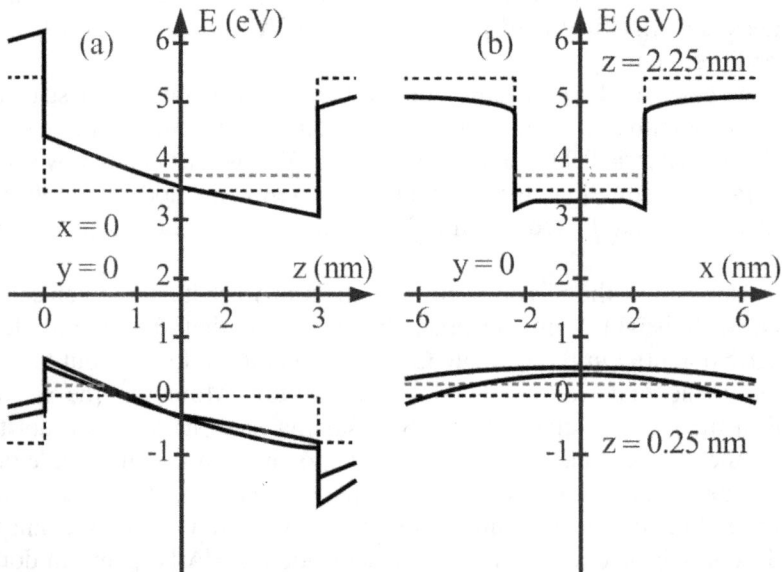

Figure 16.7. Conduction and valence-band edges along the (a) z axis and (b) x axis for wurtzite GaN/AlN quantum-dot structures with height 3 nm. The solid (dashed–dotted) lines are calculated in the presence (absence) of the strain field due to lattice mismatch. The grey curves show the electron and hole groundstate energies. Adapted from reference [7], copyright (2003) with permission from American Institute of Physics.

states towards the top of the quantum dot (high z values), where the conduction-band edge is lowest, while holes are shifted to the bottom of the quantum dot (where the valence-band edge is highest). In figure 16.8, the conduction and valence band edges of zincblende GaN/AlN quantum-dot structures (height 3 nm) in the presence (solid) and absence (dashed–dotted) of strain are plotted as function of (a) z and (b) x. The gray curves are the electron and hole groundstate energies. Evidently, for the zincblende structures, the conduction and band edges are not tilted asymmetrically with respect to the z coordinate as in the case of the wurtzite structures.

In figure 16.9, the first four electron states in a wurtzite (left panel) and zincblende (right panel) GaN/AlN quantum dot with height 3 nm are plotted. With reference to the conduction-band edge profiles of figures 16.7 and 16.8, it is evident why the electron in the wurtzite GaN/AlN QD is pushed to the quantum dot top while the electron in the zincblende GaN/AlN quantum dot is distributed over the entire quantum dot.

Figure 16.10 shows the first four hole states in a wurtzite (left panel) and zincblende (right panel) GaN/AlN quantum dot with height 3 nm. The strong piezo potential in the wurtzite quantum dot structure forces the first four hole states to be located in the wetting layer and near the bottom of the quantum dot. In the zincblende quantum dot, holes are forced towards the center of the quantum dot due to the bending and maximum of the zincblende quantum-dot valence band edge at the center of the quantum dot (refer to figure 16.8).

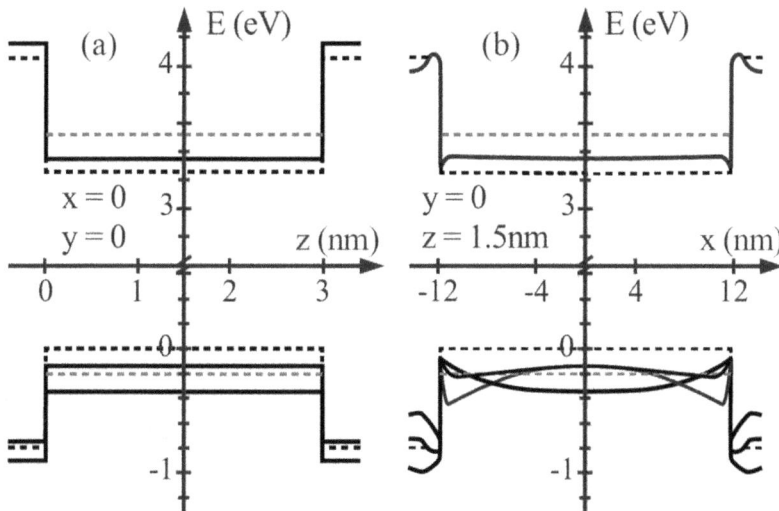

Figure 16.8. Conduction and valence-band edges along the (a) z axis and (b) x axis for zincblende GaN/AlN quantum-dot structures with height 3 nm. The solid (dashed–dotted) lines are calculated in the presence (absence) of the strain field due to lattice mismatch. The grey curves show the electron and hole groundstate energies. Adapted from reference [7], copyright (2003) with permission from American Institute of Physics.

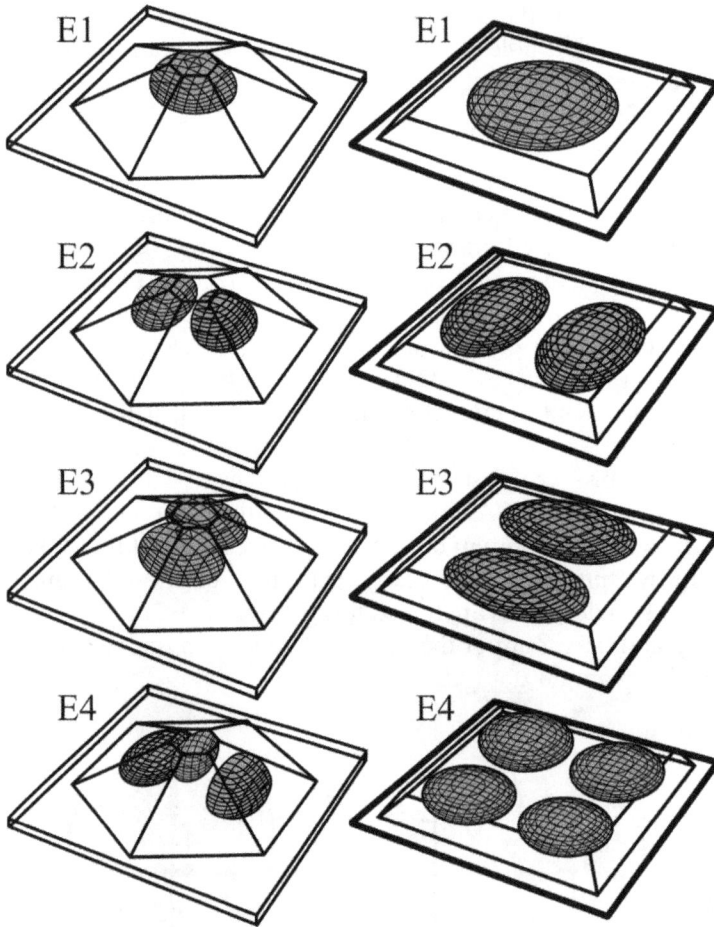

Figure 16.9. First four electron states (E1-E4) of a GaN/AlN wurtzite (zincblende) quantum dot with height 3 nm are shown in the left (right) panel. With reference to the conduction-band edge profiles of figures 16.7 and 16.8, it is evident why the electron in the wurtzite GaN/AlN QD is pushed to the quantum dot top while the electron in the zincblende GaN/AlN quantum dot is distributed over the entire quantum dot. Adapted from reference [7], copyright (2003) with permission from American Institute of Physics.

We have just emphasized the clear effects of a strong piezo potential (wurtzite quantum dots) versus a weak piezo potential (zincblende quantum dots) on electron and hole states. There is also a pronounced difference between wurtzite and zincblende GaN/AlN quantum-dot structures with respect to both single-particle and exciton energies. The differences in the location of electron and hole states in the wurtzite and zincblende GaN/AlN quantum-dot structures have significant implications on the overlap of, and momentum matrix elements between, electron and hole states in the two structures. In effect, the oscillator strength is much higher in

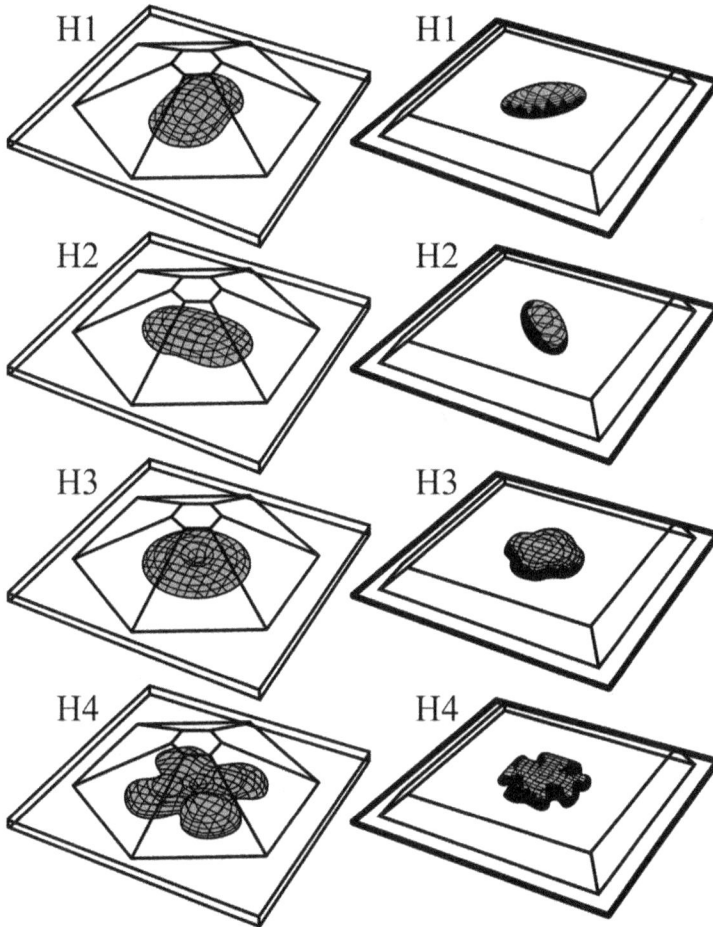

Figure 16.10. First four hole states (H1–H4) of a GaN/AlN wurtzite (zincblende) quantum dot with height 3 nm are shown in the left (right) panel. Notice the major difference in the location of the hole states between the wurtzite and zincblende quantum dots. In the wurtzite quantum dot, the strong piezo potential forces the first four hole states to be located in the wetting layer and near the bottom of the quantum dot while the first four hole states in the zincblende quantum dot are located near the center of the quantum dot due to the bending and maximum of the valence-band edge at the center. Adapted from reference [7], copyright (2003) with permission from American Institute of Physics.

the zincblende GaN/AlN quantum-dot structure compared to the wurtzite quantum-dot structure. Similarly, the radiative decay rate is much smaller in the zincblende GaN/AlN quantum-dot structure compared to the wurtzite quantum-dot structure. The combined influence on states and energies from the piezo potential in wurtzite and zincblende quantum-dot structures is evident from figure 16.11 where the radiative decay rate is shown versus GaN/AlN quantum-dot height.

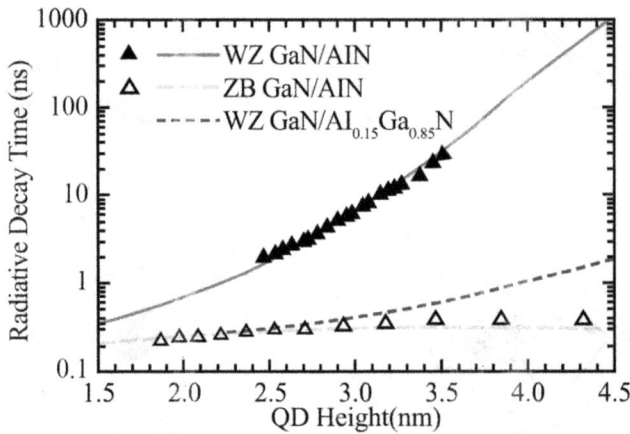

Figure 16.11. Radiative decay rate for wurtzite GaN/AlN (solid), zincblende GaN/AlN (large-dashed), and wurtzite GaN/Al $_{0.15}$ Ga $_{0.85}$N (small-dashed) quantum-dot structures as function of the quantum-dot height. Filled and empty triangles represent experimental points from reference [20, 21] for wurtzite GaN/AlN and zincblende GaN/AlN quantum dots, respectively. Adapted from reference [7], copyright (2003) with permission from American Institute of Physics.

References

[1] Willatzen M 1993 Theory of gain in bulk and quantum-well semiconductor lasers *PhD Thesis* Niels Bohr Institute, University of Copenhagen

[2] Sakurai J J 1987 *Advanced Quantum Mechanics* (Reading, MA: Addison-Wesley)

[3] Dresselhaus G, Kip A F and Kittel C 1955 Cyclotron resonance of electrons and holes in silicon and germanium crystals *Phys. Rev.* **B98** 368

[4] Luttinger J M and Kohn W 1955 Motion of electrons and holes in perturbed periodic fields *Phys. Rev.* **97** 869

[5] Kane E O 1956 Energy band structure in p-type germanium and silicon *J. Phys. Chem. Sol.* **1** 82

[6] Lew Yan Voon L C and Willatzen M 2009 *The k · p Method–Electronic Properties of Semiconductors* 1st edn (Berlin: Springer)

[7] Fonoberov V A and Balandin A A 2003 Excitonic properties of strained wurtzite and zincblende GaN/Al$_x$Ga $_{1-x}$ N quantum dots *J. Appl. Phys.* **94** 7178–86

[8] Fonoberov V A, Pokatilov E P and Balandin A A 2002 Exciton states and optical transitions in colloidal CdS quantum dots: shape and dielectric mismatch effects *Phys. Rev.* **B085310** 66

[9] Knox R S 1963 *Theory of Excitons* (New York: Academic)

[10] Taherkhani M, Willatzen M, Mørk J, Gregersen N and McCutcheon D P S 2017 Type-II quantum-dot-in-nanowire structures with large oscillator strength for optical quantum gate applications *Phys. Rev.* **B96** 125408

[11] Taherkhani M, Willatzen M, Denning E, Protsenko I and Gregersen N 2019 High-fidelity optical quantum gates based on type-II double quantum dots in a nanowire *Phys. Rev.* **B99** 165305

[12] Chuang S L and Chang C S 1996 **k · p** method for strained wurtzite semiconductors *Phys. Rev.* **B54** 2491

[13] Gil B, Briot O and Aulombard R-L 1995 Valence-band physics and the optical properties of GaN epilayers grown onto sapphire with wurtzite symmetry *Phys. Rev.* **B52** R17028

[14] Dingle R, Wiegmann W and Henry C H 1974 Quantum states of confined carriers in very thin GaAs-Al$_x$Ga$_{1-x}$As heterostructures *Phys. Rev. Lett.* **33** 827

[15] Bastard G 1981 Superlattice band structure in the envelope-function approximation *Phys. Rev.* **B24** 5693

[16] Burt M G 1987 An exact formulation of the envelope function method for the determination of electronic states in semiconductor microstructures *Semicond. Sci. Technol.* **2** 460

[17] Foreman B A 1993 Effective-mass Hamiltonian and boundary conditions for the valence bands of semiconductor microstructures *Phys. Rev.* **B4964** 48

[18] Bastard G 1988 *Wave Mechanics Applied to Semiconductor Heterostructures* (Les Ulis: Les Editions de Physique)

[19] Vurgaftman I, Meyer J R and Ram-Mohan L R 2001 Band parameters for III-V compound semiconductors and their alloys *Appl. Phys. Rev.* **89** 5815

[20] Dang L S *et al* 2003 GaN quantum dots: physics and applications *J. Korean Phys. Soc.* **42** S657–61

[21] Simon J, Pelekanos N T, Adelmann C, Martinez-Guerrero E, Andre R, Daudin B, Si Dang L and Mariette H 2003 Direct comparison of recombination dynamics in cubic and hexagonal GaN/AlN quantum dots *Phys. Rev.* **B62**

IOP Publishing

Piezoelectricity in Classical and Modern Systems

Morten Willatzen

Chapter 17

Sonoluminescence

Sonoluminescence is an exotic effect that requires a multi-physics analysis, including concepts from fluid mechanics, thermodynamics, elasticity, ultrasonics, optics, quantum mechanics, and atomic physics. This topic has recently attracted substantial interest in the general physics community. Single-bubble sonoluminescence, as presented in chapter 17, represents a different type of coupling between ultrasonics, typically generated via piezoelectric active components, and optics. For this reason, the treatment of sonoluminescence supplements well the analysis and applications of piezoelectricity presented in the other chapters of this book.

17.1 Sonoluminescense due to bubble oscillations

Sonoluminescence may emerge when bubbles oscillate in the presence of a strong external ultrasonic field or due to cavitation (figures 17.1 and 17.2). A bubble contains a gas that can be described by a van der Waal equation-of-state law combined with a polytropic assumption. When the bubble radius $R(t)$ oscillates, the volume oscillates, and therefore the gas temperature oscillates (the gas temperature $T(t)$ is assumed to be homogeneous within the bubble). If the gas temperature approaches or surpasses approximately 1000 K during bubble oscillations, visible light is emitted from the gas (note that, obviously, visible light is emitted at all temperatures, but the light intensity becomes visible to the naked eye at temperatures above approximately 1000 K). Since the bubble radius multiplied by the total light absorption coefficient at wavelength λ, α_λ, is small ($\alpha_\lambda R(t) \ll 1$), the emitted power output per wavelength is proportionally smaller than that predicted by Planck's radiation law for a black body [1], i.e., $P_\lambda(t) = \frac{4}{3}\alpha_\lambda R(t)P_\lambda^{\text{Planck}}(t)$. The polytropic coefficient is a dynamic function of the Péclet number, $Pe = R(t)|\dot{R}(t)|/\chi(t)$, where $\chi(t) = \chi(R(t), T(t))$ is the thermal diffusivity [1–4]. In the following, a simple model is aimed for by using a constant polytropic coefficient and a Planck black-body radiation assumption.

doi:10.1088/978-0-7503-5557-5ch17

Figure 17.1. When a Mantis shrimp defends itself against an enemy or attacks a prey it generates sonoluminescence by moving its claw at around 100 km h^{-1}. When the claw closes in water a bubble is generated. As the bubble implodes, the bubble gas temperature increases to several thousand kelvins whereby visible light is emitted.

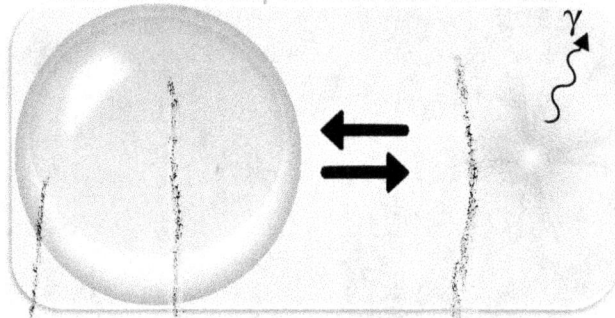

Figure 17.2. Sonoluminescence from a bubble. Visible light emission occurs as the bubble gas temperature increases to several thousand kelvins when the bubble implodes.

17.2 Incompressible fluids

From the equation of continuity, equation (12.4),

$$\frac{\partial \rho}{\partial t} + \boldsymbol{\nabla} \cdot (\rho \mathbf{v}) = 0, \tag{17.1}$$

and assuming fluid incompressibility, namely that the density of a fluid element does not change as it evolves in space and time,

$$\frac{D\rho}{Dt} \equiv \frac{\partial \rho}{\partial t} + \mathbf{v} \cdot \boldsymbol{\nabla}\rho = 0, \tag{17.2}$$

it follows that,

$$\frac{\partial \rho}{\partial t} = -\boldsymbol{\nabla} \cdot (\rho \mathbf{v}),$$
$$\frac{\partial \rho}{\partial t} = -\mathbf{v} \cdot \boldsymbol{\nabla}\rho. \tag{17.3}$$

Equating the two right-hand side expressions yields,

$$\boldsymbol{\nabla} \cdot \mathbf{v} = 0, \tag{17.4}$$

which is the condition of incompressibility of a fluid.

17.2.1 Spherical symmetry in incompressible fluids

The condition of fluid incompressibility, equation (17.4), written in spherical coordinates, is,

$$\boldsymbol{\nabla} \cdot \mathbf{v} = \frac{1}{r^2}\frac{\partial}{\partial r}(r^2 v_r) + \frac{1}{r \sin\theta}\frac{\partial}{\partial \theta}(v_\theta \sin\theta) + \frac{1}{r \sin\theta}\frac{\partial v_\phi}{\partial \phi} = 0. \tag{17.5}$$

If spherical symmetry applies, $\mathbf{v} = v_r \mathbf{e}_r + v_\theta \mathbf{e}_\theta + v_\phi \mathbf{e}_\phi = v_r \mathbf{e}_r$, then,

$$\frac{1}{r^2} \frac{\partial}{\partial r}(r^2 v_r) = 0, \tag{17.6}$$

so,

$$v_r(r, t) = \frac{F(t)}{r^2}, \tag{17.7}$$

where F is a function of time alone.

17.3 Derivation of the Rayleigh–Plesset equation

The Rayleigh–Plesset equation is a dynamic equation for the radius of a single spherical bubble $R(t)$ (t is time) contained in an infinite liquid. The liquid temperature far from the bubble, T_∞, is considered to be a constant. The liquid pressure far from the bubble, $P_\infty(t)$, may depend on time as is the case in the presence of, e.g., an ultrasonic excitation pulse. The pressure and temperature in the bubble are assumed homogeneous in space and designated $P_B(t)$ and $T_B(t)$, respectively. The pressure, temperature, and outward radial velocity in the liquid at a radial distance r from the bubble center are $P(r, t)$, $T(r, t)$, and $v(r, t)$, respectively. As such, the latter three properties are defined for $r \geqslant R(t)$. The liquid's mass density and dynamic viscosity are ρ_L and μ_L, respectively.

Assuming the liquid is incompressible, the radial liquid velocity is, from equation (17.7),

$$v_r(r, t) = \frac{F(t)}{r^2}. \tag{17.8}$$

In the case of zero mass transport across the bubble surface, continuity of the bubble velocity forces the liquid velocity at the interface between the bubble and the liquid to satisfy,

$$v_r(R, t) = \frac{dR}{dt} = \frac{F(t)}{R^2}, \tag{17.9}$$

so,

$$v_r(r, t) = \frac{R^2(t)}{r^2} \frac{dR}{dt}. \tag{17.10}$$

17.4 Momentum conservation

Assuming the fluid is a Newtonian fluid, the radial component of the incompressible Navier–Stokes equation in spherical coordinates is,

$$\rho_L \left(\frac{\partial v}{\partial t} + v \frac{\partial v}{\partial r} \right) = -\frac{\partial P}{\partial r} + \mu_L \left[\frac{1}{r^2} \frac{\partial}{\partial r} \left(r^2 \frac{\partial v}{\partial r} \right) - \frac{2v}{r^2} \right]. \tag{17.11}$$

Rearranging and introducing the kinematic viscosity, $\nu_L = \mu_L / \rho_L$,

$$-\frac{1}{\rho_L}\frac{\partial P}{\partial r} = \frac{\partial v}{\partial t} + v\frac{\partial v}{\partial r} - \nu_L\left[\frac{1}{r^2}\frac{\partial}{\partial r}\left(r^2\frac{\partial v}{\partial r}\right) - \frac{2v}{r^2}\right]. \tag{17.12}$$

Using equation (17.10),

$$-\frac{1}{\rho_L}\frac{\partial P}{\partial r} = \frac{2R}{r^2}\left(\frac{dR}{dt}\right)^2 + \frac{R^2}{r^2}\frac{d^2R}{dt^2} - \frac{2R^4}{r^5}\left(\frac{dR}{dt}\right)^2 = \frac{1}{r^2}\left(2R\left(\frac{dR}{dt}\right)^2 + R^2\frac{d^2R}{dt^2}\right) - \frac{2R^4}{r^5}\left(\frac{dR}{dt}\right)^2. \tag{17.13}$$

It should be noted that the viscous terms cancel out in equation (17.12). Integration from $r = R(t)$ to $r = \infty$ gives,

$$-\frac{1}{\rho_L}\int_{R(t)}^{\infty}\frac{\partial P}{\partial r}dr = -\frac{1}{\rho_L}\int_{P(R)}^{P(\infty)}dP = \int_{R(t)}^{\infty}\left[\frac{1}{r^2}\left(2R\left(\frac{dR}{dt}\right)^2 + R^2\frac{d^2R}{dt^2}\right) - \frac{2R^4}{r^5}\left(\frac{dR}{dt}\right)^2\right]dr,$$

$$\frac{P(R) - P_\infty}{\rho_L} = \left[-\frac{1}{r}\left(2R\left(\frac{dR}{dt}\right)^2 + R^2\frac{d^2R}{dt^2}\right) + \frac{R^4}{2r^4}\left(\frac{dR}{dt}\right)^2\right]_{R(t)}^{\infty} = R\frac{d^2R}{dt^2} + \frac{3}{2}\left(\frac{dR}{dt}\right)^2. \tag{17.14}$$

17.5 Boundary conditions

For a viscous fluid with constant density and constant viscosity, the normal stress in the liquid, T_{rr}, that points radially outward from the center of the bubble, is,

$$T_{rr} = -P + 2\mu_L\frac{\partial v}{\partial r}. \tag{17.15}$$

The total pressure P_{tot} at the bubble-liquid interface consists of three terms, the normal stress in the fluid T_{rr}, the bubble pressure at the bubble surface P_B, and the pressure due to surface tension $-2\sigma/R$ [5], pp 517–23. Hence,

$$\begin{aligned} P_{tot} &= T_{rr}(R) + P_B - \frac{2\sigma}{R} = -P(R) + 2\mu_L\frac{\partial v}{\partial r}\Big|_{r=R} + P_B - \frac{2\sigma}{R} \\ &= -P(R) + 2\mu_L\frac{\partial}{\partial r}\left(\frac{R^2}{r^2}\frac{dR}{dt}\right)\Big|_{r=R} + P_B - \frac{2\sigma}{R} \\ &= -P(R) - \frac{4\mu_L}{R}\frac{dR}{dt} + P_B - \frac{2\sigma}{R}, \end{aligned} \tag{17.16}$$

where σ is the surface tension.

Since it is assumed there is no mass transfer across the bubble-liquid interface, the total pressure must vanish ($P_{tot} = 0$), i.e.,

$$P(R) = P_B - \frac{4\mu_L}{R}\frac{dR}{dt} - \frac{2\sigma}{R}, \tag{17.17}$$

and the end result from momentum conservation becomes,

$$\frac{P(R) - P_\infty}{\rho_L} = \frac{P_B - P_\infty}{\rho_L} - \frac{4\mu_L}{\rho_L R}\frac{dR}{dt} - \frac{2\sigma}{\rho_L R} = R\frac{d^2R}{dt^2} + \frac{3}{2}\left(\frac{dR}{dt}\right)^2, \quad (17.18)$$

where the latter equation follows from equation (17.14). Finally, upon rearranging, the Rayleigh–Plesset equation is obtained,

$$\frac{P_B(t) - P_\infty(t)}{\rho_L} = R\frac{d^2R}{dt^2} + \frac{3}{2}\left(\frac{dR}{dt}\right)^2 + \frac{4\nu_L}{R}\frac{dR}{dt} + \frac{2\sigma}{\rho_L R}. \quad (17.19)$$

17.6 Adiabatic gases and perfect gas law

Assuming the bubble medium obeys a perfect gas law,

$$pV = nRT, \quad (17.20)$$

where p, V, n, R, and T are the bubble pressure, volume, number of moles per volume, universal gas constant, and absolute temperature, respectively. For a perfect gas, the internal energy U is a function of temperature only,

$$dU = C_V dT = nc_V dT, \quad (17.21)$$

where $C_V = \left(\frac{\partial U}{\partial T}\right)_V$ is the heat capacity at constant volume, and c_V is the molar heat capacity at constant volume.

From the first law of thermodynamics,

$$dU = d'Q + d'W, \quad (17.22)$$

where $d'Q$ and $d'W = -pdV$ are quasistatic heat and work differentials, respectively,

$$d'Q = dU - d'W = nc_V dT + pdV. \quad (17.23)$$

Similarly, from the definition of enthalpy, also a function of state [i.e., $H = H(T)$],

$$H = U + pV = U + nRT, \quad (17.24)$$

the heat capacity at constant pressure becomes,

$$C_p = \left(\frac{\partial H}{\partial T}\right)_p = \frac{dH}{dT} = \frac{dU}{dT} + nR = C_V + nR, \quad (17.25)$$

or, in terms of the molar heat capacities,

$$c_p = c_V + R. \quad (17.26)$$

Further, since,

$$dT = \left(\frac{\partial T}{\partial V}\right)_p dV + \left(\frac{\partial T}{\partial p}\right)_V dp = \frac{1}{nR}(pdV + Vdp), \quad (17.27)$$

using $T = \frac{pV}{nR}$ in obtaining the second equality, it follows from equation (17.23) that,

$$d'Q = nc_V dT + nRdT - Vdp = nc_p dT - Vdp, \tag{17.28}$$

where equation (17.26) was used to obtain the last equality.

17.6.1 Adiabatic processes

For an adiabatic process heat exchange between the bubble and the surrounding fluid can be neglected, i.e.,

$$d'Q = 0. \tag{17.29}$$

Equations (17.23) and (17.28) can then be written as,

$$nc_V dT = -pdV,$$
$$nc_p dT = Vdp.$$

Taking the ratio between the two expressions yields,

$$\frac{c_p}{c_V} = \frac{Vdp}{-pdV}. \tag{17.30}$$

The ratio $\frac{c_p}{c_V}$ is the heat capacity ratio and is usually denoted γ. It follows from the last expression that,

$$\frac{dp}{p} = -\gamma \frac{dV}{V}. \tag{17.31}$$

If the process is initiated in (p_0, V_0),

$$\int_{p_0}^{p} \frac{dp}{p} = -\gamma \int_{V_0}^{V} \frac{dV}{V}, \tag{17.32}$$

or,

$$\ln \frac{p}{p_0} = -\gamma \ln \frac{V}{V_0} = \ln\left(\frac{V_0}{V}\right)^{\gamma}, \tag{17.33}$$

and,

$$\frac{p}{p_0} = \left(\frac{V_0}{V}\right)^{\gamma}. \tag{17.34}$$

The latter equation is equivalent to the quasistatic adiabatic relation of a perfect gas,

$$pV^{\gamma} = p_0 V_0^{\gamma} = \text{constant}. \tag{17.35}$$

If the adiabatic equation is expressed in terms of pressure and temperature,

$$p\left(\frac{nRT}{p}\right)^{\gamma} = \text{constant,} \tag{17.36}$$

or,

$$p^{1-\gamma}T^{\gamma} = \text{constant.} \tag{17.37}$$

If the adiabatic equation is expressed in terms of temperature and volume,

$$\left(\frac{nRT}{V}\right)V^{\gamma} = \text{constant,} \tag{17.38}$$

or,

$$TV^{\gamma-1} = \text{constant.} \tag{17.39}$$

Notice that,

$$\gamma = \frac{c_p}{c_V} = 1 + \frac{R}{c_V} > 1. \tag{17.40}$$

For a single-atomic perfect gas, it holds that $c_V \approx \frac{3}{2}R$ while for a two-atomic perfect gas $c_V \approx \frac{5}{2}R$. Thus, $\gamma \approx 5/3$ for a single-atomic perfect gas and $\gamma \approx 7/5$ for a two-atomic perfect gas. Summarizing, the adiabatic T–V relation shows how the temperature increases as the volume of the bubble decreases. Based on Planck's radiation law the latter finding explains the observation that a bubble contracting heavily in the presence of, e.g., ultrasound begins to emit visible light.

17.7 Derivation of Planck's black-body radiation law

Let us proceed to derive Planck's black-body radiation law.

17.7.1 Number of electromagnetic states per volume

Firstly, the number of electromagnetic modes confined in a cube with length L defined by $0 \leqslant x \leqslant L$, $0 \leqslant y \leqslant L$, $0 \leqslant z \leqslant L$ is determined. Assuming the box has conducting walls, the electric field $\mathbf{E}$ must vanish at the walls, i.e., modal solutions take the form,

$$\mathbf{E} = \mathbf{E}_0 \sin(k_x x)\sin(k_y y)\sin(k_z z) = \mathbf{E}_0 \sin\left(\frac{l\pi}{L}x\right)\sin\left(\frac{m\pi}{L}y\right)\sin\left(\frac{n\pi}{L}z\right), \tag{17.41}$$

where $\mathbf{E}_0$ is the electric field amplitude and l, m, n are positive integers.

17.7.2 Exercise

Convince yourself that only the positive integers l, m, n must be considered when determining the number of electromagnetic modes.

From the electromagnetic wave equation,

$$\nabla^2 \mathbf{E} + \frac{\omega^2}{c^2}\mathbf{E} = 0, \tag{17.42}$$

where $\nu = \frac{\omega}{2\pi}$ is the light frequency and c is the speed of light, the dispersion relation becomes,

$$k^2 \equiv k_x^2 + k_y^2 + k_z^2 = \frac{\omega^2}{c^2}, \tag{17.43}$$

so, for modes (l, m, n),

$$k^2 = \frac{\pi^2}{L^2}(l^2 + m^2 + n^2) = \frac{\pi^2 p^2}{L^2}, \tag{17.44}$$

where,

$$p^2 = l^2 + m^2 + n^2. \tag{17.45}$$

The volume of a spherical shell of radius p and thickness dp in p space is $4\pi p^2\, dp$. Since l, m, n are positive integers and there is one k point per unit volume in (l, m, n) space, the number of k points in an octant (positive l, m, n) of the spherical shell is,

$$dN(p) = N(p)dp = \left(\frac{1}{8}\right)4\pi p^2\, dp, \tag{17.46}$$

or, with $k = \pi p/L$, $dk = \pi dp/L$,

$$dN(p) = N(p)dp = \frac{L^3}{2\pi^2}k^2 dk. \tag{17.47}$$

Since $V = L^3$ is the volume of the box and $k = 2\pi\nu/c$,

$$dN(p) = \frac{V}{2\pi^2}k^2 dk = \frac{V}{2\pi^2}\frac{8\pi^3\nu^2}{c^3}d\nu = \frac{4\pi\nu^2 V}{c^3}d\nu. \tag{17.48}$$

Finally, since there are two polarization modes per k value of the electromagnetic field, the number of states in the frequency interval ν to $\nu + d\nu$ becomes,

$$dN = \frac{8\pi\nu^2 V}{c^3}d\nu. \tag{17.49}$$

17.7.3 Boltzmann distribution and mean energy of a mode of frequency ν

To supplement the above-obtained relation for the number of states, it is necessary to determine the expected occupancy of each state. Assuming electromagnetic radiation in thermal equilibrium (black-body radiation), the probability that a single mode of frequency ν has energy $E_n = nh\nu$ is given by the Boltzmann distribution,

$$p_\nu(n) = \frac{e^{-E_n/(k_B T)}}{\sum\limits_{n=0}^{\infty} e^{-E_n/(k_B T)}}, \tag{17.50}$$

where the denominator ensures that the total probability is unity. Note that $p_\nu(n)$ is the probability that a state of frequency ν contains n photons.

The mean energy of the mode of frequency ν is, therefore,

$$\bar{E}_\nu = \sum\limits_{n=0}^{\infty} E_n p_\nu(n) = \frac{\sum\limits_{n=0}^{\infty} E_n e^{-E_n/(k_B T)}}{\sum\limits_{n=0}^{\infty} e^{-E_n/(k_B T)}} = \frac{\sum\limits_{n=0}^{\infty} nh\nu e^{-E_n/(k_B T)}}{\sum\limits_{n=0}^{\infty} e^{-E_n/(k_B T)}}. \tag{17.51}$$

To simplify the calculation, let us substitute $x = e^{-h\nu/(k_B T)}$. Then,

$$\bar{E}_\nu = h\nu \frac{\sum\limits_{n=0}^{\infty} nx^n}{\sum\limits_{n=0}^{\infty} x^n} = h\nu \frac{x + 2x^2 + 3x^3 + \cdots}{1 + x + x^2 + \cdots} = h\nu\, x \frac{1 + 2x + 3x^2 + \cdots}{1 + x + x^2 + \cdots}. \tag{17.52}$$

Since,

$$\frac{1}{1-x} = 1 + x + x^2 + x^3 + \cdots,$$

$$\frac{1}{(1-x)^2} = 1 + 2x + 3x^2 + \cdots,$$

the mean energy of a mode of frequency ν is,

$$\bar{E}_\nu = \frac{h\nu\, x}{1-x} = \frac{h\nu}{x^{-1}-1} = \frac{h\nu}{e^{h\nu/(k_B T)}-1}. \tag{17.53}$$

17.8 The Planck formula

Planck's radiation formula is finally obtained as the energy of radiation in the frequency range ν to $\nu + d\nu$, i.e.,

$$u(\nu)d\nu = \bar{E}_\nu dN = \bar{E}_\nu \frac{8\pi\nu^2 V}{c^3} d\nu$$

$$= \frac{8\pi h\nu^3 V}{c^3} \frac{1}{e^{h\nu/(k_B T)}-1} d\nu, \tag{17.54}$$

where equations (17.49) and (17.53) are used to obtain the second and third equality, respectively.

17.9 Stefan–Boltzmann law

Let us compute the energy density (energy per unit volume) over the whole spectrum using the Planck formula,

$$u = \frac{1}{V} \int_0^\infty u(\nu) d\nu = \frac{8\pi h}{c^3} \int_0^\infty \frac{\nu^3 d\nu}{e^{h\nu/(k_B T)} - 1}. \tag{17.55}$$

Substituting $x = h\nu/(k_B T)$ and $dx = h/(k_B T)d\nu$,

$$u = \frac{8\pi h}{c^3} \left(\frac{k_B T}{h} \right)^4 \int_0^\infty \frac{x^3 dx}{e^x - 1}. \tag{17.56}$$

The latter integral is a standard integral given by,

$$\int_0^\infty \frac{x^3 dx}{e^x - 1} = \frac{\pi^4}{15}, \tag{17.57}$$

thus,

$$u = \frac{8\pi^5 k_B^4}{15 c^3 h^3} T^4 = aT^4, \tag{17.58}$$

which is one form of the Stefan–Boltzmann law of black-body radiation. Upon inserting the constants, the coefficient a takes the value,

$$a = 7.566 \times 10^{-16} \mathrm{J\,m^{-3}\,K^{-4}}. \tag{17.59}$$

17.10 Derivation of Wien's displacement law from Planck's black-body radiation law

Recasting the Planck formula in terms of wavelength, using $u(\nu, T)d\nu = u(\lambda, T)d\lambda$,

$$u(\lambda, T) = \frac{2hc^2 V}{\lambda^5} \frac{1}{e^{hc/(\lambda k_B T)} - 1}, \tag{17.60}$$

and differentiating with respect to wavelength,

$$\frac{\partial u}{\partial \lambda} = 2hc^2 V \left(\frac{hc}{k_B T \lambda^7} \frac{e^{hc/(\lambda k_B T)}}{(e^{hc/(\lambda k_B T)} - 1)^2} - \frac{1}{\lambda^6} \frac{5}{e^{hc/(\lambda k_B T)} - 1} \right) = 0, \tag{17.61}$$

yields,

$$\frac{hc}{\lambda k_B T} \frac{e^{hc/(\lambda k_B T)}}{e^{hc/(\lambda k_B T)} - 1} - 5 = 0. \tag{17.62}$$

Defining,

$$x \equiv \frac{hc}{\lambda k_B T}, \tag{17.63}$$

the equation becomes,

$$\frac{xe^x}{e^x - 1} - 5 = 0, \tag{17.64}$$

with solutions $x = 0$, $x = 4.965\,114$. The second solution for x specifies the wavelength where the energy density is at a maximum,

$$\lambda_{\max} = \frac{hc}{k_B T x} = \frac{2.897\,772 \text{ mm} \cdot \text{K}}{T}. \tag{17.65}$$

This expression is known as Wien's displacement law.

17.11 Numerical solution of the Rayleigh–Plesset equation, i.e., bubble radius versus time for a given ultrasonic pulse

Including the effect of liquid compressibility in deriving the Rayleigh–Plesset equation, an additional term, $\frac{R}{c}\frac{d}{dt}p_B$, must be added to equation (17.19), i.e.,

$$\rho\left(R\frac{d^2R}{dt^2} + \frac{3}{2}\left(\frac{dR}{dt}\right)^2\right) = p_B(t) - P_\infty - P_{\text{exc}}(t) - 4\mu_L\frac{dR/dt}{R} - \frac{2\sigma}{R} + \frac{R}{c}\frac{d}{dt}p_B, \tag{17.66}$$

and the addition of an ultrasonic pressure is assumed,

$$P_{\text{exc}}(t) = -P_a \sin(\omega t), \tag{17.67}$$

so as to generate the bubble dynamics. For a perfect gas law, the bubble pressure $p_B(t)$ is determined from the bubble radius $R(t)$ using the adiabatic relation,

$$p_B(t)V(t)^\gamma = p_0 V_0^\gamma,$$
$$p_B(t) = \frac{p_0 R_0^{3\gamma}}{(R(t))^{3\gamma}}, \tag{17.68}$$

where R_0 is the static bubble radius in the absence of ultrasonic excitation, and $V_0 = \frac{4}{3}\pi R_0^3$. It then follows that,

$$\frac{d}{dt}p_B = p_0 R_0^{3\gamma}(-3\gamma)(R(t))^{-3\gamma-1}\frac{dR}{dt}. \tag{17.69}$$

If instead a van der Waal's model for the equation-of-state of the gas in the bubble is adopted,

$$p_B(t) = \left(P_\infty + \frac{2\sigma}{R_0}\right)\left(\frac{R_0^3 - h^3}{R^3(t) - h^3}\right)^\gamma, \tag{17.70}$$

where h is the characteristic hard-core radius of the gas inside the bubble, and γ is the ratio between the specific heat at constant pressure and at constant volume, i.e., the adiabatic coefficient, then the derivative of the bubble pressure becomes,

$$\frac{d}{dt}p_B = -3\gamma p_B \frac{R^2\dfrac{dR}{dt}}{R^3 - h^3}. \tag{17.71}$$

17.11.1 Exercise

Prove equation (17.71) from equation (17.70).

In figure 17.3 the bubble radius is shown as a function of time using the parameters in table 17.1 and the van der Waal's gas law in equation (17.70).

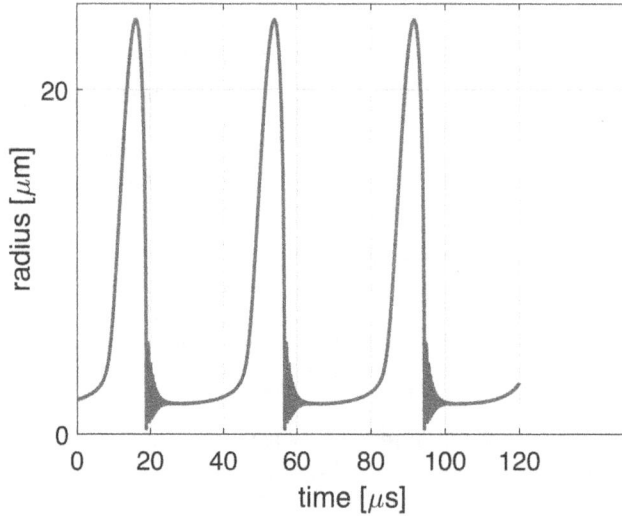

Figure 17.3. Bubble radius oscillations versus time determined by solving the modified Rayleigh–Plesset equation. The ultrasonic excitation is a sinusoidal pressure of frequency 26.5 kHz and amplitude 1.40 atm. The initial bubble radius and temperature are 2 μm and 300 K, respectively.

Table 17.1. Parameters used in the numerical evaluation of bubble dynamics based on the modified Rayleigh–Plesset equation.

Parameter	Value	Unit
Surface tension σ	72.8×10^{-3}	N m^{-1}
Density (water) ρ	1000	kg m^{-3}
Adiabatic index (argon) γ	5/3	
Speed of sound (water) c	1500	m s^{-1}
Dynamic viscosity (water)	1.002×10^{-3}	Pa s
Ambient pressure P_∞	1.00	atm
Static radius R_0	2.0×10^{-6}	m
Hard core (argon) h	$R_0/8.86$	m
Ultrasound frequency $\omega = 2\pi f$	$f = 26.5$	kHz
Ultrasonic amplitude P_a	1.40	atm

Results are obtained by solving the modified Rayleigh–Plesset equation [equation (17.66)] assuming the bubble gas is argon. The initial bubble radius oscillates in a nonlinear fashion in the presence of a harmonic ultrasonic pressure field of frequency 26.5 kHz and amplitude 1.40 atm. A close inspection of the data reveals that the minimum bubble radius is 0.3 μm.

17.11.2 Numerical solution of $T(t)$ versus time using the adiabatic relation

Assuming the gas is a perfect gas and adiabatic conditions when converting the bubble radius dynamics to bubble temperature dynamics, the gas temperature oscillates nonlinearly in time as shown in figure 17.4. Input to the calculation is the bubble radius dynamics shown in figure 17.3. The dynamics show that the maximum bubble gas temperature is 13 000 K corresponding to the minimum bubble radius 0.3 μm.

17.11.3 Numerical solution of λ_{max} (EM wavelength at the peak intensity) versus time using Wien's displacement law

In figure 17.5, the light wavelength λ_{max} associated with the maximum energy density in the Planck black-body spectrum is shown as a function of time. Note that λ_{max} reaches a minimum value of 220 nm (corresponding to peak emission in the near-ultraviolet range) when the temperature reaches 13 000 K corresponding to the minimum bubble radius 0.3 μm.

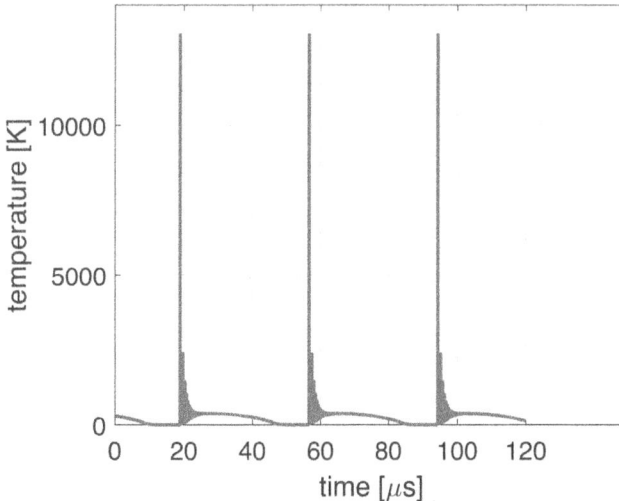

Figure 17.4. Temperature oscillations of the bubble gas versus time determined by solving the modified Rayleigh–Plesset equation. Assuming a perfect gas law and adiabatic conditions, the bubble gas temperature is calculated from the bubble radius dynamics shown in figure 17.3. The ultrasonic excitation is a sinusoidal pressure of frequency 26.5 kHz and amplitude 1.40 atm. The initial bubble radius and temperature are 2 μm and 300 K, respectively.

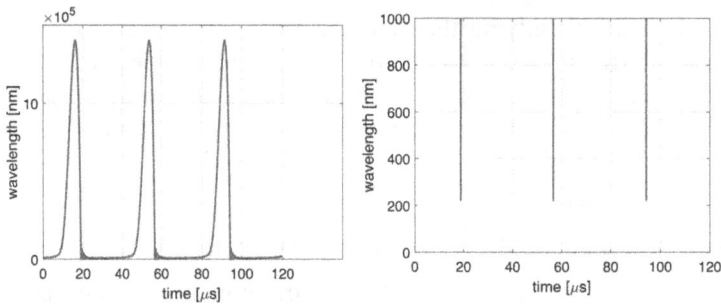

Figure 17.5. Light wavelength λ_{max} versus time associated with the maximum energy density in the Planck black-body spectrum. The plot is obtained assuming the gas in the bubble behaves as a black body using Wien's displacement law [equation (17.65)] on the temperature data in figure 17.4. The right panel is a zoom-in of the left panel at low wavelengths.

Figure 17.6. Matlab program for solving the modified Rayleigh–Plesset equation. Bubble radius versus time, bubble gas temperature versus time, and the light wavelength at maximum energy density versus time.

17.11.4 Matlab program for solving the Rayleigh–Plesset equation

In figure 17.6, the MATLAB code for solving the modified Rayleigh–Plesset equation is shown. The bubble radius, bubble gas temperature, and the light wavelength at maximum energy density versus time are plotted as output.

17.11.5 Exercises

(a) Write a Matlab program for solving the Rayleigh–Plesset equation assuming a perfect gas law instead of a van der Waal's model. Use the parameters in table 17.1.

Calculate the bubble radius as a function of time. Then, change the frequency of the ultrasound to $f = 30$ kHz and calculate the bubble radius as a function of time.

(b) Use the parameters in table 17.1. Calculate and plot the minimum pressure in time as a function of the ultrasonic pressure amplitude P_a in the range 0.5–2 atm.

References

[1] Hilgenfeldt S, Grossmann S and Lohse D 1999 A simple consistent explanation of sonoluminescence light emission *Nat. Suppl. Inf.* **398** 402

[2] Prosperetti A 1977 Thermal effects and damping mechanisms in the forced radial oscillations of gas bubbles in liquids *J. Acoust. Soc. Am.* **61** 17

[3] Hilgenfeldt S, Grossmann S and Lohse D 1999 A simple consistent explanation of sonoluminescence light emission *Nature* **398** 402

[4] Hirschfelder S, Curtiss C F and Bird R B 1954 *Molecular Theory of Gases and Liquids* (New York: Wiley)

[5] Landau L D, Lifshitz E M and Pitaevskii L P 1984 *Course of Theoretical Physics* Vol. 5 (Oxford: Butterworth-Heinemann)

Part IV

Appendices

IOP Publishing

Piezoelectricity in Classical and Modern Systems

Morten Willatzen

Appendix A

Stiffness tables

Table A.1. Values given are c^E.

Material	Stiffness constants c_{IJ} (10^{10} N m^{-2})§						
	c_{11}	c_{33}	c_{44}	c_{66}	c_{12}	c_{13}	c_{14}
INSULATORS							
Barium titanate[†]	27.5	16.5	5.43	11.3	17.9	15.1	
Barium titanate[†‖]	15.0	14.6	4.4		6.6	6.6	
Lead titanate-zirconate (PZT-2)[†‖]	13.5	11.3	2.22		6.79	6.81	
Lead titanate-zirconate (PZT-5H)[†‖]	12.6	11.7	2.30		7.95	8.41	
Quartz[†]	8.674	10.72	5.794		0.699	1.191	1.791

§ Only the independent constants are given.

[†] Piezoelectric.

[‖] Poled ceramic. The stiffness matrix has the same form as for the hexagonal crystal system, with Z along the poling axis.

Table A.2. Values given are c^E.

Material	Stiffness constants c_{IJ} (10^{10} N m^{-2})§						
	c_{11}	c_{33}	c_{44}	c_{66}	c_{12}	c_{13}	c_{14}
SEMICONDUCTORS							
Cadmium selenide[†]	7.41	8.36	1.317		4.52	3.93	
Cadmium sulfide[†]	9.07	9.38	1.504		5.81	5.10	
Gallium arsenide[†]	11.88		5.94		5.38		
Gallium phosphide[†]	14.12		7.047		6.253		
Germanium	12.89		6.71		4.83		
Indium antimonide[†]	6.72		3.02		3.67		
Indium phosphide[†]	10.22		4.60		5.76		
Silicon	16.57		7.956		6.39		
Zinc oxide[†]	20.97	21.09	4.247		12.11	10.51	
Zinc sulfide[†]	12.04	12.76	2.28		6.92	6.20	

§ Only the independent constants are given.
[†] Piezoelectric.
‖ Poled ceramic. The stiffness matrix has the same form as for the hexagonal crystal system, with Z along the poling axis.

Table A.3. Values given are c^E.

Material	Stiffness constants c_{IJ} (10^{10} N m^{-2})§						
	c_{11}	c_{33}	c_{44}	c_{66}	c_{12}	c_{13}	c_{14}
METALS							
Aluminium, crystal	10.80		2.85		6.13		
Aluminium, polycrystal	11.1		2.5				
Gold, crystal	18.6		4.20		15.7		
Gold, polycrystal	20.7		2.85				
Iron, crystal	23.7		11.6		14.1		
Iron, polycrystal	27.7		8.2				
Silver, crystal	11.9		4.37		8.94		
Silver, polycrystal	13.95		2.7				
Tungsten, crystal	50.2		15.2		19.9		
Tungsten, polycrystal	58.1		13.4				

§ Only the independent constants are given.
[†] Piezoelectric.
‖ Poled ceramic. The stiffness matrix has the same form as for the hexagonal crystal system, with Z along the poling axis.

Appendix B

Piezoelectric constant tables

Table B.1.

Piezoelectric strain constants $d_{iJ} = d_{Ji}$ (10^{-12} C/N)§							
Material	Symmetry	d_{x1}	d_{x4}	d_{x5}	d_{y2}	d_{z1}	d_{z3}
INSULATORS							
Barium titanate	Tetr. $4mm$			392		-34.5	85.6
Barium titanate	Uniaxial[†]			260		-78	190
Lead titanate-zirconate (PZT-2)	Uniaxial[†]			440		-60	152
Lead titanate-zirconate (PZT-5H)	Uniaxial[†]			741		-274	593
Lithium niobate	Trig. $\bar{3}\frac{2}{m}$			68	21	-1	6
Quartz	Trig. 32	-2.3	-0.67				

§ Only the independent constant sare given.

† Piezoelectric ceramics are poled during preparation and are isotropic in the plane transverse to the poling (or Z axis). The piezoelectric strain matrix is of the form $\begin{bmatrix} 0 & 0 & 0 & 0 & d_{x5} & 0 \\ 0 & 0 & 0 & d_{x5} & 0 & 0 \\ d_{z1} & d_{z1} & d_{z3} & 0 & 0 & 0 \end{bmatrix}$.

Table B.2.

Piezoelectric strain constants $d_{iJ} = d_{Ji}$ (10^{-12} C/N)§							
Material	Symmetry	d_{x1}	d_{x4}	d_{x5}	d_{y2}	d_{z1}	d_{z3}
SEMICONDUCTORS							
Cadmium sulfide	Hex. $6mm$			-14		-5	10.3
Gallium arsenide	Cub. $\bar{4}3m$		2.6				
Zinc oxide	Hex. $6mm$			-11.34		-5.43	11.67
Zinc sulfide	Hex. $6mm$			-2.8		-1.13	3.23

§ Only the independent constants are given.

† Piezoelectric ceramics are poled during preparation and are isotropic in the plane transverse to the poling (or Z axis). The piezoelectric strain matrix is of the form $\begin{bmatrix} 0 & 0 & 0 & 0 & d_{x5} & 0 \\ 0 & 0 & 0 & d_{x5} & 0 & 0 \\ d_{z1} & d_{z1} & d_{z3} & 0 & 0 & 0 \end{bmatrix}$.

doi:10.1088/978-0-7503-5557-5ch19

Table B.3.

Material	Symmetry	e_{x1}	e_{x4}	e_{x5}	e_{y2}	e_{z1}	e_{z3}
Piezoelectric stress constants $e_{iJ} = e_{Ji}$ (C/m²)§							
INSULATORS							
Barium titanate	Tetr. $4mm$			21.3		−2.74	3.70
Barium titanate	Uniaxial†			11.4		−4.35	17.5
Lead titanate-zirconate (PZT-2)	Uniaxial†			9.8		−1.9	9.0
Lead titanate-zirconate (PZT-5H)	Uniaxial†			17.0		−6.5	23.3
Lithium niobate	Trig. $\bar{3}\frac{2}{m}$			3.7	2.5	0.2	1.3
Quartz	Trig. 32	0.171	−0.0436				

§ Only the independent constants are given.

† Piezoelectric ceramics are poled during preparation and are isotropic in the plane transverse to the poling (or Z axis). The piezoelectric stress matrix is of the form $\begin{bmatrix} 0 & 0 & 0 & 0 & e_{x5} & 0 \\ 0 & 0 & 0 & e_{x5} & 0 & 0 \\ e_{z1} & e_{z1} & e_{z3} & 0 & 0 & 0 \end{bmatrix}$.

Table B.4.

Material	Symmetry	e_{x1}	e_{x4}	e_{x5}	e_{y2}	e_{z1}	e_{z3}
Piezoelectric stress constants $e_{iJ} = e_{Ji}$ (C/m²)§							
SEMICONDUCTORS							
Cadmium sulfide	Hex. $6mm$			−0.21		−0.24	0.44
Gallium arsenide	Cub. $\bar{4}3m$		0.154				
Gallium phosphide	Cub. $\bar{4}3m$		−0.10				
Indium antimonide	Cub. $\bar{4}3m$		0.71				
Zinc oxide	Hex. $6mm$			−0.48		−0.573	1.32
Zinc sulfide	Hex. $6mm$			−0.0638		−0.0140	0.272

§ Only the independent constants are given.

† Piezoelectric ceramics are poled during preparation and are isotropic in the plane transverse to the poling (or Z axis). The piezoelectric stress matrix is of the form $\begin{bmatrix} 0 & 0 & 0 & 0 & e_{x5} & 0 \\ 0 & 0 & 0 & e_{x5} & 0 & 0 \\ e_{z1} & e_{z1} & e_{z3} & 0 & 0 & 0 \end{bmatrix}$.

IOP Publishing

Piezoelectricity in Classical and Modern Systems

Morten Willatzen

Appendix C

Permittivity tables

Table C.1.

Relative permittivity constants for piezoelectric materials§							
Material	Symmetry	$\epsilon^S_{xx}/\epsilon_0$	$\epsilon^S_{yy}/\epsilon_0$	$\epsilon^S_{zz}/\epsilon_0$	$\epsilon^T_{xx}/\epsilon_0$	$\epsilon^T_{yy}/\epsilon_0$	$\epsilon^T_{zz}/\epsilon_0$
INSULATORS							
Barium titanate	Tetr. $4mm$				2920		168
Barium titanate	Uniaxial[†]	1115		1260	1450		1700
Lead zirconate-titanate (PZT-2)	Uniaxial[†]	504		260	990		450
Lead zirconate-titanate (PZT-5H)	Uniaxial[†]	1700		1470	3130		3400
Lithium niobate	Trig. $3\frac{2}{m}$	44		29	84		30

§ Only the independent constants are given. $\epsilon_0 = 8.854 \times 10^{-12}$ F m^{-1}.
[†] Poled ceramic material. The permittivity matrix has the same form as for the hexagonal crystal system with Z along the poling axis.

Table C.2.

Relative permittivity constants for piezoelectric materials§							
Material	Symmetry	$\epsilon^S_{xx}/\epsilon_0$	$\epsilon^S_{yy}/\epsilon_0$	$\epsilon^S_{zz}/\epsilon_0$	$\epsilon^T_{xx}/\epsilon_0$	$\epsilon^T_{yy}/\epsilon_0$	$\epsilon^T_{zz}/\epsilon_0$
SEMICONDUCTORS							
Cadmium sulfide	Hex. $6mm$	9.02		9.53	9.35		10.3
Gallium arsenide	Cub. $\bar{4}3m$				12.5		
Gallium phosphide	Cub. $\bar{4}3m$				11.1		
Indium antimonide	Cub. $\bar{4}3m$				17.7		
Indium phosphide	Cub. $\bar{4}3m$				12.35		
Zinc oxide	Hex. $6mm$	8.55		10.2	9.16		12.64
Zinc sulfide	Hex. $6mm$				8.7		8.7

§ Only the independent constants are given. $\epsilon_0 = 8.854 \times 10^{-12}$ F m^{-1}.
† Poled ceramic material. The permittivity matrix has the same form as for the hexagonal crystal system with Z along the poling axis.

Table C.3.

Relative permittivity constants for non-piezoelectric materials§				
Material	Symmetry	ϵ_{xx}/ϵ_0	ϵ_{yy}/ϵ_0	ϵ_{zz}/ϵ_0
INSULATORS				
Diamond	Cub. $\frac{4}{m}\bar{3}\frac{2}{m}$	5.67		
Sapphire	Trig. $\bar{3}\frac{2}{m}$	9.34		11.54
SEMICONDUCTORS				
Germanium	Cub. $\frac{4}{m}\bar{3}\frac{2}{m}$	15.8		
Silicon	Cub. $\frac{4}{m}\bar{3}\frac{2}{m}$	11.7		

§ Only the independent constants are given. $\epsilon_0 = 8.854 \times 10^{-12}$ F m^{-1}.

www.ingramcontent.com/pod-product-compliance
Lightning Source LLC
Chambersburg PA
CBHW080515220326
41599CB00032B/6096